Iron Making during the Migration Period

The case of the Lombards

Vasco La Salvia

BAR International Series 1715
2007

Published in 2016 by
BAR Publishing, Oxford

BAR International Series 1715

Iron Making during the Migration Period

ISBN 978 1 4073 0159 4

BAR Publishing is the trading name of British Archaeological Reports (Oxford) Ltd. British Archaeological Reports was first incorporated in 1974 to publish the BAR Series, International and British. In 1992 Hadrian Books Ltd became part of the BAR group. This volume was originally published by Archaeopress in conjunction with British Archaeological Reports (Oxford) Ltd / Hadrian Books Ltd, the Series principal publisher, in 2007. This present volume is published by BAR Publishing, 2016.

Printed in England

BAR
PUBLISHING

BAR titles are available from:

BAR Publishing
122 Banbury Rd, Oxford, OX2 7BP, UK
EMAIL info@barpublishing.com
PHONE +44 (0)1865 310431
FAX +44 (0)1865 316916
www.barpublishing.com

Ai miei cari Zaroui, Arechi e Giulia Sofia, dolci impegni quotidiani
saldi di fronte al calore della fiamma di Apollo
ed agili danzatori al flauto di Bacco

Acknowledgements

This work has been thought and prepared as a dissertation to be submitted to the Department of Medieval Studies of the CENTRAL EUROPEAN UNIVERSITY of *Budapest* (Hungary) in candidacy for **Ph.D.** degree in Medieval Studies. Thus, I have to thank first of all the Members of the *Examination Committee* (J. LASZLOVSZKY, I. BARBIERA, A. CHOYKE, N. CHRISTIE, T. VIDA) and the *External Readers* (R. FRANCOVICH, J. GÖMÖRI, A. AUGENTI). As a matter of fact, their comments and suggestions, about which I was advised during my successful defense (4th March 2006), have all proven to be extremely useful for the present publication of this study.

A Ph.D. dissertation marks, in my opinion, an important point in scholarly activity, a sort of research tank in which most of the studies and the ideas collected during the entire University life are channeled. Therefore, many people in different ways and to a different degree contributed to the final shape of my work. Thus many people have to be thanked for my achievement. Some of them are friends and/or University colleagues, some other University Professors (and some, after long lasting acquaintance, became both things together). First of all I feel that I owe the deepest gratitude and credits to the late Prof. ANNA MARIA GIUNTELLA (Chair of Medieval Archaeology University of *Chieti*, Italy), the late Professor RICCARDO FRANCOVICH (Chair of Medieval Archaeology University of *Siena*, Italy), the late Prof. ISTVÁN BÓNA (ELTE Universiy, *Budapest*) with whom I had the honor and the pleasure to work. I will never forget their kindness, their passion for teaching, in other words their love for Archaeology. This study is for them, dedicated to their memory hoping that they would have appreciated my scholarly efforts.

Moreover, in the preparation of this work, several people have been of great help and assistance. A special thanks goes again to the late Prof. ISTVÁN BÓNA who kindly granted me the permission to study and analyze unpublished evidence from his excavations. In addition, I would like to thank JÁNOS ÓDOR of the Direction of the Museum of *Szekszárd* (Hungary), who generously allowed the sampling of the iron items from *Kajdacs-Homokbánya* and *Tamási-Csikólegelő* (Hungary) necropolises which are discussed in the present work and Dr. ÁRPÁD KOVÁCS and Dr. BÉLA TÖRÖK, from the Department of Physical Metallurgy of the University of *Miskolc* (Hungary) who carried out, using a *Scanning Electron Microscope* (SEM), the metallographic analyses on these objects. Also Dr. JÁNOS GÖMÖRI, Director of the Museum of *Sopron* (Hungary) played an important role as the main Hungarian connection for Archaeometallurgy. I owe credits to Dr. CORNELIA RUPP for granting me the permission of using her yet unpublished data on *Nocera Umbra* swords.

I also would like to thank Prof. Dr. LUBOMIR MIHOK of the Faculty of Metallurgy of the Technical University of *Kosice* (Republic of Slovakia), under whose supervision I had been introduced to archaeometallurgical analyses, for its numerous and unrivalled advices concerning many of the technical aspects of the discussion related to metallography.

I have to thank my friend and colleague Dr. FRANCESCA ZAGARI for introducing me in the 'wonderful world' of medieval agricultural technology.

I would like to underline as well the crucial help I received from Dr. ALICE CHOYKE of the Medieval Studies Department of the **C**entral **E**uropean **U**niversity of *Budapest*. In my case her work has not been limited to improving the English academic writing, an improvement that ended up in the 'dissolution' (or almost) of my splendid Italianized English (or Anglicized Italian). I actually found in her a disposition to discussion (basically, long distance e-mail quarrel), a sound support which was fundamental for clarifying my ideas, my concepts and the way to express them properly.

In addition, I owe credits to my beloved CSILLA DOBOS: without her help I would not have completed my work. She has been my 'super supervisor.' I also have to thank all the people of the staff of the Medieval Studies Department of the **C**entral **E**uropean **U**niversity of *Budapest* for their consistent assistance and encouragement, particularly ANNABELLA PAL and DOROTTYA 'DODO' DOMANOWSZKY.

Last but not least, my thesis supervisor Prof. JOZSEF LASZLOVSZKY of the Department of Medieval Studies of the **C**entral **E**uropean **U**niversity of *Budapest* deserves a special mention particularly for his patience and willingness to wait for so long a time for the final version of my work.

Moreover, during my University years first as a student and, later as a faculty member of the Dipartimento di Scienze dell'Antichità dell'Univeristà di *Chieti* (Italy), many people had contributed to my work.

First, I would like again to mention my dearest Prof. ANNA MARIA GIUNTELLA (Chair of Medieval Archeology, University of *Chieti*) who kindly and warmly accepted me in her working team.

Prof. RICCARDO FRANCOVICH (Chair of Medieval Archeology, University of *Siena*) has allowed me to maintain a constant and fruitful scholarly dialogue with him and his collaborators, as well as given me the opportunity to "work" for and with archaeology.

Prof. LETIZIA ERMINI PANI (Chair of Medieval Archaeology, University of *Rome* I) has always followed my work with great care and attention, demonstrating a rare sensibility towards my studies related to material culture.

I have to thank Prof. ANDREA AUGENTI (Chair of Medival Archaeology, University of *Ravenna*, Italy) to whom I am indebted since 1996 when my M.A. thesis was published. He has been a good elder friend of mine during my youth and I am extremely happy and proud to have found him again as a "master" in medieval archaeology. Probably words are not enough to express how much I feel indebted to him, as he has been always so patient with me since the time we were only youngsters.

Prof. GIORGIO STABILE (Chair of History of Science, University of *Rom*e I), my professor, THE professor. If I have done anything as a scholar I owe it to his teaching, to his clear explanations, to his compassion, and to his precious friendship.

I would like also to thank Prof. SOFIA BOESCH GAJANO (Chair of Medieval History, University of *Rome* III). It is difficult to say how and how much a person that knows you since childhood can contribute to the scholarly formation of that person. However, the fact that I became a medievalist expresses more than any words of gratitude. I have to thank as well the *maestro* DANIELE DE LUCA and his 'disciples,' MASSIMO MENCHETTI and JACOPO BRUTTINI. Working side by side with them I learned what it means to read 'for real' archaeological features on the field. I owe gratitude to all my colleagues from the University of *Siena*, who tolerated my archaeometallurgical frenzy, not to mention others, particularly Dr. MADDALENA BELLI, Dr. FRANCESCA GRASSI, Dr. LUISA DALLAI, Dr. CARLO CITTER and Dr. ROBERTO FARINELLI. I feel indebted also to my colleagues from the University of *Chieti*: Prof. MARIA CARLA SOMMA for her constant support, Dr. OLIVA MENOZZI and Dr. DOMENICO FOSSATARO for the wonderful 'confetti' and not only for them; then my Department roommates Dr. SONIA ANTONELLI and Dr. MARZIA TORNESE and all my students.

Now let's turn to Budapest and CEU. I thank my friends, my MA-year classmates Martin, Tiho, Mitko, Sasha, Zsolt, Adrian, Jucika, Marina, Agneska and many others, those of my Ph.D. years Cristian, Levan, Gusti, Lenka, Natasha, Vladimir…

FINALLY, MY FAMILY, MY WIFE ZAROUI AND MY KIDS ARECHI AND GIULIA SOFIA. WITHOUT THEIR SMILES AND LOVE MY ENTIRE LIFE IS NOTHING.

List of Figures

Table of Contents

0.0 Introduction

Pochi temi della storia occidentale hanno avuto la capacità di colpire non solo l'interesse degli studiosi ma anche l'immaginazione popolare con la stessa fortissima carica di suggestione delle migrazioni (o, dal punto di vista dei romani delle invasioni) barbariche, cioè di quel vasto fenomeno di spostamenti a catena, tra l'Asia e l'Europa, di popolazioni eterogenee che, a partire dal IV-V secolo d.C., finirono per stabilirsi in sedi diverse da quelle originarie, spesso sul suolo che era già appartenuto all'impero romano.[1]

The present study seeks to explore the contribution of the peoples of the *Barbaricum* to the shaping of early medieval technology in Europe, with a particular reference to iron-making. Within this general cultural framework, the case of Lombards will be analyzed in more detail, tracing the way their iron-making technological heritage developed: first, during their settlement on the Lower *Elbe* (first centuries AD) characterized by a Western Germanic technical culture, then, in Central Europe (AD 3rd/4th-6th), where they came into contact with a Celtic and provincial Roman sub-stratum, and finally in Italy (second half of AD 6th to 8th). At this stage, Lombard craftsmen, who possessed the full range of technical-artisanal skills of iron-production that were integral to western Germanic culture, would have come into contact with practitioners embodying the technical knowledge of the Mediterranean heritage. This encountering of material cultures seems to have resulted in reshaping of the entire economic structure of the peninsula.

This economic reshaping involved important developments in iron production as well. This kind of changes will be analyzed through the archaeometry of metallurgical production processes (from ore exploitation to the finished product), that is to say through *archaeometallurgy.* In fact, archaeometallurgical analysis of various iron artifacts, performed on Migration Period objects, are an integral part of this work and therefore, **Chapter 1** is devoted to clarifying the terminology and methodology of archaeometallurgy. Briefly, through archaeometallurgical investigation, the methods of natural sciences are applied for studying material culture. They provide the scientific tools –such as metallography, X rays analyses— necessary for the diachronic analysis of the "evolutionary processes" of human societies and their interactions with natural resources and not merely the means to establish typology and chronology of objects. Such an approach is extremely important since, through the analyses of the production processes, it is possible to gain information about the organization of production, the social strategy related to the appearance of standardization and craftsmanship specialization, and on the general economic framework, including distribution and consumption of the different artifacts, of a given society. [2] Moreover, through the archaeometallurgical analysis of the metal objects possessed and used by a given society, it is possible to fix the boundaries of the material patrimony this populace had produced, distributed and used and, thus, to attempt to understand the different levels of wealth and status within the very same human group. Thus, the focus of this research method is always on the evaluation of the different aspects of the constant and direct relationship between men and objects and, through them, between men and society. [3] This study will, thus, particularly emphasize the key role of technology and economy in reconstructing the social organization of pre-industrial civilizations.[4] However, this method of analysing human societies should not lead into the dangerous quagmire of reductionism and, thus, of promoting hasty generalizations.[5]On the contrary, it should help analyze the complex lines of development of pre-industrial societies.

In addition, since this study is concerned with interaction between cultures (material cultures and identity/ethnicity), migration, technology transfer and acculturation, basic terminology must inevitably be defined. Therefore, **Chapter 2** will lay the theoretical groundwork on how to define ethnic identities based on the analysis of the different artifacts that constitute a given material culture. Using this general theoretical

[1] C. Azzara, *Le invasioni barbariche* (Bologna: Il Mulino, 1999), 7.

[2] C. Lynne Costin, "Craft Specialization: Issues in Defining, Documenting and Explaining the Organization of Production," in M.B. Schiffer ed., *Archaeological Method and Theory* (Tucson: The University of Arizona Press, 1991), 1, 2.

[3] See, V. La Salvia, "Archaeometallurgy as a Source for the History of Changes and Developments within Medieval Daily Life," in G. Jaritz ed., *History of Medieval Life and the Sciences* (Vienna: Austrian Academy of Sciences), 113-14.; M. Shanks and Chr. Tilley, *Social Theory and Archaeology* (Cambridge: Polity Press, 1987), 79; M.C. Nelson, "The study of technological organization," in M.B. Schiffer ed., *Archaeological Method*, 88; J. Chapman and H. Hamerow, "Introduction," in J. Chapman and H. Hamerow eds., *Migrations and invasions in archaeological explanation* (Oxford: BAR, I.S. 664, 1997), 2.

[4] G. Gibbon, *Anthropological Archaeology* (NewYork: Columbia University Press, 1984), 180-81: "Technology refers to the material equipment and technical knowledge components of a techno-economic system. This is the way in which resources, technology and work are combined to satisfy the material requirements of human beings and social groups ... Regardless of the ultimate importance of the techno-economic subsystems, the collection of techno-economic-ecological data remains a basic task for archaeologists for several more mundane reasons. First, it is essential to reconstruct the techno-economic subsystem. Second, given the close interaction between socio-cultural subsystem and the reliance of archaeologist on indirect measures, techno-economic-ecological data have proven essential in many instances in modeling social and political organisations, settlements, interactions, exchange networks, and patterns of socio-cultural change. For instance, identifying the presence of particular social elements and political organisations may depend on the recognition of craft specialization, the redistribution of goods, the subsistence emphasis, wealth and other techno-economic traits."

[5] B.C. Burnham and J. Kingsbury, "Introduction," in B.C. Burnham and J. Kingsbury eds., *Space, hierarchy and society: interdisciplinary studies in social area analysis* (Oxford: BAR, I.S. 59, 1979), 1; M. Godelier, *L'Idéel et le Matériel* (Paris: Fayard, 1984), 71: "... matérialisme 'grossier,' l'économisme, qui ravale tous les rapports sociaux au rang d'épiphénomènes accompagnant des rapports économiques, eux-mêmes réduits à un ensemble de techniques d'adaptation à l'environnement naturel et biologique."

framework, the Chapter will proceed to the specific case of defining the identity/ethnicity of the Germanic peoples, then, more precisely that of the Lombards and the role of material culture within human groups. Here, the most recent scholarship on the relationship between Roman and the so-called Barbarian civilizations will be discussed and the contribution of archaeology of production to this debate will also be evaluated. In doing so, the chapter will suggest a shift in emphasis from the model of peaceful acculturation[6] to the more complex one of cultural integration/osmosis. This model of cultural interaction corresponds better, in my opinion, to the events that, between AD 5th up to 7th centuries, involved the inhabitants of the former Roman soil and the various people from the *Barbaricum*. It was, in fact, a framework of relationships (i.e. economic, military, needs for daily life) that cannot, to any extent, be regarded as a one way process as it is for acculturation. Acculturation, in fact, implies a 'simple' cultural change which occurs when one group adopts the culture of another group, among whom it lives.[7] Actually, even the long-lasting movement of goods and ideas, which had taken place since the Bronze Age and before, between the Mediterranean regions and northern Europe was not uniform, nor only conditioned by the interests of the supposedly 'culturally stronger,' usually southern, part of Europe.[8] As noted by K. RANDSBORG:

> The centuries of Classical Antiquity display many examples of rejections of Mediterranean norms on the part of, for example, the Celtic and, in particular, the Germanic region of the continent. The rise of particular Barbarian art-styles is a strong testimony to this phenomenon, which is also indicated by the nature of Roman imports, which in the Germanic parts of Europe were almost exclusively artefact types fitting readily into the local culture.[9]

In order to follow the path of these cultural exchanges this study will emphasize the role of the archaeological evidence, that is to say, material culture, and will not rely only on written sources:

> However, while it might be said that the written sources of the age tend to focus on issues such as honour and beliefs, they are virtually about values and, in particular, the needs of daily life. Settlement-archaeology and material culture relating the character and intensity of agriculture as well as the social and economic dimensions are seldom described in the written sources. This is especially the case in the Germanic world and other non-Roman realms of the Migration Period. Only Late Roman sources voice the concerns of the state in term of taxation, revenue and landed interest, throwing light on the economics, if only, in an indirect way.[10]

As a matter of fact, the presence of "Barbarians," especially as intruders and invaders, is actually too often invoked as the simplistic explanation for an otherwise seemingly inexplicable problem such as the breakdown of the Classical civilization.[11] Their role in the many developments within the Old Continent between Late Antiquity and the Early Middle Ages in technology and economy has been often underestimated, if not ignored. Human mobility has always been a major contributor to cultural development and change and migrations and invasions are, thus, important processes with potential explanatory status.[12] However, migration/invasion as explanatory model does not represent simplistic explanation since migration is a complex social strategy and not an automatic demographic scheme responding to crowding.[13]

Accordingly, this study will stress the importance of analyzing the different material cultures patrimonies: those typical of the *Barbaricum* and those of Mediterranean origin. These are the core elements that make it possible to trace the strategies for inter-cultural interactions and technological transfer between these two different worlds. For this purpose, as already suggested forty years ago by R.S. LOPEZ, *homo faber* and his manufacturing activities need to re-take a central position in the evaluation of the period between Late Antiquity and the early Middle Ages.[14] In this respect, the analysis of the iron production technology typical of the western Germanic populace, as a fundamental portion of its more general patrimony of material culture, will help to establish the 'original' technical background of each culture, the different levels of interaction, their impact on the development of technological skills and production strategies as well as on their integration into new economic and social contexts. Therefore, the reconstruction of the complex "scenario" comprising bi-lateral interactions and influences between Roman and Germanic cultures (for the Italian case the Lombards), is a more representative scholarly 'target' pursued in this work. In this way, the material basis for the general socio-economic environment of the so-called Roman-Barbarian kingdoms, finally formed in the interval between Late Antiquity and the early Middle Ages, becomes evident.

---

[6] Strong criticism on contemporary historiography's views on peaceful acculturation are in B. Ward-Perkins, *The fall of Rome and the end of civilization* (Oxford: University Press, 2005), 1-10, 169-83.

[7] For the definition of acculturation see, K.R. Dark, *Theoretical Archaeology* (London: Duckworth, 1995), 194-95.

[8] K. Randsborg, "The Migration Period: Model History and Treasure," in R. Hodges and W. Bowden eds., *The sixth century: production, distribution and demand* (Leiden/Boston: Brill, 1998), 61-62.

[9] Ibid.

[10] Ibid.

[11] R. Hodges, *Early Medieval Archaeology* (Gwynedd (UK): Headstart History, 1991), 11.

[12] J. Chapman and H. Hamerow , "Introduction," 1.

[13] H. Hamerow, "Migration theory and the Anglo Saxon identity crisis," in J. Chapman and H. Hamerow eds., *Migrations and invasions*, 35; D. Anthony, "Prehistoric Migration as social Process," in J. Chapman and H. Hamerow eds., *Migrations and invasions*, 21-22.

[14] R.S. Lopez, "Discorso inaugurale," in *Artigianato e tecnica nella società dell'altomedioevo occidentale. Atti della Settimana di Studio del Centro Italiano di Studi sull'Alto Medioevo, 18.(Spoleto 2-8-IV-1970)* (Spoleto: CISAM, 1971), 16. Lopez's advice, as noted again more recently by B. Ward-Perkins, should not only be considered an old fashion and useless suggestion since there is a trend in contemporary Late Antiquity studies to primarily interpret this period as a spiritual and mental world, almost to the exclusion of the secular and material one. About this topic see, B. Ward-Perkins, *The fall of Rome*, 171. On material culture and archaeology see, K.R. Dark, *Theoretical Archaeology*, 117-42; M. Shanks and Tilley Chr., *Social Theory*, 79-117; L. Lavan and W. Bowden eds., *Theory and Practice in Late Antique Archaeology* (Leiden/Boston: Brill, 2003), 113-32.

Through the lenses of archaeology of production other thorny concepts such as ethnicity and culture will also be thus scrutinized without dismissing all anthropological theories about ethnicity because of contemporary political correctness and/or concerns about unpleasant socio-political circumstances.[15] In the same vein, the current trend in archaeology which emphasizes the active symbolic role of material culture in building social relations and social strategies,[16] ignoring its primary economic and technical function, will be viewed with a critical eye. For example, the study of ethnogenesis is actually too often focusing only on the search, within the archaeological record, for signs/symbols of status, rank and social stratification. Thus, ethnicity is often used as a simple and convenient marker of a supposedly entirely symbolic process that, mediating the relations between the economic and political spheres, forms socially ordered differences in wealth, rank and status. Surprisingly, the traces of this symbolic strategy should be searched for and found within the archaeological record. Hence, artifacts (i.e. archaeological evidence) are mainly considered for their symbolic content which is often equated with their stylistic (aesthetic) features. In this way, the analysis of the symbolic/semiotic features of a given human society becomes the prevailing tool for interpreting their structures and developments. The entire realm of human economic and social evolution is, accordingly, restricted to the invention of meaningful strategies for reinforcing and enhancing the level of self-perception and ethnic cohesion. In this approach, style turns into a sort of iconography purposefully and consciously used by artisans mainly for generating a symbolic code which communicates to people information about ethnic affiliation and identity and which represents the main stream within the processes of social formation.[17] However, these beliefs are not necessarily true and, as a consequence, would need, in my opinion, fundamental revisions. First, the complicated functioning of the human socio-economic systems cannot be simplistically explained by reduction to a single component. The economic and social life of any human group, actually, consists of a complex set of interrelationships that, "as an extra somatic adaptive system," integrates a society and its ecological/natural environment and other socio-cultural systems. Consequently, culture is a structure based on the combination of people, things and habitats. The dynamic and articulated relationship between these different fields constitutes the real focus of the process of cultural formation. In this sense, any culture, being a system of relationships, is not simply shared by people but it is participated in by them.[18] Secondly, within such an environment, the origin of a given style should not be searched in the conscious creation of a symbolic code which would serve to identify the presence of specific self-conscious social groups within a broader social context. In contrast to this, any ethnic style is apparently generated by a much deeper source of human behavior which has more to do with the real process of manufacturing craft products (i.e. the relationship between men, tools and working matter) than with the invention of social and ethnic symbols. As a matter of fact, a given style is formed exactly when the choices of craftsmen in the processes of production become consistent and specific, selecting among various available technical options definite strings of technical operation. These technical preferences are largely based on the craft traditions the artisans grew up in as members of a distinct social group. Thus, the specific patterns of stylistic variations within material culture can be considered as markers of specific "ethnic units" since it is exactly this pattern of variations that is perceived as style.

Accordingly, ethnic style does not have to be regarded as a positive hidden/immanent force that, permeating the entire life of a human group, can be invoked as "deliberately" producing "meaningful" technologies and objects. On the contrary, the expression of any ethnic style may very well be essentially ordinary and omnipresent within material culture since stylistic attributions can be assigned to any trait of a given object regardless of whether it had been produced primarily for a non-utilitarian or an utilitarian use. Consequently, it is only the specific choice of a distinct technical option, among a range of equally available possibilities, that define a stylistic trend. As a matter of fact, this choice is essentially dictated by the prescriptions of a craft tradition that, in its turn, is strictly bound to the social group within which craftsmen have been and, eventually, continue to be culturally involved. Besides, there is a strong trend among artisans to conform to and perpetuate the technical options already selected by their craft traditions. This conservative attitude, then, fits with the general tendency to conform to and to perpetuate the entire cultural framework within which also craft traditions are cultivated and it permeates all aspects of social and cultural life. In this respect, this specific technical behavior has its own symbolic and iconic value as well. It is, however, an iconography of the commonplace mainly acquired unconsciously since it is customarily taught by insinuation and employed automatically. Craftsmen, in fact, do not need to intentionally and/or consciously invest their products with ethnic symbolism any more than they usually do for their gestures, eating habits or managing their family relationships. Actually, craft traditions are maintained through regular consistency during manufacturing processes, that is to say, through the constant respect of

[15] M. Banks, *Ethnicity: Anthropological Constructions* (London: Routledge, 1996), 50-51; see as well, E.R. Wolf, "Perilous Ideas: Race, Culture and People," *Current Anthropology* 35.1 (1994): 2; R. Narroll, "On Ethnic Unit Classification," *Current Anthropology* 5.4 (1964): 288: "A society is an actual system of social interaction … Social practice in most if not in all societies departs somewhat from social theory."

[16] This 'symbolic' trend is represented among the others by I. Hodder, "Economic and Social Stress in Material Culture Patterning," *American Antiquity* 44 (1979): 446-54; Id., *Symbols In Action* (Cambridge: Cambridge University Press, 1982); Id. ed., *Symbolic and Contextual Archaeology* (Cambridge: Cambridge University Press, 1982); Id., *The Present Past* (London: B. T. Batsford, 1982); P. Wiessner, "Style or Isochrestic Variation? A Reply to Sackett," *American Antiquity* 50.1. (1985): 160-66.

[17] As rightly pointed out by J.R. Sackett, "Style and Ethnicity in the Kalahari: A Reply to Wiessner," *American Antiquity* 50.1 (1985): 154.

[18] J.R. Sackett, "Style and Ethnicity," 205.

the rules prescribed by the craft tradition.[19] In a production environment as such and at a social level entirely built around verbal/oral communication rather than written tradition, the act of acquiring and conserving technical knowledge is tied with mechanisms that are certainly socially complex such as apprenticeship and discipleship. Learning takes place through observation and practice and verbal explanations would be quite limited since this practical and "operational" knowledge does not conceive of technical discourses as a separate moment different from "doing." Therefore, the constant presence of the "master," the only depositary of the secrets of the craft, is indeed indispensable: he is the only one capable of ensuring the continuation of the proper sequences of both individual operations and general technological procedures. Thus, technical traditions and crafts do last through time until the last craftsman leaves the workshop or dies without transmitting his own knowledge to others. Without this continuous transmission, the artisan's knowledge and experience, in fact, become lost. Actually, skills and technical tasks that are deeply tied to workshop organization should enjoy a continuity of existence beyond an individual life-span. The consistent transmission of knowledge through generations was, thus, the only way to keep workshops alive and active, maintain their tools and, most of all, the technical practices that represent the very core of a craft and the very essence of its persistence.[20]

Last but not least, it is also essential to remember that archaeologists remain (and should remain) dependent on material culture as the primary source of evidence and social organization is more difficult to interpret from the surviving evidence than either subsistence or settlement because it has fewer tangible manifestations: social organization does have fewer direct connections with material things. Actually, symbols, indicating positions within and membership in a given human group, should first be archaeologically identifiable; "unfortunately," archaeologists face both methodological and practical difficulties in this endeavor.[21]

To sum up, the first two chapters of this study will be devoted to methodological issues in order to establish the general framework for the entire research. The remaining Chapters will, in their turn, place this theoretical framework into practice, illustrating how production processes, craft traditions, social stratification and ethnic identities are in effect interconnected.

Chapter 3 will include a 'demonstration' that Lombards were an integral part of the Western Germanic *isotechne* [22] and how their traditional iron-making technologies were enriched through various contributions during their migration within central-northern Europe. The process of the birth of this craftsmen tradition will be followed through the analyses of the archaeological evidence, consisting of tools and workshops beside artifacts from grave goods. This approach will enable to establish the 'original' technical background of each culture, the stages of interactions, their impact on the development of technological skills and production strategies as well as on their integration into a new economic and social context. The core of this methodology will also make a great use of archaeometric analysis, in this way, helping to reconstruct virtually beyond a shadow of doubt, the technical strategies and the production processes used for iron manufacturing.

Finally, **Chapter 4** contains an exploration of the ways portions of this Western Germanic technical knowledge penetrated the Mediterranean areas of Europe, introducing instruments and modes of production that were alien to the Graeco-Roman world. This process as related to iron technologies will be studied in particular as it evolved in Italy. Here, the contribution of material culture of Lombard origin or which reached the peninsula through the mediation of Lombards seems to have contributed to the formation of a new production system which was too complex to be explained as simply the result of a juxtaposition of Germanic and Late Latin elements. Rather, it underwent a real reconstruction/restructuring based on new criteria after the profound economic crisis in the long period between the end of the AD 5th and the first half of the 7th century.[23]

[19] J.R. Sackett, "Style and Ethnicity," 155, 157, 158; see also K. E. Lloyd, "Behavioral anthropology: A review of Marvin Harris' Cultural Materialism," *Journal of Experimental Analysis of Behavior* 43.2 (March 1985): 279–87.

[20] V. La Salvia, "Gap or Continuity? Mining in Early Medieval Italy," in G. Magnusson ed., *The Importance of Ironmaking: Technical Innovation and Social Changes. Papers presented at the Norgberg Conference on May 8-13, 1995* (Stockholm: Jernkontorets Bergshistorika Utskott, 1995), 262, 266; Id., "La fabbricazione delle spade delle Grandi Invasioni. Contributo alla riflessione sulla storia del *processo diretto* nella lavorazione del ferro," *Quaderni Medievali* 44 (1997): 31-32, 50-54; Id. "*De la Pirotechnia* of Biringuccio. A Key to the Craftsmen Production. Technical Questions and Aspects of Organisation of Production," in M. Lette and M. Oris eds., *Technology and Engineering. Proceedings of the XX International Congress of History of Science (Liège, 20-26 July, 1997)*, Vol. 7 of *De Diversis Artibus, Collection of Studies from the International Academy of the History of Science*, by E. Poulle and R. Halleux (Turnhout: Brepols, 2000), 13-25; M.Cima, *Archeologia e Storia dell'industria di una valle* (Florence: All'Insegna del Giglio, 1981), 16 and H. Bartlett Wells, "A contribution to the history of the heat treatment of steel," *Journal of Arms and Armour Society* 6.8 (dec. 1969): 231 about the death of the former Toledo swordsmith *Dionisio Corriente* in 1733 resulted in a gap of the manufacture of blades by the Toledo method for about hundred years.

[21] O. Bryony. *Anthropology for archaeologists: an introduction* (London: Duckworth, 1981), 135. S.J. Pierpoint, "Who's who in the Northern British Bronze Age," in B.C. Burnham and J. Kingsbury eds., *Space, hierarchy and society*, 35; B.F. Williams, "Of Straightening Combs, Sodium Hydroxide, and Potassium Hydroxide in Archaeological and Cultural-Anthropological Analyses of Ethnogenesis," *American Antiquity* 57.4 (Oct., 1992): 608-12.

[22] For the meaning of *isotechne* see J. David, *L'outil*, in *Typologie des sources du Moyen Age occidental*, 78 (Turnhout: Brepols, 1997), 58.

[23] A.M. Romanini, "Il concetto di classico e l'altomedioevo," in M.G. Arcamone *et al.*, *Magistra Barbaritas* (Milan: Garzanti/Scheiwiller, 1984), 668-69 where the early medieval period is regarded as an *unicum* and not an interlude between the Roman and the Medieval (post AD 1000) civilizations. Its own character, thus, does not stem from a simple overlapping of Germanic and late Latin elements but from the presence of positive features specifically consistent with the economic, juridical, political and cultural situation of the given period. About the economic crisis between AD 5th and 7th see, B. Ward-Perkins, *The fall of Rome*, 41, 87, 95, 117, 128; specifically on the crisis of agricultural production

This crisis, nevertheless, should not have lead to an interpretation of the early Middle Ages as a simple and static period. Nothing is farther from the truth. As an example, archaeological investigation of the early medieval countryside in Italy has demonstrated the existence of an increasing complexity in the mode of exploiting natural resources.[24] Therefore, in my opinion, the debate focused on *Continuity vs. Disruption*, apparently, is no longer an appropriate model for discussing the changes and developments between Late Antiquity and the early Middle Ages. In contrast, it would be better to distinguish between the forms of concrete and particular technological/cultural and/or economic continuity (i.e. regional and/or contingent) and the real and meaningful scale of this continuity or discontinuity within a given archeological context.[25]
In this respect, the investigation of the development trajectories of iron-making technology, through the use of archaeometallurgical analyses can contribute to a better understanding of the dynamics of interaction between these different material cultures.

However, notwithstanding a massive program of archaeometallurgical investigation, involving analysis of iron artifacts from the La Têne period to the Middle Ages, the time span between Late Antiquity and the Migration Period remains still unexplored. Consequently, the blacksmithing techniques employed in these epochs are not fully understood.[26] The aim of this study is, thus, to fill in some of these gaps in our understanding of blacksmithing across all societies of this time period. As has recently (1997) been pointed out by R. STUPPERICH *et al.* for non-ferrous metallurgy, such an approach is crucial for understanding the different technological paths that defined the early medieval milieu of metal production. It is also important to understand, how much of this production variability was related to commercial exchange of objects, technology transfer or migration of craftsmen and how much to variability in strategies of raw material exploitation.[27]

see R. Hodges, *Early Medieval Archaeology*, 12; Id., "Rewriting the Rural History of Early Medieval Italy: Twenty-five Years of Medieval Archaeology Reviewed," *Rural History* 1 (1991): 31-32; G. Barker and J. Llyod eds., *Roman Landscapes. Archeological Survey in the Mediterranean Region* (London: British School at Rome, 1991); R. Hodges, *Wall-to-Wall History: The Story of Roystone Grange* (London: Duckworth, 1991), illustrating the developments in later Roman Agriculture in the West.

[24]R. Hodges, "Rewriting the Rural History of Early Medieval Italy," 31.

[25]L. Lavan, "Late Antique Archaeology: an Introduction," in L. Lavan and W. Bowden eds., *Theory and Practice*, VII. A good example of this analysis of complexity within a specific geographical and economic framework is in R. Francovich and R. Hodges, *From Villa to Village. The Transformation of the Roman Countryside in Italy, c. 400-1000* (London: Duckworth, 2003).

[26] R. Pleiner, "Spätkaiserzeitliche und Völkerwanderungszeitliche Stahlklinge aus Nordwestböhmen im Lichte der Metallographie," *Jahrbuch der Römisch-Germanischen Zentralmuseums Mainz* 35.2 (1991): 602: "Die europäische Archäologie verfügt heute über Tausende von metallographischen Untersuchungen an Eisenwerkzeugen aus verschiedenen geschichtlichen Period seit der Laténezeit bis ins Mittelater. Eine Forschungslücke ist jedoch in der jüngeren römischen Kaiserzeit und in der Volkerwanderungszeit in Mitteleuropa zu beobachten." and 606: " ... so ist die Situation für die Völkerwanderungszeit noch schlimmer als für die unmittelbar vorausgehende jüngere römische Kaiserzeit. Zu wenig ist uns über die Schmiedetechnologie jener Zeit bekannt. In Böhmen wurden z.B. nur einige Lanzenspitzen und ein Schieldbuckel aus der Körpergräberfeld in Klučov untersucht (Pleiner 1962, 119-121, T. XXXVIII-XL). Nur einfache Erzeugnisse aus Schmiedeeisen oder Weichstahl wurden festgestellt. Auch in Polen wurde dieser Periode bisher nur geringere Aufmerksamkeit gewidmet. Im Falle eines Messers aus Lubieszewo in Pommern wurde Eisen-Stahl-Verbund beobachtet (Piaskowski 1976, 183). Die Gesamtlage des Schmiedegewerbes bleibt jedoch noch unübersichtlich."; on the same matter see, E. Pernicka, "Förderschwerpunkt Archäometallurgie –Eine Bilanz," *Archäologisches Nachrichtenblatt* 3.1 (1998): 82.

[27]R. Stupperich, "Export oder Technologietransfer? Beobachtungen zu römischen Metallarbeiten in Germanien," in C. Bridger and C. von Carnap-Bornheim eds., *Römer und Germanen –Nachbarn über Jahrhunderte* (Oxford: BAR, I.S. 678, 1997), 19-24; P. Hammer, H.-U. Voß, "Metallkundliche Untersuchungen an römischen und germanischen Funden des Elbegebietes" in C. Bridger and C. von Carnap-Bornheim eds., *Römer und Germanen*, 25-28; W.R. Teegen, "Zur Metallversorgung germanischer Buntmetallschmiede am Beispiel des Pyrmonter Brunnenfundes und des Moorfundes von Strückhausen" in C. Bridger and C. von Carnap-Bornheim eds., *Römer und Germanen*, 29-35.

Chapter 1
Archaeometallurgy: methods and problems

L'archeologia, che alla storia fornisce ciò che i documenti non sono in grado di testimoniare, vive di ipotesi e di modelli dove le conferme sperimentali spesso cozzano contro l'esigenza di rigorose diacronie. La storia dei popoli senza scrittura è per molti versi anche una storia senza tempo. L'archeologia si interessa per definizione agli oggetti, che in gran parte sono oggetti tecnici, e per questo motivo non si può guardare alla storia dell'uomo senza distogliere lo sguardo dalle cose, le quali risultano intimamente connesse con lo sviluppo sociale dell'uomo.[28]

Il faut prendre conscience du fait que l'accession aux métaux représente une étape technique majeure, comparable à l'accession à la vapeur, ou à l'életricité, c'est-à-dire un état qui peut entraîner des modification capitales dans le rapports de l'homme et du milieu naturel. Étant donné que les plus vieux métallurgistes nous apparaisent comme des agriculteurs-éleveurs, on peut se demander si, avant eux, agriculture et élevage étaient pratiqués.[29]

Archaeology is not a natural science nor do its roots lie in it. Archaeology belongs in the realm of historical inquiry. Deriving historical information through excavations is a complex process and far from being a simple attempt to structure chronologies through seriation of objects. In this respect, the use of natural scientific methodologies plays a crucial role in reconstructing a more complete picture of past societies. [30] Methods borrowed from the natural sciences represent important tools for fully extracting all possible information from the excavation data. [31] The combination of archaeology with the natural sciences is known as archaeometry. Using these methods, archaeological research can begin to grasp the history of both the production and consumption of artifacts and their ongoing impact on the environment.[32]

Archaeometallurgy as a branch of archaeological and historical research can be primarily regarded as the archaeometry of metallurgical producing processes (from ore exploitation to the finished product). Through this kind of research, thermomechanical, thermical (quenching and tempering), and thermochemical (*Cementation* or *carburization*) treatments as well as types of welding can be evaluated. Moreover, archaeometallurgy also subsumes the history of the metallurgical crafts (craftsmen's rank and status, the location of workshops and their organization). Archaeometallurgical research also concerns the "economy of metals." This refers to the investigation of the geographical distribution of the evidence and its quantitative analysis in order to evaluate the commercial network, the use of woods (essential in charcoal production) and water, and the economic and ecological impact of the metallurgical production in a given area and period. In other words, archaeometallurgy is a branch of archaeology concerned with production, circulation, and consumption of metal goods, with a final goal of reconstructing the economic and social history of metal production, on the basis of the investigation of the different forms of means and organization of production.[33]

Metals and metal alloys have always represented the principal means of achieving prosperity and military power in pre-industrial societies. Accordingly, it is not surprising that during the Middle Ages, there existed basically two main indicators of prosperity beside the bellowing of a cow, the sound of a plough (an iron tool), and the metallic din of hammers working in the smithy.[34] Iron in particular, thanks to its wide geological distribution and to the fact that tools can be produced more cheaply from it than from other metal raw materials, enabled the diffusion of metal objects among all social strata. Thus, the evaluation of the general level of metal producing processes reflects the efficiency of the products themselves. As a matter of fact, highly specialized production processes create the most functional tools. Investigations into the quality of the metal equipment used by peasants, craftsmen, and warriors, permit a general picture of the quality of life of a given populace to be drawn.[35]

Modern science has already refined several scientific methods such as physical, chemical, biological, and metallographic analyses, developed as tools for checking and improving new technologies. These methods can easily be applied to the evaluation of pre-industrial artifacts since the vast majority of these objects share the raw materials of their production with contemporary ones. Thus, the main connection between contemporary scientific research and the study of the products from past ages consists of the continuity of use of the same raw material obtained through the exploitation of natural resources. [36] Any raw material (or any archaeological object) includes and preserves within itself the traces of the working processes it has passed through during its manufacture; information about its geographical origin, that is to say the region where its raw material originated as well as chronological information which is completely

[28] V. Marchis, "*Homo faber*. Le tecniche della produzione nel mondo antico," in V. Castronovo ed., *Storia dell'economia mondiale*, 1, *Permanenze e mutamenti dall'antichità al Medioevo* (Rome-Bari: Laterza, 1996), 27.

[29] A. Leroi-Gourhan, *Milieu et Techniques* (Paris: Éditions Albin Michel, 1973), 316.

[30] E. Pernicka, "Förderschwerpunkt Archäometallurgie," 79.

[31] R. Francovich, "Premessa," in T. Mannoni and A. Molinari eds., *Scienze in Archeologia* (Florence: All'Insegna del Giglio, 1990), 8.

[32] A. Molinari, "Il ciclo di lezioni sulla ricerca applicata in Archeologia ed un progetto di laboratorio," in T. Mannoni and A. Molinari eds., *Scienze*, 12.

[33] E. Pernicka, "Förderschwerpunkt Archäometallurgie," 80: "Das Forschungziel der Archäometallurgie ist die Aufklärung der Geschichte der Metalle, ihrer Gewinnung, Verarbeitung und Verbreitung, sowie deren Auswirkungen auf Umwelt und Gesellschaft. In diesen grossen Rahmen lassen sie praktisch alle Disziplinen der Natur- und Ingenieurwissenschaften sowie Altertums-und Geschichtwissenschaften integrieren."

[34] E. Power, *La vita nel Medioevo* (Turin: Einaudi 1966), 11.

[35] V. La Salvia, "Archaeometallurgy as a Source," 105-15; R. Delort, *La vita quotidiana nel Medioevo* (Rome/Bari: Laterza 1997), 15.

[36] T. Mannoni, "Introduzione all'Archeometria," in T. Mannoni and A. Molinari eds., *Scienze*, 31.

independent from the context of its discovery. This is known as absolute dating. It is, thus, possible to reconstruct the production processes and the strategies used in choosing raw materials but not the nature of apprenticeship or the mentality of the craftsmen.

By focusing on the borderline between nature and culture, that is, on the methods historically used by human societies to exploit natural resources, archaeometallurgy is forced to use a multidisciplinary approach from the very beginning. In fact, in order to control and manage all the different competences (i.e. geology; metallurgy, archaeology, history) involved in this field of archaeological investigation it is essential to organize procedures for working teams.[37]

The metallurgical inheritance of the Middle Ages from both the Classical world and the Germanic technical patrimony included various metallurgical techniques and tools. During this period some of these were made more efficient, some were invented *ex novo*, some were forgotten and not all were equally efficient. Particularly during the Early Middle Ages, the results of their application were not at all geographically or technically homogeneous. However, the principal means of exploiting many natural resources were always based on the use of fire and iron tools.[38] Thus, many questions can be studied using this method of evaluating the archaeological evidence pertaining to the Middle Ages and the Migration Period in particular. These include the problem of the gap or the continuity with respect to technology between Late Antiquity and the Early Middle Ages, the types of organization of production (whether home-produced, craftsmanship, or pre-industrial production), and circulation of products (for example differences between towns and the countryside).[39]

Within the framework of this study, the results of various archaeometallurgical analyses carried out on iron objects will be presented. A full understanding of this archaeometric methodology is, thus, essential for grasping the final goal of the entire study. It is, therefore, necessary to briefly explain the ironmaking technology of the period and the scientific methods through which modern scholars can reconstruct it.

The archaeometallurgical methodologies used within this investigation are:

a) Metallographic examinations under a *Scanning Electron Microscope* (*SEM*) and/or an *Optical Metallographic Microscope.*

b) X-ray analyses.

Such methodologies allow the study of the internal composition and structure of the metal, the form of which is evidently the result of the working processes it was subject to as it was being manufactured. Whereas microscopic analyses imply a partial destruction of the artifacts through the cutting off a small sample, X-ray investigation is a non-invasive method that leaves the object intact.[40]

c) In order to evaluate the quality of an iron artifact it is important to be able to measure its hardness. The hardness of a metal is measured by its resistance to indentation. The metal is indented under a known load using a small steel bar (as in the Brinell test) or a square-based diamond pyramid (as in the Vickers test). In the Vickers test the result is given as the *Diamond Pyramid Number* (DPN) or H_V. For the other test there is a Brinell Hardness Scale (H_B).[41]

As far as iron production technology is concerned, the principal aspects of this complex process are the following:

During the period between Late Antiquity and the Early Middle Ages, iron was produced through the *direct method*, that is to say, through reduction of the oxides (between 800 and 1150° C). This process made it possible to melt veinstones and inclusions, such as siliceous gangue, and obtain a spongy *bloom* composed of a compound of iron oxides and slag. This *bloom* had to be worked further before it was possible to have a mass of useful solid iron. However, the conditions of reduction resulting from the lower temperature and the CO/CO_2 ratio, implies that there was still a large quantity of iron (FeO) within the slag, often more than 40%, and that there was no melting of metallic iron. The chemical principle of reduction is based on the reducing reaction of the carbon oxide (CO) on the iron oxide at high temperatures. The combustion of charcoal, aided by the action of an air draft, produces carbon dioxide (CO_2) which produces carbon oxide at temperatures of more than 1000° C,[42] thanks to the great quantity of charcoal. Once in contact with the mass of iron oxides, this carbon oxide acts as an energy-reducing factor, thus:

$Fe_2O_3 + CO \text{ ----> } 2\ FeO + CO_2$

$2Fe_2O_3 + CO \text{ ----> } 2\ Fe_3O_4 + CO_2$

$Fe_3O_4 + CO \text{ ----> } 3\ FeO + CO_2$

$FeO + CO \text{ ----> } Fe + CO_2$

The reducing action of oxygen on the iron sesquioxide begins at a temperature of approximately 200°C and usually produces iron monoxide which is always reduced by carbon oxide at temperatures of approximately 850°C. Small quantities of iron sesquioxide produce ferrous-ferric oxide which is reduced at a starting temperature of 600°C. (**see figure 1**).[43] A radical change in iron making

[37] D. Morin "Sauvegarde et protection des sites archeologique miniers. L'exemple de la France de l'est," in G. Magnusson ed., *The Importance of Ironmaking*, 311, 314.

[38] V. La Salvia, "Archaeometallurgy as a Source," 105-15.

[39] A. Molinari, "Il ciclo di lezioni," 16.

[40] J.I. Goldstein, H.Yakowitz, *Practical scanning electron microscopy, electron and ion microprobe analysis* (New York: Plenum Press, 1975); F. Borile, C. Dondolato, "Il microscopio elettronico a scansione e le sue applicazioni in metallurgia," *La metallurgia italiana* 7/8 (1973): 425-34; V. La Salvia, "Archeometallurgia," in R. Francovich and D. Manacorda eds., *Dizionario di archeologia* (Rome-Bari: Laterza), 18-24

[41] D.A. Scott, *Metallography and Microstructure of Ancient and Historic Metals* (Singapore: Tien Wah Press Ldt, 1991), 2, 77.

[42] Burning of charcoal produce mostly CO and somewhat CO_2 Afterward, CO_2 reacts with C from charcoal: $CO_2 + C$ 2CO, known as Boudoard reaction.

[43] M. Cima, *Archeologia del ferro* (Brescia-Turin: Edizioni Grafo-Edizioni Nautilus, 1991), 17; J.F. Healy, *Mining and Metallurgy in the Greek and Roman World* (London: Thames and Hudson, 1978), 182; G. Sperl, "Il sentiero del ferro," in *Dal basso fuoco all'altoforno. Atti del I*

occurred during the Early Modern period when cast iron was obtained through the use of blast furnaces (the reducing conditions were improved by construction elements and intensive blast) and later of coke (instead of charcoal). In this way, the melting point of iron was finally reached.[44]

After the successful reduction of iron through the *direct method*, another problem to be resolved was certainly enhancement of the low carbon content of this soft bloomery iron as well as to create uniformity in the carbon distribution in the iron's structure. These actions were taken in order to improve the quality and the efficiency/functionality of the final product increasing its capacity to properly resist mechanical stresses. The process to which iron was subjected in order to produce steel, a carbon-iron alloy, is known as *carburization* or *Cementation*. Moreover, in order to enhance the homogeneity of iron with respect to carbon diffusion, other procedures such as *annealing, quenching*, and *tempering* were mastered on swords, knives and many other iron tools.

Carburization or *Cementation*: By reheating, always below melting point, and hammering wrought iron at a high temperature and keeping its surface in close contact with charcoal and/or with another material able to liberate carbon (any carbonaceous matter) when heated and through slow cooling, it is possible to combine iron and carbon, giving a true steel structure to iron. At this stage, the structure of the metal may be composed of the following three elements depending on the quantity of carbon diffused into the iron: 1) *Ferrite* (a solid solution of carbon inside the iron) which at ambient temperature comprises not more than 0.02% rich in carbon. It is somewhat fragile and relatively soft (90 Brinnell). Its micrographic feature consists of relatively large polyedric grains; 2) *Cementite* (corresponding to the phase Fe_3C) is remarkable for its high level of hardness (840 Brinell) and, consequently, for its brittleness. The micrographic feature of this alloy appears as needles on the borders of grains inside a *ferritic* and *pearlitic* structure; 3) *Pearlite,* a lamellar compound of the previous two structures with a carbon content of 0.8% and a hardness of 200 Brinell. Steel with this very carbon content is called eutectoide; steel with a lower carbon content is hypoeutectoid and that with a higher one is hypereutectoid. It is clear that the presence of *Ferrite* depends on the level of carburization which leads to the fact that there is no *Ferrite* in the hypereutectoid stage with the result that a completely *pearlitic* strucuture is formed. This metal would be hard but also britlle.

The process of this complex operation depended greatly on the intuitive instincts of the skilled craftsmen, certainly made keener by repeated observation of the phenomena and, therefore, by a continuous accumulation of empirical experience.[45] This experience consisted basically of maintaining the constant action of the carbon on the iron by regulating the temperature appropriately and by controlling the length of time the iron was subjected to heat treatment.[46]

Annealing: Work-hardening is accomplished by hammering iron at room temperature, an action that increases the metal's hardness value but decreases ductility, that is to say the metal's capacity to be drawn into threads or hammered into sheets. To avoid this problem the metal must be annealed, that is to say, heated up to bring about recrystallization. This heat-treatment process is usually carried out to soften the material and to allow further deformation.[47] Practically, this heat treatment consists of the heating of a metal or alloy to a predetermined temperature, holding it at this temperature for a certain time, and then allowing it to cool slowly to room temperature. This method was first discovered around 5000-4500 BC and used in the working of copper. The main goal of this working procedure is to improve ductility and reduce brittleness. Thus, the structure of brittle steel can also be modified by this method and given a degree of softness. In the case of iron, the necessary temperature in order to master annealing ranges from between 700°C to 900°C, depending on the type of steel required. Moreover, this process was mastered intermittently during the working of an artifact to restore ductility lost through repeated hammering or other treatments. Annealing also relieves internal stresses in the structure of a given metal. During Antiquity and the Middle Ages, the success of annealing depended exclusively on the skill of the craftsmen: there were no carefully regulated modern annealing furnaces.[48]

Quenching and Tempering: It is possible to change the quality of metal by re-heating. At room temperature after carburization the iron's structure is still not very homogeneous with regard to the carbon distribution. But when the metal is re-heated at a high temperature, while the iron is still solid, a moment occurs during which the carbon dissolves entirely into the iron. This alloy is called *Austenite*. When slow cooling occurs, the *Austenite* decomposes and *Ferrite, Cementite* and *Pearlite* take up their basic position in the iron structure once more. In contrast, if a quick and immediate cooling occurs, there is insufficient time for the *Austenite* to be dissolved, and in

Simposio Valle Camonica 1988: La siderurgia nell'antichità, published in *Sibrium*, 20 (1989): 17-18; and L. Salvi, "La fusione del ferro all'uso catalano in Campania," in *Dal basso fuoco all'altoforno*, 318; concerning the bloomery process to obtain iron, see also H. Cleere and D. Crossley, *The Iron industry of the Weald* (Leicester: University Press, 1985), 31-51; R. Pleiner, *Iron in Archaeology* (Prague: Institute of Archaeology of the Czech Academy of Sciences, 2000), 131-40.

[44]E. Tholander, "Did the blast furnace occur before the 16th Century?," in *Dal basso fuoco all'altoforno*, 111 and ff.; M. Cima, "Il cannecchio bresciano tra forno a manica ed altoforno," in *Dal basso fuoco all'altoforno*, 275 and 278; and G. Tanelli, M. Benvenuti and I. Mascaro, "Aspetti giacimentologici dei Minerali estratti in Età Preindustriale," in R. Francovich ed., *Archeologia delle Attività Estrative e Metallurgiche* (Florence: All'Insegna del Giglio, 1993), 272.

[45]C. Panseri, *Ricerche metallografiche sopra una spada da guerra del XII sec. Quaderno 1* (Milan: Centro per la Storia della Metallurgia, 1954), 19; V. La Salvia, "La fabbricazione delle spade," 50-54.

[46] F. Wewer, "Das Schwert im Mythos und Handwerk," *Arbeitsgemeinschaft für Forschung des Landes Nordrhein Westfalen* 91 (1961): 36.

[47]D.A. Scott, *Metallography*, 3, 137.

[48]J.F. Healy, *Mining and Metallurgy*, 231.

this way the iron saturation of carbon (0.8% eutectoid) is maintained. The resulting needle-like structure is *Martensite*. This kind of worked metal is very hard (710 Brinnell). When cooling is not quick enough, the *Austenite* is partially dissolved, resulting in a lamellar and thin nodular compound of *Ferrite* and *Cementite* called *Troostite* (350 Brinnell). *Sorbite* is another decomposition product of *Martensite* and consists of fine particles of *Cementite* in *Ferrite* matrix. However, this structure may also consist of rounded *Cementite* not necessarily deriving from *Martensite*. Moreover, it is clear that depending on the original carbon content of the re-heated metal (that is, whether it is hypo- or hypereutectoid), the final structure of the metal will change. It will be more or less hyper-saturated (with carbon) and consequently, more or less hard. In wrought low-carbon steels, *Widmanstätten* structure may occur as a result of the working process or deliberate heat treatment during manufacturing.
The aim of this whole process is the attempt to fix the various properties possessed by steel at different temperatures in the cold structure of iron tools. At this stage, steel is very hard but still brittle: by subjecting the quenched metal to consistent heat treatment (between 220 and 400° C), that is *Tempering*, it is possible to reduce hardness while retaining strength. This same operation is controlled by observing the steel's surface color during the re-heating and cooling until the required color appears (for swords it is purple at 270-290 C). Good steel was obtained by alternating the *Quenching* and *Tempering* processes on the given object.[49]

As already mentioned, all these processes leave specific 'signs' in the structure of iron tools and, thus, can be detected through metallographic analyses. Therefore, these archaeometallurgical investigations allow evaluation of the level of the working processes used during the manufacturing and of their effect on the final products. In this way, they provide crucial information for understanding the strategies involved in developing blacksmithing technology in the period between Late Antiquity and Early Middle Ages. Moreover, since the history of metals and the history of civilization are inseparable, each depending on the other for its development, the investigation of the level of metal production is extremely important in evaluating the overall level of technology in a given place and/or time.[50] This is particularly true for the Early Middle Ages when iron production, and particularly weapons' and swords' production, was regarded as primary and essential.[51] The best metallurgical knowledge of these periods was, thus, evident in the work of the best blacksmiths.[52]

[49] J.F. Healy, *Mining and Metallurgy*, 31-36; J.F. Finó, "Notes sur la production du Fer et la fabrication des Armes en France au Moyen Âge," *Gladius* 1 (1964): 52-53; E. Salin, *La Civilisation merovingienne d'apres les Sepoltures, les Textes, les Laboratoires*, 4 vols. (Paris: Edition A. et J. Picard et C., 1949-1959), vol. 3, 9-11; J. Bronowski, *L'ascesa dell'uomo* (Milan: Fratelli Fabbri Editori, 1976), 125-26 discusses the hypersaturation of iron thanks to the diffusion of carbon into it. See also D.A. Scott, *Metallography*, 20-21, 33, 35-36.

[50] A. J. Wilson, *The Living Rock* (Cambridge: Woodhead Publishing Ldt., 1994), xiii.

[51] An interesting discussion on this matter with specific bibliography is in M. Arnoux, *Mineurs, férons et maitres de forge* (Paris: Editions du CTHS, 1993), 11-26.

[52] E. Salin, *La Civilisation merovingienne*, 5; C. Panseri, *Ricerche metallografiche*, 9; Id., *La tecnica di fabbricazione delle lame d'acciaio presso gli antichi. Quaderno 2* (Milan: Centro per la Storia della Metallurgia, 1957), 5. See also I. Bóna, *The Dawn of the Dark Ages: The Gepids and the Lombards in the Carpathian Basin* (Budapest: Corvina Press, 1976), 9: "... duel between warlords or the death of one of them decided the outcome of a battle - sometimes even the fate of a whole people."

Chapter 2
Vanishing Barbarians versus everlasting Romans. An introduction to the study of Migration Period Material Culture.

Vorrei solo osservare che questa nuova presentazione dei problemi non sembra essere meno ideologicamente condizionata di quella che faceva riferimento ai contrapposti principi di romanità e germanesimo quali costanti strutturali della civiltà europea e criteri di spiegazione della sua dinamica storica. Mi pare infatti che la storiografia recente tenda a descrivere i fenomeni del passato in modo consentaneo ai problemi dell'odierna società occidentale avanzata, che si avvia a diventare multietnica e si preoccupa di attenuare il peso dei contrasti di civiltà, esaltando invece le ipotesi e le speranze di integrazione fra gruppi etnici e culturali posti a contatto dalle nuove migrazioni.[53]

The importance of the first of these, the barbarian influence, has been far too little understood even by those who have dabbled in the history of technology. Students of the fine arts have only recently led the way towards an appreciation of the essential unity and originality of that vast northern world of so-called 'barbarians' which, in ancient times, had its focal point on the plains of Russia and of Western Siberia, but which extended from the Altai mountains to Ireland: we are beginning to learn how profoundly it affected the aesthetic expression of the Middle Ages. But even before the Germanic migrations, these barbarians had begun to influence Roman technology, and in later centuries they contributed many distinctive ingredients to medieval life.[54]

2.1.0 Material Culture, Ethnic Identity, Migration and Technological Transfer: the Contribution of the "Vanishing" Barbarians to European Early Medieval Technology. Some Preliminary and Methodological Remarks.

In this study, the term 'Lombard(s)' refers to one specific portion of this larger human group which was comprised of diverse peoples whose culture falls under the common denominator of Western Germanic. Characteristics of this greater human group cannot and should not be analyzed and defined exclusively through the ethnogenetic interpretative model which is based mainly on analyses of the written sources and considers archaeological data only through the lens of such written sources. In fact, archaeological evidence cannot be directly connected to ethnogenetic matters.[55] On the contrary, the diachronic development of a culture should be studied primarily through the archaeological analysis of features from settlements and villages. The same approach should be used in order to reconstruct the main lines of development in animal breeding, agricultural and crafts production which constitute the economic base of any human group, that is to say, the material basis for any culture.[56] In this way, the 'Lombard' migration should

[53]P. Delogu, "L'editto di Rotari e la società del VII secolo," in J. Arce and P. Delogu eds., *Visigoti e Longobardi* (Florence: All'Insegna del Giglio, 2001), 331.

[54]L. White Jr., "Technology and Invention in the Middle Ages," in Id., *Medieval Religion and Technology. Collected Essays* (Berkely, Los Angeles, London: University of California Press, 1978), 4.

[55]As recently written by W. Pohl, "Le identità etniche nei Ducati di Spoleto e Benevento," in *I Longobardi dei Ducati di Spoleto e Benevento. Atti del XVI Congresso internazionale di studi sull'alto medioevo. (Spoleto-Benevento 20-27-X- 2002)* (Spoleto: CISAM, 2003), 83: "In conseguenza non abbiamo un accesso diretto alla definizione di identità storiche attraverso criteri oggettivi, ma dobbiamo trovare una via indiretta, un'ermeneutica dei segni, percezioni, azioni individuali e collettivi ..."

[56]I. Rouse, "The place of 'peoples' in prehistoric research," *Journal of the Royal Anthropological Institute of Great Britain and Ireland,* 95.1 (1965): 7: "I do not believe that one can properly reconstruct a people's culture until one has determined when and where the people lived, for one will not be able to understand the culture until has acquired some knowledge of its physical and cultural environment." Moreover, R. Francovich has recently discussed the evolution of the Italian and European cultural landscape and rural settlement during the early medieval period from a historical-technological, or more precisely the archaeology of a production-based point of view. He suggests that archaeological data permits diachronic explanation without the need to impose other paradigms on them developed from other branches of historical research. These are more often than not informed by written sources. However, their conclusions cannot be directly imposed on archaeological evidence. Archaeology, which uses data which are generated in very different ways must be allowed to create its own interpretative, more appropriate, models of the historical process. See R. Francovich, "Villaggi dell'altomedioevo: invisibilità sociale e labilità archeologica," in M. Valenti, *L'insediamento altomedievale nelle campagne toscane. Paesaggi, popolamento e villaggi fra VI e X secolo* (Florence: All'Insegna del Giglio, 2004) x: "... evidenti segnali di afasia fra storici ed archeologi, questi ultimi non propensi a delineare quadri interpretativi generalizzanti, partendo dai loro singoli momenti di approfondimento, e gli altri, soprattutto nell'ultimo trentennio, propensi ad offrire un quadro talvolta contraddittorio, ... e 'derivante' sostanzialmente da un assetto tardo romano."; xii-xiii "Appare chiaro che l'archeologo che appiattisse un inquadramento dei dati materiali prodotti dal proprio lavoro sul campo entro modelli costruiti sulle fonti scritte si priverebbe di strumenti essenziali, tali da escludere interpretazioni innovative anche a livello storiografico, e si priverebbe degli strumenti indispensabili per individuare i contesti e le strategie per le indagini future. L'unica strategia possibile per accrescere la conoscenza ... è quella di costruire e mettere alla prova paradigmi interpretativi, rimanendo disponibili a modificarli e a superarli sulla base delle nuove conoscenze acquisite, e la verifica non può che ripartire dalla lettura delle fonti: chi le ha usate infatti non necessariamente si è confrontato con sufficienti strumenti critici alle fonti materiali. ... la logica di conservazione della materialità della storia è ben diversa dalla logica di conservazione delle fonti scritte. In particolare per l'altomedioevo dobbiamo aver chiaro che ormai gli scavi hanno prodotto ... documenti che investono qualità e quantità di dati assai superiori ai pochi documenti privati superstiti." On the strategy through which archaeology should build up its own paradigms see, A. Bietti and A.M. Bietti Sestrieri, "Problemi di teoria e di metodo in Archeologia preistorica," in M. Liverani, A. Palmieri and R. Peroni eds., *Studi di paletnologia in onore di Salvatore M. Pugliesi* (Rome: Università di Roma "La Sapienza," Dipartimento di scienze storiche, archeologiche e antropologiche dell'antichità, 1985), 13-29, particularly 26: "1) Paradigma di tipo antropologico, cioè basato in ultima analisi sul concetto di cultura come interazione fra uomo ed ambiente. 2) Costruzione di teorie e modelli per lo studio delle culture preistoriche e del loro sviluppo nel tempo basati essenzialmente (anche se non esclusivamente) su un confronto critico fra elaborazione teoriche e i risultati di ricerche specifiche in campo archeologico e antropologico. 3) Importanza fondamentale del momento sperimentale della ricerca (raccolta dei dati, analisi e classificazione di tutte le variabili, sia "archeologiche" che naturalistiche) e della elaborazione e sperimentazione delle tecniche relative (ad esempio tecniche dell'analisi territoriale e dello scavo, con distinzione fra analisi di tipo funzionale e "comportamentale," da applicare a contesti sincronici, e analisi di tipo "normativo" e dei processi di trasformazione culturale, da applicare a contesti non identificabili per piani sincronici e alla correlazione fra contesti diversi, contemporanei e non; tecniche di analisi quantitativa e statistica, in particolare analisi multivariate per lo studio globale delle

not be considered as a series of definable and complete transfers of populations from an ancestral land around the lower and middle course of the *Elbe* all the way to *Pannonia* and finally to Italy, but rather as a medium- to long-term process of definition and a continuous relocation of people along with the natural dynamics of development of Western Germanic culture. It seems that for Lombards this process got underway during the period of the Marcomannic wars in AD 166.

However, the attribution of the cinerary urns fields and their associated funerary goods from this area of Northern Germany (i.e. the basin of the lower *Elbe*) dating to the *Seedorf* period (late Early Roman Iron Age) to the Lombards, has been based on the assumption that these objects can be linked to evidence from Antique written sources. These sources, actually, have consistently located the *gens Langobardorum* within this region. The evaluation of this stage of Lombard history is extremely important since it is exactly at this point that their *Formengut* was being defined, a set of forms and/or shapes of artifacts that are characteristic of this particular portion of Western Germanic culture which can be recognized and traced archaeologically as 'index fossil' which independently of the location of the discovery, always reflects the same basic data. Thus, it is this *Formengut* that guides us through the long history of 'Lombard' settlements which stretches over a period of around 700 years, and which can be divided into two main periods: the period of settlement along the *Elbe* and the period of migration, or more precisely of partial departure towards the Danubian *limes*. [57] As rightly pointed out by H. SCHUTZ, in order to get information on this matter one has to use:

> "diagnostic analyses and involve geography and archeology –settlement patterns in identified regions, the consistent appearance and distinction of forms and designs in closed complexes of artifacts and the known presence of a named people in that region. If in such a closed complex the pottery, the armaments and ornaments all showed particularly consistent stylistic characteristics, and if these were significantly different from neighboring inventories, and if such a closed complex could be convincingly connected with a tribal name, then the link between the material evidence and ethnicity can be firmly suggested. Thereafter a pincer fibula can be identified as a Thuringian brooch regardless where it is found, though, of course, it reveals nothing about the identity of the wearer."[58]

Thus, ethnicity is not explained as a biological characteristic of any human group but as a cultural feature attached to a particular portion of a human mass. In consequence, as noted by J. OKELY, it is not either correct or any more plausible to consider the language supposedly spoken by a given human group as synonymous with culture and in turn to equate it with a given biologically defined race.[59] In this sense, according to CH. BLACKMORE, M. BRAITHWAITE and I. HODDER, ethnicity only occurs if we use our ethnic group affiliation and culture as a framework for social interaction and as a mechanism for supporting economic and political strategies and competitions. It is therefore, necessary to provide large amount of independent evidence to support any supposed display of ethnicity. Fortunately, the one thing of which the archaeologist often does have a lot is material culture, so this is an approach which is often feasible. This approach is, however, not deterministic. On the contrary, it simply suggests that archaeologists can examine the relationship between cultural patterning and social and economic stress, using the underlying concept of ethnicity. [60] Consequently, if group formation is strong, the style of artefacts should show distinct blocks, if it is weak it should follow geographical limitations.[61]Thus, the culture of a people is represented chiefly by its material culture. These objects are, thus, representative of the intellectual development and of the identity of a given ethnic group.[62] However, the most recent orientation in historiography refuses to consider funerary goods and customs – the single greatest resource of archaeology for the paleoethnology of Germanic peoples – absolute indicators of definite and distinct ethnic identity. Grave goods consisting of the weapons and jewelry of the deceased are thought to provide information on his/her social rank rather than as indicators of ethnicity. For the same reason, one may doubt the "ethnicity" of many findings in the sense that our rigid distinctions can never correspond to a reality revealed by archaeological documentation which points to a strong cross-penetration between various Germanic cultures. These suggestions, then, fit into a larger debate on the usefulness or correctness of the label "Germanic" which is applied to the culture of the Barbarian peoples who invaded the provinces of the Western Roman Empire.[63]
An emblematic example of such an approach is quoted below:

> In the fluid political landscape of the Migration Period, material culture took on a new identity-creation function ... became the symbol of the new communities of political groups or "nations," but it was also

interrelazioni fra le variabili che costituiscono la cultura nel suo complesso, in senso sia sincronico che diacronico; tecniche di analisi dei contenuti simbolici della documentazione archeologica)."

[57]M. Menke, "Archeologia Longobarda tra la bassa Elba e l'Ungheria," in S. Gasparri and P. Cammarosano eds., *Langobardia* (Udine: Casamassima, 1990), 35-106.

[58]H. Schutz, *Tools, Weapons and Ornaments* (Leiden/Boston/Köln: Brill, 2001), xxvi-vi, see also the entire "Foreword" and 17-19 specifically on the Lombards.

[59]J. Okely, "An Anthropological contribution to the History and Archaeology," in B.C. Burnham and J. Kingsbury eds., *Space, hierarchy and society*, 81-92.

[60]Ch. Blackmore, M. Braithwaite and I. Hodder, "Social and Cultural patterning in Late Iron Age England," in B.C. Burnham and J. Kingsbury eds., *Space, hierarchy and society*, 93-94

[61]S.J. Pierpoint, "Who's who in the Northern British Bronze Age," 41; J.R. Sackett, "Style and Ethnicity," 208: "To put it another way, the tradition is seen in continuity in those formal attributes which vary with the social context of manufacture exclusive of the variability related to the use of the item."

[62]H. Kilbride-Jones, *Celtic craftsmanship in bronze* (London: Trowbridge & Esher, 1980), 258.

[63]S. Gasparri, "I Germani immaginari e la realtà del regno. Cinquanta anni di studi sui Longobardi," in *I Longobardi dei Ducati di Spoleto e Benevento*, 26 and 27: "Al tempo stesso, però, la risoluta negazione dell'esistenza di un'obbiettiva identità etnica germanica, e, in questo caso longobarda, ci porta a revocare in dubbio l'esistenza stessa di un'autentica 'cultura tradizionale' longobarda."

apparently used to mark both the warrior aristocracy and the new kingships, just as the display gear of warrior did[64]

Nonetheless, in my opinion, such an assumption should not remain uncontested. First, the brazen metaphor with which the author describes goods buried centuries ago that suddenly acquire a precise meaning demonstrates how these objects are taken into consideration exclusively for their semiotic/symbolic aspects, while their context is narrowed down only to social issues, thus leaving aside economic and, even more, production related questions. However, we do not know (or rather, we are not told) how, when and, more importantly, for what reason at which point in their existence objects become symbols. If we take into consideration only the semiotic aspect of objects, i.e. when those are explained as symbols first and foremost, then a fibula is no longer used for fastening clothes, a sword is no longer for fighting or a chair for sitting on. Their practical functions become secondary and it seems that the main purpose of the material culture of the Migration period is that of providing modern scholars with signs/meanings/symbols of hypothetical contemporary social values.[65] This approach completely ignores the point of view of the archaeology of production which reveals the fact that these objects before representing any idea or social value are the result of a process of production. Moreover, this approach fails to take into consideration that in order to be so up-to-date and sophisticated in responding to commissioners' tastes (and thus to correspond to a semiotic-semantic function attributed to them *a posteriori*) this process of production had to be rather refined and sophisticated. And finally, it does not make any distinction between different categories of products based on their respective functions, diverse strings of technical operations necessary for production and the context of their use and consumption, since these objects are studied only for their meaning. This is a particularly serious drawback when dealing with objects that are classified as working tools. This methodology of analyzing material culture has justly been coined as "pansemiotic" by the Italian anthropologist A. M. CIRESE, meaning that the entire reality of human action has been reduced to the single dimension of signs/symbols. The purpose of the study of material culture, however, should not only be to identify the meaning of objects or that of their production processes. Neither should it abstractly define their "ethnic" characteristics trying to establish an absolute correspondence between the denomination of a people and specific types of objects.[66]

Denominations of human groups are always bound to be approximate definitions whose heuristic function is to provide a basis for organizing research material in an orderly manner and, consequently, to make it easier to trace the development of certain techniques and instruments along a time/space axis (geographical location and historical periods). These denominations are far from corresponding exactly to diverse ethnic identities which characterize any human group and the attribution of a name is far from corresponding to the attribution of identity.[67] In fact, neither the specific form of an object

[64]L. Hedeager, "The creation of Germanic identity. A European origin myth," in P. Brun, S. van der Leeuw and C.R. Whittaker, *Frontieres d'empire. Nature et signification des frontières romaines. Actes de la Table Ronde Internationale de Nemours. 21-22-23 Mai 1992* (Nemours: A.P.R.A.I.F, 1992), 123; a similar approach is in S. Jones, *The Archaeology of Ethnicity: constructing identities in the past and present* (London: Routledge, 1997), 118, 125, 126.

[65] On the 'difficult relationship' between archaeological evidence, symbolism and social values see, O. Bryony. *Anthropology for archaeologists*, 135: "According to archaeological tradition, social organisation is more difficult to interpret from the surviving evidence than either subsistence or settlement, because it has less tangible manifestations. The tradition is based on fact, and social organisation does have fewer direct connections with material things than do food and dwellings. This makes it all the more necessary for archaeologists to make themselves aware of the manifold ways in which pre-industrial societies may organise themselves."; F.R. Hodson, "Inferring status from burial in Iron Age Europe. Some recent attempts," in B.C. Burnham and J. Kingsbury eds., *Space, hierarchy and society*, 25: "Of course, as for all prehistoric contexts, the degree of relationship between the disposition of material remains and social structure must remain a matter of judgment."; S.J. Pierpoint, "Who's who in the Northern British Bronze Age," 35: "The question of symbolism is very large and complex … Most groups use symbols to indicate position within and membership of that group. One of the key features is that the symbols must be recognisable and this is not something that archaeologists can readily consider."; J. Shephard, "The social identity of the individual isolated barrows and barrow cemeteries in Anglo-Saxon England," in B.C. Burnham and J. Kingsbury eds., *Space, hierarchy and society*, 52: "There are varying interpretations of status in archaeological terms, but no general consensus has been achieved. Clearly, status and wealth are not commensurate. Status has more to do with social persona (sensu Saxe), and wealth with intrinsic value of personal possessions."; and, finally, M. Vidale, *Che cos'è l'etnoarcheologia* (Rome: Carocci, 2004), 20: "Nel contesto archeologico, però, non sono accessibili le dimensioni cognitive e simboliche che, in quello originario, hanno accompagnato la creazione, l'uso e la trasformazione di un manufatto o di uno strato."

[66]A.M. Cirese "Homo Faber, Homo loquens. Testimonianza per Andrè Leroi-Gourhan," in A. Leroi-Gourhan, *Ambiente e Tecniche* (Milan: Jaca Book, 1994), 295: "C'è però un rischio, non sempre evitato: quello del pansemiotismo. È il rischio di pensare, più o meno inavvertitamente, che l'operare dell'uomo nel mondo si riduca al suo produrre e governare i segni (o i simboli). Si dimentica l'ordine che l'uomo produce col suo *fare* extra simbolico e fabbrile. L'uomo si riduce a essere solo *homo loquens*." See also A.M. Cirese, *Segnicità, Fabrilità, Procreazione* (Roma: CISU, 1984), 18: "E allora il mondo intero … viene cambiando faccia: sfuma e svanisce la distinzione fra dare o aver un significato e fare o essere un prodotto; tutto il reale tende a ridursi o si riduce alla sola sfera dei segni (significanti di significati, o significati di significanti); l'universo (o la natura e simili) appare umanamente ordinabile e ordinato solo per via di semiosi e non anche di fabbrilità. Si leva sempre più fitta le nebbia del pansemiotismo." See as well, M. Shanks and Tilley Chr., *Social Theory*, 85 and 95-97 where the authors advise that material culture should not be regarded merely as a reflection of cognitive systems and social practices but actively involved in the formation and structuring of those practices. Therefore, the entire realm of material culture should not be seen as constituting a symbolic and active communicative field otherwise the information concerning technology and production processes would be underestimated and neglected.

[67] A. Leroi-Gourhan, *Milieu et Techniques*, 305: "L'esprit, pour construire une image cohérente du monde, a besoin de lui prêter une certaine stabilité: cette stabilité est artificielle. Lorsqu'on présente l'histoire d'un group humain, on prête à ce groupe une personalité moyenne, artificielle, et on le condamne ainsi à perdre cette personalité peu en deçà et peu au-delà du point moyen. ..."; 308: "L'éthnie sur ce plan, est moins un passé qu'un devenire."; 309: "C'est donc sur un terrain doublement instable que nous avons à conisidérer l'évolution des technique materiélles. Terrain instable parce que les éléments qui constituent l'unité ethnique ne convergent que sur un point conventionnel et mènent en deçà et au-delà une existence indépendante, instable parce qu'il n'y a pas d'unité ethnique fixe, mais des divenir successifs qui ôtent a l'ethnologie la sécurité que possède l'Historire generale, en usent de noms de peuples, de dates et de lieux

nor the particular context in which it was found allow attribution of the object to a specific ethnic group. Rather, the study and the recognition of the social context of object use (always synchronic with the object and not with the archaeologist), the corporal gestures that compose the strings of technical operations of its production, the type of approach to available natural resources as raw materials for creation of the object are all factors that actually bring researchers much closer to understanding the close cultural milieu of the artisan producing these goods. Therefore, any artifact cannot simply be defined as the result of the different qualities of its production material(s) or through the description of its various possible functional uses. Both these dimensions actually contribute to shape the "functional identity/role/position" of any object within a given culture. Therefore, artifacts become the main mediators between human groups and the surrounding environment, providing the necessary set of tools for their subsistence and reproduction. In this sense, objects and their production processes became full actors in the formation of a culture. They, thus, contribute in the essential partition of the physical/material world as well as in that of the social environment. Therefore, the production of material objects and the successive creation of their synchronical social attributions are consistently connected even if they are, undoubtedly, two distinct processes. Consequently, the identity of a culture consists of this multi-layered relationship. [68] Accordingly, technology can be defined as an efficient and traditional human action, [69] where efficiency refers to the real manipulation of raw materials and tradition to the various aspects of the continuous transmission of the technical knowledge necessary to afford the permanence of a given craft and style.

In sum, it is the specific technical style of producing an object that defines it as ethnically compatible within each human group. Thus, technical behavior, to use the felicitous terminology of the French anthropologist A. LEROI-GOURHAN, is defined by variables from *internal* and *external* milieu specific for each human group. The first term refers not only to biological characteristics and abilities of any human being, but to the intellectual and material (in terms of technological know-how and preferences) heritage of any (often not fully) circumscribed human group in each historical period, a complex of mental traditions and specific technical gestures which themselves vary from person to person and that evolve over time at different rates. *External milieu* refers to everything that surrounds human beings: the geological, climatic, plant, animal and cultural environments. In addition, it includes other material items or ideas that may come from other human groups. [70] While the synchronic context of the use of a specific object is not always possible to reconstruct directly, most aspects of the *chaine opératoire* of technical operation of production and respective technical gestures are, as we shall see below, possible to reconstruct thanks to some techniques of archaeometric analysis.

Technology *does*. It does *not mean*. If anything, it is the object produced through this technology that in specific historical and social circumstances may assume a significance that goes beyond their proper function. To put it simply, the study of material culture from the point of view of archaeology of production focuses on elucidating and separating diverse technical-artistic heritages/backgrounds, corresponding production

d'événements. Nous ne nous affranchirons pas complètement de ces reères commodes: il restera indispensable de dire que telle technique est connue des Patagons au XVIIIe siècle, mais une telle précision est accesoire ..."; 320, 324: "Pour organiser les matérériaux, il faut convenir d'abord des divisions humaines. Nous nous abstiendrons de préciser le valeurs de ces divisions: ce serait supposer résolu le problème de la personalité ethnique. Les noms qu'il faut pour désigner certains groupements d'hommes ... nous les choisirons vagues, sans tenir trés fermement à leurs limites, puisqu'il est impossible de definir le 'chinois,' l' 'eskimo,' sans faire intervenir une infinitè de nuances qui nous forceraient à charger d'à-côtés, a chaque pas, notre recherche. Les quelques noms qui vont suivre n'ont que la valeur d'abréviaitons commodes, ils correspondent en gros à des distinctions ethniques ..." On the same argument see also J.V. Bromlej, *Etnos e Etnografia* (Rome: Editori Riuniti, 1975), 166: "In quanto sistema dinamico stabile l'etnos è caratterizzato non soltanto da una successione dello sviluppo, ma anche da modificazioni nel tempo ..." and, specifically on Germanic identity, H. Schutz, *Tools*, 3: "Hence it should be considered that the term 'Germanic' is merely a preferred, though imprecise term, of convenience. ... It is difficult to determine an acceptable designation with which to differentiate the groups of multi-ethnic peoples. In one sense, a word such as 'Germanic' misconstrues the identity of the tribal groups since they were neither ethnically nor linguistically uniform. It is a debatable term in Archaeology where the inventory of anonymous objects cannot readily be likened conclusively with tribal names as such, but only by deriving interdisciplinary conclusions from known settlement histories. It is, however, a specialized term reserved by the philologists to describe an aspect of Indo-European etymology as it applies to all those participants in the First Germanic Consonant Shift. ... Nevertheless, restricted to Late Antiquity and the early Middle Ages, the term 'Germanic' will be applied without prejudice of biology or chronology and with the understanding that as an ethnic term 'Germanic' implies a comprehensive, polyethnic, multilingual and multicultural designation applied to the groups of mixed population originating in the regions beyond the Roman Empire's northern frontiers."

[68] On this question see, M. Akrich, "La construction d'un système socio-technique," *Anthropologie et Sociétés* 13 (1989): 31-54; Id., "Comment decrier les objects techniques?" *Techniques et Culture* 5 (1987): 49, 51; B. Latour and P. Lemonnier, "Introduction: genèse sociale des techniques, genèse technique des humains," in B. Latour and P. Lemonnier eds., *De la Prehistoire aux Missiles ballistiques* (Paris: La Découvert, 1994), 14; M.-C. Mahias, "Façonage de l'argile et de la société en Inde," in B. Latour and P. Lemonnier eds., *De la Prehistoire aux Missiles ballistiques*, 189, 199; P. Lemonnier, "Choix techniques et representation de l'enferment chez le Anya de Novelle-Guinée," in B. Latour and P. Lemonnier eds., *De la Prehistoire aux Missiles ballistiques*, 254.

[69] See M. Mauss, "Les Techniques du corp," in *Sociologie et Anthropologie* (Paris: PUF, 1936), 392; Id., "Les Technique et la Technologie," in *Œuvres* (Paris: Minuit, 1941), 252.

[70] See, A. Leroi-Gourhan, *Milieu et Techniques*, 333-334; Id., *Il gesto e la parola* (Turin: Einaudi, 1977), 323-27, 349-63; A.G. Haudricourt, "Contribution à la géographie et à l'ethnologie de la voiture," in Id., *La technologie science humaine* (Paris: MSH, 1987), 141: "L'objet est le résultat du travail de l'homme, et le travail de l'homme est un ensemble de mouvements: une technique est donc un système de mouvements musculaires traditionnels (c'est a-à-dire mouvements qui ne sont ni naturels, ni instinctifs); et étudier un object au point de vue technique c'est d'abord lui trouver sa place dans un certain nombre de ces systèmes, c'est ensuite expliquer comment et par quels dispositifs l'objet fabriqué replit sa fonction." See as well, M.C. Nelson, "The study of technological organization," 61: "... constrains are determined by the environment as well as by the levels of concepts of technological expertise and social interaction."

relationships and their diachronic development within the history of technology. Only after such analysis has been carried out, can one eventually discuss issues of continuity and rupture within history of production, the distribution and consumption of specific categories of objects especially in the light of relevant events that brought into contact different (neither better nor worse) heritages of technical knowledge, as well as finally, the forms such contacts have assumed from a social point of view.[71] The objects and their subsequent archaeometric analysis are at the core of such a scholarly approach since the former are the result of a production process which the latter allows reconstruction of. Throughout this thesis the following themes will repeat themselves:

"l'object n'est pas une illustration du passé. Il en faut partie intégrante...pour l'historien il constitue donc un source importante, irremplaçable, dentant plus qu'il n'a, en général, pas l'intention de relater, d'expliquer ou d'excuser, son teimognage est involontaire...il peut donner un image totale d'une société. Les technique, créations de l'homme pour l'homme, sont un des éléments qui constituent la culture et sont donc, évidentement, situées dans un lieu et dans un temp, on en autres termes, dans une société."[72]

Within this structure a technical tradition is created and a human group recognizes itself as a "technical" group, i.e. a group that carries its specific and "exclusive" heritage of technical knowledge. In this sense archaeological finds from a necropolis, as the product of a specific *chaine opératoire* determined by the availability of raw materials, the type of instruments used and the specific series of appropriate technical gestures (gestures that we judge to be neither instinctive nor natural, but that have been learned in a precise socio-cultural, "ethnic" context) represent a good part of social identities in a human group. Certainly, this exclusivity is only relative since the laws of physics that dictate the behavior of materials used for production, or more generally the entire natural context that influences the technical behavior of a group —the *external milieu*—impose choices or limitations that are obligatory for humans who produce, reducing to a finite number technical possibilities/choices of an artisan with respect to the raw material he has to use. At the same time, the *internal milieu* of a human group as a whole has an equally important regulatory function between individuals within social entities with respect to the implementation strategies for the particular production activity. Thus, the continuous use of a specific stylistic pattern of decoration should not be considered an identity marker from a sociological point of view (i.e. a sign of belonging to a more or less definite social group), but rather should be recognized as an element which can allow researchers to identify a specific group as its producer through the particular method of using tools in working available raw materials, thus, creating a well-defined technology typical of a particular group.[73] Thus, the continuous use of these technologies creates a tradition through which each group accumulates its continuously re-modeled fund of technical knowledge. From this point of view, the word "traditional" referred to knowledge built up through continuous working practice can be useful.[74] To fully evaluate an object and its ethnic context, it is necessary to determine its function, its shape, its dimensions and weight, its constituent material(s), its way(s) of operating on matter(s) and its use. Only after this analysis should an object be studied in a wider context which may include sociological implications as well.[75]

Therefore, as was noted twenty years ago by WHITE, the identity of the people of the *Barbaricum* was not a "written" identity, but a "silent" and traditional identity, typical of societies that remained without a written culture for a long time. But the absence of writing cannot and should not be equated with the absence of socially, culturally and technologically defined structures:

"From its beginnings until very recently, written history has been a history of the upper classes and for the upper classes. Literacy was the prerequisite of small ruling groups. The human record normally has been confined to the interest of and activities of those who recorded it, and its interpretation has been both constricted and tinctured by their values and concerns."[76]

---

[71] See, J. David, *L'outil*, 47-48; J.D. Prown, "The truth of material culture: history or fiction?," in S. Lubar and W.D. Kingery eds., *History from things. Essays on material culture* (Washington/London: Smithsonian Books, 1993), 1-19

[72] J. David, *L'outil*, 135.

[73] A. Leroi-Gourhan, *Milieu et Techniques*, 333-334: according to this author, the meaning of "milieu exterieur" and "milieu interieur" (external and internal environment) are clear: the first refers to all material aspects that surround humans – geological environmental, climatic, the flora and fauna; the second. In addition, it can include material artifacts and ideas that may have come from other human groups. The second term embraces all that is beyond a human being as a natural creature at his/her birth, i.e. the intellectual/material heritage that each defined (often not completely) human group possesses. This refers to a complex of mental traditions and specialized technical gestures. On the subject see also, M. Fiorini, "Postfazione," in A. Leroi-Gourhan, *Ambiente e Tecniche*, 301: "Inoltre, secondo Leroi-Gourhan, il livello tecnico costituisce il principale documento d'identità del gruppo, e l'insieme dei gesti e oggetti che caratterizzano una tecnica è in armonia con l'integra unicità della cultura alla quale esso appartiene. ... È in virtù di quest'ultima convinzione che Leroi-Gourhan considera mutilante la concezione secondo la quale è possibile accantonare, nella scienza dell'uomo, la conoscenza delle competenze e delle attività tecniche rispetto a quelle simboliche, e attribuire a questa soltanto un ruolo di secondo piano; lo stesso piano delle tecniche contribuisce alla comprensione dei rapporti sociali e culturali di una collettività umana alla stesso titolo, anche se in maniera diversa, dello studio di aspetti propriamente 'segnici.' ... gli oggetti tecnici, strumenti e prodotti, sono gli unici testimoni culturali che, grazie alla materialità che ne favorisce la conservazione, si offrono alla ricerca con una continuità nel tempo, pressoché totale, che va dagli esordi dell'attività umana sino alla soglia della contemporaneità. ... le tecniche rappresentano l'unico campo all'interno del quale la storia si svolge in modo continuo attraverso i millenni, senza che sia necessario apporre alcuna cesura tra le prime selci tagliate e le macchine automatiche."; 302: "... l'etnologia può manifestarsi come scienza storica in particolar modo attraverso lo studio della vita materiale dell'uomo." See as well, M. Godelier, *L'Idéel et le Matériel*, 61: "... c'est que chaque système économique et social détermine un mode spécifique d'exploitation des ressources naturelles et de mise en oeuvre de la force de travail humaine, et par conséquent détermine des normes spécifique du 'bon' et du 'mauvais' usage de ces ressources et de cette force, c'est-a-dire une forme spécifique, originale de rationalité économique *intentionnelle*."

[74] J. David, *L'outil*, 55, 56-57.

[75] J. David, *L'outil*, 69.

[76] L. White Jr., "The life of the silent majority," in Id., *Medieval Religion and Technology*, 133.

"We must write – and write from scratch – the history of all mankind including the hitherto silent majority, and not merely that of the tiny vocal fraction which dominated the rest"[77]

"Voltaire to the contrary, history is a bag of tricks which the dead have played upon historians. The most remarkable of these illusions is the belief that the surviving written records provide us with a reasonably accurate facsimile of past human activity. 'Prehistory' is defined as the period for which such records are not available. But until very recently the vast majority of mankind was living in a sub history which was a continuation of prehistory. Nor was this condition characteristic simply of the lower strata of society. ... If historians are to attempt to write the history of mankind, and not simply the history of mankind as it was viewed by the small and specialized segments of our race which had the habit of scribing, they must take a fresh view of the records, ask new questions of them, use all the resources of archeology, iconography, and etymology to find answers when no answers can be discovered in contemporary writings. ... Since, until recent centuries, technology was chiefly the concern of groups which wrote little, the role which technological development plays in human affairs has been neglected."[78]

2.1.1. Romans and Barbarians

In recent years, the tendency among scholars to emphasize continuity without a break in certain forms typical of Late Roman social organization – for the Italian case before and after the Lombard conquest of Italy – has been revived. This re-evaluation has especially affected the study of the fate of cities, which, according to this interpretation, maintained their role as focal points of territorial organization and remained centers of organized social activity, as seats of government and administration and religious centers. In the same vein, it is believed that the structure of great landed estates did not change either, subsequently leaving intact structures of rural settlement and work. Roman-type social, legal and cultural practices, even if those are more difficult to trace, also continued to exist and have been attested all the way to the AD 8th century. These structures were supposedly inherited directly from Roman Late Antiquity. These theses have already been contested elsewhere and quite successfully, by historians of law at the end of the 19th and beginning of the 20th centuries.[79]

However, another new element in this school of thought, and its most important contribution, was to definitely break away from a clearly racist assumption that any archaeological culture is equal to a definite racial/ethnic entity living, crystallized, in a precise geographical context,[80] and consequently to question the ethno-cultural homogeneity of Germanic tribes in general and of the Lombards in particular as carriers of a civilization antagonistic to that of Rome. This theory stands in open contradiction to a historical-philosophical tradition deeply rooted in the European nation-building process of 18th and 19th centuries (represented by figures such as HUME, SISMONDI, CHATEUBRIAND, MONTLOSIER, MALÒBY, TURNER, FICHTE, CLAUSEWITZ, COLERIDGE, BURKE, among others). As recently pointed out by an Italian scholar, it was believed at that time that:

"all'origine della storia delle nazioni europee c'erano le invasioni di epoca medievale, c'erano degli atti di brutale conquista di un popolo ai danni di un altro ... lo scontro che dà origine alla lotta per la libertà nazionale, originariamente non è né uno scontro politico, né uno scontro sociale ma è un conflitto etnico, o – per usare i termini impiegati dai nostri autori – un conflitto di razze ..."[81]

Together, these diverse research hypotheses converge in emphasizing a substantial survival of Late Roman social customs and traditions. In fact, a number of scholars have recently stressed the idea that, for example, 'Lombard' domination did not alter significantly the economic structure and activities characteristic of the Roman epoch. Most importantly, supposedly, it did not change the central role of cities in the organization of the territory, in economic functions and in social life-styles.[82] However, a major drawback in this whole strain of thought is that, as a conclusion, it completely underestimates the "non-Roman" contribution to the formation of early medieval

[77] L. White Jr., "The life of the silent majority,"134.

[78] L. White Jr., *Medieval Technology and social change* (Oxford: University Press, 1964), v.

[79] A. Solmi, *Le associazioni in Italia avanti le origini del comune* (Modena: Società tipografica Modenese, 1898); A. Gaudenzi, "Le società delle Arti in Bologna nel sec. XIII. I loro statuti e le loro matricole," *Bollettino dell'Istituto storico italiano* 21 (1899): 7-126; G. Volpe, *Medioevo italiano* (Florence: Vallecchi, 1928), 248, 254. F. Valsecchi, *Le corporazioni nell'organismo politico del medioevo* (Milan: Edizioni Alpes, 1931); F. Monti, *Le corporazioni nell'evo antico e nell'alto medioevo* (Bari: Laterza, 1934), 164-65, 169.

[80] A. Melucco Vaccaro, *I Longobardi in Italia* (Milan: Longanesi, 1988), 17-18: "Ma presto avanza il nazionalismo pangermanista, che ha il suo funesto culmine, precorritore del nazismo, nell'età Guglielmina, e che si annuncia negli studi di C. Hegel, figlio del famoso filosofo, e del Planitz, ... Nel clima torbido che segue in Germania al Trattato di Versailles, il Kossinna, che sarà l'ispiratore dell'archeologia nazista, comincia subito ad affinare le sue armi in una disputa, solo in apparenza accademica, per affermare le origini germaniche della Polonia e della Slesia. Dalle sue teorizzazioni sulla razza deriva in line diretta la Siedlungsarchäologie, l'archeologia di insediamento, che, vera stampella del regime, con intense campagne di scavo alla foce dell'Elba e nel medio corso del fiume, nelle zone baltiche, materializza nei reperti delle necropoli dei campi di urne la teoria della stirpe e del suo destino fin dall'origine delineato; il metodo seguito consisteva nell'identificare territorio e razza; attribuendo un carattere culturale e politico chiuso e definito a nuclei tribali, in realtà ancora estremamente fluidi e in formazione ... Ma fin dal 1933 i rapporti fra archeologia e regime erano apparsi strettissimi; la Società Tedesca di Preistoria si trasforma in Lega del Reich per la Preistoria tedesca, cioè diviene un'associazione nazionalsocialista riconosciuta; la presidenza di Himmler assicura cospicui finanziamenti per scavi e ricerche alla Deutches Ahnenerbe (1937), l'associazione per lo studio dell'eredità dei progenitori tedeschi, di cui Jahnkuhn è l'uomo di punta; e si noti che secondo la tradizione di studi tedesca l'età delle migrazioni rientra nella ricerca pre- e protostorica e ne costituisce in questo periodo la parte essenziale. ... La ricerca storica percorreva vie analoghe ... la cultura tedesca non ha mai fatto un bilancio critico di quanto fu prodotto in età nazista; oggi, come ai tempi di Adenauer, la questione viene sistematicamente elusa, anche per la ben nota e indisturbata continuità delle presenze accademiche ..." On the puzzling relationship between archaeology, nationalism and building of identities see, among many others, S. Jones, *The Archaeology of Ethnicity*, 2-5; B. Arnold, "The past as propaganda: Totalitarian Archaeology in Nazi Germany," *Antiquity* 64 (1990): 464-78; J.R. Naess ed., *Arkeologi og Etnisitet* [*Archaeology and Ethnicity*] (Stavanger: Arkeologisk museum i Stavanger, 1985); B. Olsen, "Norwegian Archaeology and People without (Pre-)History: Or How to Create a Myth of a Uniform Past," *Archaeological Rewiev from Cambridge* 5 (1986): 25-42; B. Scott, "Archaeology and National Identity: The Norwegian Example," *Scandinavian Studies* 68.3 (1996): 321-42; J. Voss, "Antiquity Imagined: Cultural Values in Archaeological Folklore," *Folklore* 98.1 (1987): 80-90; R. Hodges, *Early Medieval Archaeology*, 3-4.

[81] A.M. Banti, "Le invasioni barbariche e le origini delle nazioni," in A.M. Banti and R. Bizzochi eds., *Immagini della nazione nell'Italia del Risorgimento* (Roma: Carocci, 2002), 22, 25.

[82] A. Cameron, *The Mediterranean world in late antiquity. AD 395-600* (London, New York: Routledge, 1993), 43.

European technical, economic and cultural spheres. This theory also assumes that Germanic peoples represented a simple conglomeration of various "illiterate" groups of tribes speaking closely-related Indo-European dialects and thus, having a very modest level of material culture. For this reason, they were also unable to "have an identity" and were constantly subject to the influences of the stronger social models of both Romans and *Steppenvölker*.[83] In this world view, the Germanic world was peripheral both to the Roman Imperial system and to that of the Eurasian steppes. The primary purpose of Barbarians, then, was to maintain the old Roman social and economic structures in place so as to be able to reap maximum benefits from them, given that the Barbarians themselves did not have their own, alternative socio-economic models, different from those of the Mediterranean world. Having been for centuries only a peripherical component of the Roman Empire, they now only wished to take a full advantage of benefits offered by the surviving Roman structures. Always according to this school, the Germanic peoples did not import any new ways of exploiting land resources but simply tried to manage the slow degradation of the old Roman agriculture as well as that of the great urban centers of the Antiquity. Thus it was posited, their creative impact on this world was minimal. [84] In this case, refuting any precise ethnic identity to the peoples of the *Barbaricum* corresponds to giving them less weight in the formation of the medieval world, which, as a consequence, would be firmly anchored in its Roman roots.[85] The idea of Germanic civilization versus that of Rome is also placed under a, more or less, radical scrutiny: the Barbarians are considered with more willingness as an organic element integrated into the complex world of Late Antiquity, if not the creation of Roman "political genius."[86]

[83]S. Gasparri, "Tra antichità e Medioevo: i modelli sociali ed economici dei popoli barbarici e il loro impatto con il mondo mediterraneo," in V. Castronovo ed., *Storia dell'economia mondiale*, 318, 322, 323, 329, 333.

[84]S. Gasparri, "I Germani immaginari," 3-28, especially 4-5, 6, 23, 25, 28; W. Pohl, "Le identità etniche," 79-103 particularly 81, 82, 83, 89, 92, 10; the last two contributions are very important for they contain the most recent bibliographic references on the subject; moreover, see W. Pohl, *Le origini etniche dell'Europa* (Roma: Viella, 2000); L. Hedeager, "The creation of Germanic identity.," 121-31, specifically 122-23.

[85]As noted, based on a quote of G. Cracco and L. Cracco Ruggini, by G. Traina this type of conclusion is the result of a historiographic cliché, rather typical of western medieval studies, which believes that: "L'Occidente o resta Roma e continuità di Roma, o è poca cosa; e questa poca cosa non può venire compensata da Bisanzio, che è solo una Roma destinata al naufragio." See, G. Traina, "La fine del regno d'Armenia," in *Atti dei Convegni Lincei, 201. La Persia e Bisanzio. Roma 14-18 ottobre 2002* (Roma: Accademia Nazionale dei Lincei, 2004), 366. See also R. Francovich and R. Hodges, *From Villa to Village*, 28: Plainly, we must all begin to question whether this 'continuity' or 'elongation' is now justified or, as Liebeschuez has suggested, the product of a kind of contemporary political correctness resisting Edward Gibbon's view of decline and fall in the fourth century. See as well, J.H.W.G. Liebeschuez, "The uses and abuses of the concept of "decline" in later Roman history, or was Gibbon politically incorrect?" in L. Lavan ed., *Recent Research in Late-Antique Urbanism*, in *Journal of Roman Archaeology Supplementary Series* 42 (2001): 233-45.

[86]See also, W. Goffart, *Barbarians and Romans. AD 418-584. The techniques of accommodation* (Princeton: Princeton University Press, 1980), 30, 31.

However, in my opinion, even without going back to schools of thought inherited from the nineteenth century which presented the idea of a Germanic civilization versus that of the Roman world, through the lens of archaeology of production, the hypothesis of discontinuity (which does not mean a sudden gap or disruption) for the passage between Late Antiquity and the Early Middle Ages is still plausible.[87] Actually, for the economic sphere, archaeological research is able to provide sufficiently strong data to support the thesis of a profound change not only in the cultural landscape in the Early Middle Ages but also a profound restructuring of production cycles.[88] The end of Roman-type settlements

[87]P. Delogu, "L'editto di Rotari," *passim* but particularly 329-31, 345, 348; R. Hodges, *Early Medieval Archaeology*, 12, 15: "Modern excavations show beyond any doubt that the Dark Ages represent an important stage separating Antiquity from the High Middle Ages." See also the recent B. Ward-Perkins, *The Fall of Rome*, 167.

[88]As a matter of fact, other craft activities besides metallurgy such as pottery also underwent profound changes in its production structures. On the character of pottery production in Italy during the Lombard period see, P.M. De Marchi, "La ceramica longobarda. Osservazioni," in R. Fiorillo and P. Peduto eds., *III Congresso Nazionale di Archeologia Medievale, Castello di Salerno, Complesso di Santa Sofia (Salerno, 2-5-X-2003)* (Florence: All'Insegna del Giglio, 2003), 18: "Gli studi finora condotti sulle ceramiche longobarde sono concordi nell'indicare tre aree produttive che si differenziano per qualità, struttura e apparato decorativo. In Piemonte e Lombardia occidentale abbiamo prodotti più rozzi, talvolta dalle forme irregolari e asimmetriche e con motivi decorativi meno variati, in Lombardia orientale prodotti tendenzialmente più raffinati, di forma slanciata ed elegante. Anche all'interno di questi ampie suddivisioni, si individuano prodotti di qualità molto differenziata, ad esempio in area alto milanese, la fiasca di *Castellanza* priva di confronti, nonostante l'asimmetria del corpo, ha una varietà rara di motivi ornamentali disposti entro un campo definito da due linee incise regolari, indizio della capacità dell'artigiano nel gestire le fasi di rifinitura dei prodotti, come è stato osservato per un gruppo di ceramiche di Brescia/S. Giulia. Abbiamo, quindi, atelier che tendono alla standardizzazione del prodotto, con una buona organizzazione dei cicli di lavorazione, accanto a produzioni più domestiche, che in base all'elevato numero delle località di ritrovamento e alla loro distribuzione, sembrano polverizzarsi sul territorio con artigiani più o meno abili e legati alle tradizioni originarie, cui corrisponde, allo stato attuale delle conoscenze, un raggio di diffusione del prodotto piuttosto ristretto appena più ampio in area bresciana, con influenza che si estende al Veronese e al Veneto meridionale. Il ritrovamento di Brescia/S.Giulia (composto da 583 frammenti) rimane, finora, l'unico di cui si possano individuare le caratteristiche dell'insediamento, impostato su una *domus* palaziale romana in un'area della città densa di edifici pubblici e con tradizioni produttive, che persistono nell'altomedioevo. Un'area urbana rimasta pubblica nella prima età longobarda e poi trasmessa al patrimonio regio, con un insediamento, datato a partire dal 569 e che perdura fino alla metà circa del VII secolo, frequentato da una popolazione servile a caratteri antropologici ibridizzati, soggetta alla corte ducale, per la quale svolgeva attività di servizio e artigianali, relative alla lavorazione dei metalli e dell'osso. La vicinanza tra l'area di ritrovamento di due fornaci per ceramiche altomedievali all'insediamento longobardo, la compresenza nei due scavi di ceramiche romane e longobarde, aventi i medesimi impasti, indica da un lato che gli artigiani appartenevano ad un ceto sociale dipendente, all'interno del quale era difficile distinguere romani e longobardi, dall'altro che le modalità produttive legate all'attività dipendente, se non servile, dei lavoratori sono molto mutate rispetto all'età romana quando si operava in autonomia. Questa organizzazione del lavoro secondo gerarchie rigide è attestata anche nella Germania altomedievale, dove le botteghe non lavoravano in proprio, sempre che sia appurata l'interdipendenza tra l'insediamento di S. Giulia e l'atelier produttivo poco distante. Il ritrovamento bresciano, relativo ad una città ducale, aperta agli scambi in ogni direzione geografica, rimane al momento l'unico indagato e non può essere, allo stato attuale delle ricerche generalizzato, anche perché si conosce troppo poco sulle strutture produttive. … Nel complesso la ceramica

in the countryside and the introduction of new types of residential buildings, carrying with it a radical change in architectural standards, can be traced throughout continental Europe between the AD 5th/6th and 7th/8th centuries, without distinction between former Roman territories and those in the *Barbaricum.* [89] The contemporary development of these changes and the insertion of the Germanic technical and economic background into the former Imperial territory might well not be coincidental, but rather a clear indication of an interaction between groups with different material cultures.

longobarda attesta la scarsa mobilità degli stampi o degli artigiani, dovuta forse anche alla loro mancanza di autonomia, sono, infatti, molto pochi i confronti puntuali tra recipienti rinvenuti in aree tra loro lontane. A differenza di quanto riscontrato per altre classi di materiali (gli scudi da parata, e recipienti in bronzo, armi, oggetti in osso e stampi per lamine a sbalzo, agemine), la ceramica longobarda, salvo il caso di Brescia, non sembra giungere ad una produzione standardizzata distinta da una circolazione ad ampio raggio, ma concentrarsi negli insediamenti rurali, prevalentemente indicati da ritrovamenti funerari con corredi d'arme di ricchezza media e di potere economico non elevatissimo, probabilmente relativi a proprietari terrieri di 'ceto medio,' corrispondenti ad una nobiltà rurale, tradizionalista e conservatrice, in qualche modo 'periferica' e meno aperta all'interscambio culturale, che tende a mutare ideologia funeraria intorno alla metà del VII secolo per l'assimilazione delle usanze locali e del cristianesimo ortodosso. Un dato interessante, di cui non è possibile al momento verificare la reale portata, è costituito dalla mancanza di ceramica "etnica" in contesti funerari di elevato potere economico, ne sono, ad esempio, privi i corredi dei cinque nobili cavalieri di Trezzo ai quali si debbono attribuire – per le caratteristiche dei doni funebri – alte funzioni nelle gerarchie del regno e contatti 'internazionali' (*La necropoli longobarda* 1986). In questi casi forse prevaleva la necessità di rappresentare 'il cosmopolitismo' dei defunti, attraverso prodotti preziosi che connotassero le relazioni sociali intrattenute in vita. ... È però certo che un'ampia circolazione non è favorevole a prodotti fragili che non presentano nessuna utilità nel trasporto di alimenti in quantità utile." On the very same subject see as well, M. Vitali, "La ceramica longobarda," in G.P. Brogiolo ed., *S. Giulia di Brescia: gli scavi dal 1980 al 1992: reperti preromani, romani e alto medievali* (Florence: All'Insegna del Giglio, 1999), 192, 207; F. Rossi ed., *Nuove ricerche sul* Capitolium *di Brescia. Scavi, studi e restauri* (Milan: Et, 2002), 449-558; G.P. Brogiolo, "Trasformazioni urbanistiche nella Brescia longobarda: dalle capanne di legno al monastero regio di S. Salvatore," in C. Stella and G. Brentegani eds., *S. Giulia a Brescia. Archeologia, arte, storia di un monastero dai Longobardi al Barbarossa*, (Brescia: Grafo, 1992), 179-210; and Id., "Considerazioni sulle sequenze altomedievali nella zona monumentale della città romana," in F. Rossi ed., *Carta Archeologica della Lombardia. V. Brescia. La città* (Modena: F.C. Panini 1996), 257-63; A. Guglielmetti, "Ceramica di età longobarda dall'area del Capitolium: analisi di una struttura produttiva," in F. Rossi ed., *Carta Archeologica della Lombardia. V. Brescia. La città* (Modena: F.C. Panini, 1996), 265-83; O. Von Hessen, "A proposito della produzione di ceramica nel periodo delle migrazioni nell'Europa centrale e meridionale," in *Artigianato e tecnica*, vol. 2, 760-61; G.P. Brogiolo and S. Gelichi, *Le ceramiche altomedievali (fine VI-X sec.) in Italia settentrionale: produzione e commerci* (Mantova, Padus, 1996), 221-24; L. Paroli, "La cultura materiale nella prima età longobarda," in J. Arce and P. Delogu eds., *Visigoti e Longobardi*; J. Werner, "Eine langobardische Schild von Ischl an der Alz, ge. Seeon (Oberbayern)," *Bayerische Vorgeschichte Blätter*, 18-19 (1974): 45-58; P.M. De Marchi, "Gli scudi da parata longobardi in Lombardia: luoghi e centri di potere," in P. Baj ed., *Studi in memoria di Carlo Mastorgio* (Gavirate: Nicolini, 2002), 61-84; M. Ricci, "La produzione di merci di lusso e di prestigio a Roma da Giustiniano a Carlo Magno," in M.S. Arena ed., *Roma dall'antichità al Medioevo. Archeologia e storia* (Milan: Electa, 2001), 79-87; C. Giostra, *L'arte del metallo in età longobarda* (Spoleto: CISAM, 2000), 107-12.

[89] M. Valenti, *L'insediamento altomedievale nelle campagne toscane*; L. Pani Ermini, "Città fortificate e fortificazione delle città italiane fra V e VI secolo," and Ead. "Il recupero dell'altura nell'Alto Medioevo," in L, Pani Ermini, *'Forma' e Cultura della città altomedievale*, eds. A.M. Giuntella and M. Salvatore, (Spoleto: CISAM, 2001), 45-58 and 59-110; R. Hodges, "Rewriting the Rural History of Early Medieval Italy," 26; R. Francovich and R. Hodges, *From Villa to Village*, 21-22: "It seems beyond doubt that the estate system of the *villa rustica* – predominant in Roman times – was in the early middle ages replaced by a new basic unit of production, that is the by peasant households cooperating as villages. Despite dozens of intensive field surveys in Italy, not a single case of continuity of a Roman estate (the farm and its managed landscape) into the early middle ages has been identified ... the rupture ... should be ascribed to the sixth or seventh century, when an extreme structural change of the economic system occurred."; see also, R. Hodges, *Early Medieval Archaeology*, 12.

2.2.0 Roman-Barbarian Relationships between Late Antiquity and Migration Period: a Critical Approach to Acculturation.

> *I confini settentrionali dell'Impero Romano, dopo il fallimento del tentativo augusteo di conquistare i territori fino all'Elba, si attestarono sulla linea costituita dai fiumi Reno e Danubio, dal Mare del Nord fino al Mar Nero; a parte qualche modifica, tali confini rimasero sostanzialmente immutati per un lunghissimo periodo di tempo. Per secoli il continente europeo si trovò così diviso stabilmente in due grandi aree: il territorio sotto il controllo diretto romano, e quello che invece ne rimase escluso, genericamente denominato Barbaricum. Questo nome non designa un insieme di territori preciso, ma piuttosto definisce in termini negativi tutto ciò che non era Impero Romano; concretamente il Barbaricum comprenderebbe tutta l'Europa centrosettentrionale e orientale, compresa la penisola scandinava e la parte europea della Russia.* [90]

Within the last few years, numerous studies have been dedicated to the archaeology of production in Italy during the Lombard period. These researches have attempted not only to define the fundamental elements of Late Antique economic development in Italy but also to identify the technological background that lay at its base. They have attempted to analyze its diverse components, or, more precisely, the contribution of different cultures which made up, in the French anthropologist A. LEROI-GOUHRHAN's words, "the general technological environment" of the period under examination.[91]

[90] T.M. Lucchelli, *La Moneta nei rapporti tra Roma e l'Europa barbarica: aspetti e problemi* (Florence: La Nuova Italia, 1998), 1.

[91] For early medieval Italy see, L. Paroli, "Aspetti e problemi dell'archeologia della produzione in età longobarda," in M. S. Arena and L. Paroli eds., *Le arti del fuoco in età longobarda* (Rome: Museo dell'Alto Medioevo, 1994), 11-18; L. Paroli, "La cultura materiale," 257-304; R. Francovich and G. Noyè eds., *La storia dell'Alto Medioevo italiano alla luce dell'archeologia* (Florence: All'Insegna del Giglio, 1994); C. Giostra, *L'arte del metallo in età longobarda*; G.P. Brogiolo ed., *S. Giulia di Brescia: gli scavi dal 1980 al 1992*; G.P. Brogiolo and L. Castelletti, *Il territorio tra tardo antico e altomedioevo. Metodi di indagine e risultati* (Florence: All'Insegna del Giglio, 1992); S. Lusuardi-Siena ed., *Ad Mensam. Manufatti d'uso da contesti archeologici fra tarda antichità e medioevo* (Udine: Del Bianco, 1994); M. Rotili, "Le necropoli di tradizione germanica," *Archeologia Medievale* 10 (1983): 143-174; C. Citter, "I corredi funebri nella Toscana longobarda nel quadro delle vicende storico-archeologiche del popolamento," in L. Paroli ed., *L'Italia centro-settentrionale in età longobarda* (Florence: All'Insegna del Giglio, 1997), 185-211; Id., "I corredi nella Tuscia longobarda: produzione locale, dono o commercio? Note per una storia delle attività produttive nella Toscana altomedievale," in G.P. Brogiolo ed., *Sepolture tra VI e VIII secolo. 7° Seminario di Monte Barro, Gardone Rivera 1996. Documenti di*

The present work, thus, will contribute to the study of history of iron technologies,[92] which are connected with and directly influenced the development of agriculture, arms and many other branches of craftsmanship.[93] Such data will be useful for understanding the development of the production system and of the *direct method* of iron production in Italy during the Early Middle Ages. This system underwent complete and profound structural transformations between the AD 5th and 8th centuries, from large-scale intensive iron production (a typical model of Imperial economic organization) to an economic system that tended to be regional (marked by the dissemination of production centers in the territory and a certain subordination of metallurgical activities to agricultural production – agriculture maintaining its leading economic role). [94] In analyzing this 'transformation,' which is full of elements of discontinuity, the study of the Late Antique heritage should not be limited to macroeconomics and the exploration of the general process of economic development. It is equally, if not more important, to analyze the material structures of production and its instruments in Late Antiquity on the one hand, and to explore in detail the technological heritage of Germanic populations on the other hand. The interrelations between these two modes of production need to be explored in detail.[95] To achieve this task, the tools of archaeometry can be key since this science is able, if not to specify an exact chronological horizon, at least to provide exact technical-productive information and consequently provide evidence for the various modes of technical transfer and transformation of technical knowledge.[96]

When discussing production activities within traditional and pre-industrial societies it is more appropriate to speak of "discontinuity" rather than of "rupture/disruption". In such a context rupture/disruption has too much of a negative connotation, often implying a cessation, a disappearance or at least a loss of technical knowledge. Such a vision does not take into account the formation of traditional artisans' know-how which was practical and not at all theoretical. Thus, artisans know-how required continuity in craft traditions and stability in the workshop organization not only and simply through maintaining the material structures of production that "contain" the place and working instruments, but, more importantly, the uninterrupted execution and transmission of those technical instructions which comprised the true content of a craft and assured its persistence. Such continuity was also indispensable for pinning down technical procedures aimed at modifying products to fit the tastes of commissioners.[97] For the period of our interest there is no indication or even the slightest archaeological evidence that suggests there might have been a uniform and homogeneous disappearance of all Graeco-Roman technological heritage, unless the decided decline in production volume might be considered the same as technical regression. [98] Thus, the continuity vs.

Archeologia 13 (Mantova: SAP, 1998), 179-95; M. Ricci, "Relazioni culturali e scambi commerciali nell'Italia centrale romano-longobarda alla luce della Crypta Balbi in Roma," in L. Paroli ed., *L'Italia centro-settentrionale*, 239-73; F. Zagari and V. La Salvia, "Aspetti della produzione metallurgica longobarda," in *Paolo Diacono e il Friuli alto medievale (Secc. VI-X). Atti del XIV Congresso internazionale di studi sull'Alto Medioevo. (Cividale del Friuli-Bottenicco di Moimacco 24-29-IX-1999)* (Spoleto: CISAM, 2001), 865-86; V. La Salvia and F. Zagari, "Cultura materiale e tradizione tecnica: la metallurgia del ferro dei Longobardi in Italia," in *I Longobardi dei Ducati di Spoleto e Benevento*, 945-1008.

[92]Concerning iron production in Italy during the Lombard Kingdom and the early Middle Ages see, V. La Salvia, "Gap or Continuity?," 262-71; Id., "La fabbricazione delle spade," 28-54; Id., *Archaeometallurgy of Lombard Swords* (Florence: All'Insegna del Giglio, 1998); Id., "L'artigianato metallurgico dei Longobardi alla luce delle fonti archeologiche con particolare riferimento alla lavorazione del ferro. Suggerimenti e problemi," *Archeologia medievale* 25 (1998): 7-26; Id., "Notes on Early Medieval Ironmaking in Italy," in J. Gömöri ed., *Traditions and Innovations in the early Medieval Iron Production* (Sopron/Somogyfajsz: Dunaferr-Somogy Archaeometallurgical Foundation/ Soproni Muzeum, 1999), 83-87.

[93]See, V. La Salvia, "Archaeometallurgy as a Source," 106-07; M.S. Mazzi and S. Raveggi eds., *Gli uomini e le cose nelle campagne fiorentine del Quattrocento* (Florence: Olschki, 1983), 156, 160-61; M.-Chr. Bailly-Maitre, "Les methodes de l'Archéologie Miniere," in R. Francovich ed., *Archeologia delle Attività Estrattive,* 237; J.C. Edmondson, "Mining in the later Roman empire and beyond: continuity or disruption?," *The Journal of Roman Studies* 79 (1989): 84: "At all times iron remained the metal most in demand for arms and armour, constructional and agricultural tools and building materials such as nails, hinges and fittings."

[94]R. Farinelli and R. Francovich, "Potere e attività mineraria nella Toscana altomedievale," in R. Francovich and G. Noyè eds., *La storia dell'Alto Medioevo italiano*, 445; Edmondson, "Mining in the later Roman empire," 85: "It will be suggested that mining did not cease, but that it underwent a restructuring or reorganization of production. Such a restructuring seems consistent with an overall decline in centrally controlled large-scale mining but it also indicate that mining on a smaller scale still had a role to play within the local economies of the Empire." Moreover, on the same subject see 91 and 93-97; V. La Salvia, "L'artigianato metallurgico," 7, 23-24 and Id., *Archaeometallurgy*, 29; M. Cima, *Archeologia del Ferro*, 121 and Id., "Metallurgia in ambiente rurale al sito alto-medievale di Misobolo," *Archeologia Medievale* 13 (1986): 189.

---

[95]L. White jr., "Technology and Invention," 4-6.

[96]V. La Salvia, "L'artigianato metallurgico," 24, T. Manoni and E. Giannichedda, *Archeologia della Produzione* (Turin: Einaudi, 1996), 49-54.

[97] Concerning the long lasting question about the dichotomy continuity/disruption within the history of technology of pre-industrial societies see L. White Jr., "Medieval roots of modern technology", in L. White Jr., *Medieval religion and technology*, 76: "Any discussion of continuity and changes runs grave risks of setting up false dichotomies. Those who stress change sometimes seem to forget that it takes a river to make a waterfall, as well as a cliff to tumble over. Those who tend to dwell on continuity are so entranced by the river that they disregard the cliff. This sort of nonsense can best be avoided if we distinguish between change and discontinuity in cultural history. Discontinuity occurs when an item or set of items is borrowed from outside a culture and when that borrowing alters the whole style of the relevant activity in the recipient culture." See as well 83-84. On the organization of production of craftsmanship see V. La Salvia, "Gap or Continuity?," 266-67; Id., "La fabbricazione delle spade," 50-54; C. Panseri, *Ricerche metallografiche*, 13, 16-17, 21 and note 10; Id., *La tecnica di fabbricazione*, 5, 23-24; E. Salin, *La Civilisation merovingienne*, 3; G. Stabile, "La torre di Babele: confusione dei linguaggi e impotenza tecnica," in J.-C. Maire Vigueur and A. Paravicini Bagliani eds., *Ars et Ratio. Dalla torre di Babele al ponte di Rialto* (Palermo: Sellerio, 1990), 246, 258-59, 261-64, 268-70, 272-73; M Cima, *Archeologia e Storia*, 15-16; A. Leroi-Gourhan, "Notes pour une histoire des aciers," in A. Leroi-Gourhan, *Le fil du temps* (Paris: Fayard, 1983), 35; F. Wewer, "Das Schwert," 54; G. Angioni, *Il sapere della mano* (Palermo: Sellerio, 1986), 95; W. Ong, *Oralità e scrittura* (Bologna: Il Mulino, 1986), 27, 72.

[98]C. Wickham, "L'Italia e l'Alto Medioevo," *Archeologia medievale* 15 (1988): 121; U. Monneret De Villard, "L'organizzazione industriale nell'Italia longobarda durante l'Alto medioevo," *Archivio Storico*

rupture/disruption dichotomy fits poorly with attempts at reconstructing technological-productive structures of pre-industrial societies. These must be presented in all their complexities rather than in terms of simple models and definitions.[99]

Contemporary scholarship recognizes that it is difficult to consider the collapse of the Roman Empire a simple and sudden military event brought about by Barbarian invasions.[100] As a matter of fact, the penetration of the 'Barbarians' inside the borders of the Empire took place over many centuries: the settlement of Barbarians of Germanic and so-called 'Scythian' origin, sanctioned by Imperial authority was already common at the time of the Emperor *Marcus Aurelius* and continued until the reign of Emperor *Honorius*. In the three dioceses of Gaul in the AD 5th century there were 18 of these ethnically different communities within the Romanized population. It is important to stress here that these settlements, organized as military colonies, were already a separate body from the general Roman State organization. In fact, within these colonies, the population preserved their customs and did not observe Roman law.[101]

However, the contacts between Romans and Germanic peoples were not only limited to warfare or frontier military struggles, nor were they confined to commercial exchanges along the *limes* where the two communities were closer. Not all Barbarians were veterans placed along the frontiers or serving in the Imperial army. Many of them worked in the Imperial administration and many others were deeply scattered and diffused through society and in every day life as part of the domestic and agricultural workforce. Thus, there was no annihilation of a flourishing Roman civilization by hordes of Barbarians. Neither were the new established Roman-Germanic kingdoms, organized on former Imperial soil, the result of a military victory that implied a loss of freedom and real estates for the Romanized population and the spread of the "primitive traditions" of the new barbaric lords. The Germanic peoples settling in the Mediterranean areas were, in fact, always and everywhere a transitory, minority superstratum, only thinly plastered over a broadly based and long-term indigenous substratum. Moreover, the historical evidence suggests that ethnic isolation, which, as anthropologists tell us, normally promotes linguistic and cultural persistence if not continuity, was rarely, if ever, a real alternative for the Germanic Mediterraneans in Late Antiquity. From this point of view, the various relationships between the Barbarians and both the Imperial administration and the Romanized population were apparently established before the period of flux and struggle known as the Migration Period. The Roman-Germanic kingdoms of the AD 5th century seem to have been the natural conclusions of long lasting contacts and movements of peoples which had already begun long before the period of the "decline and fall" of the Empire in the late AD 5th century.[102] From this point of view, integration rather then acculturation seems to be a more suitable concept for describing such a situation.

Moreover, the idea of the Imperial border as an impermeable wall between two different worlds cannot be considered valid any longer since the archaeology of northern Europe has shown what intense exchange went on between the two sides of the *limes*. Actually, excavations in areas far from the Roman border have also yielded relevant evidence, especially in the richest Germanic graves, for the diffusion of artifacts produced in Imperial workshops.[103] According to the WHITTAKER model, as suggested in his 1990 contribution: 1) Roman Frontiers were not fixed lines but broad zones which by their nature were open to cultural economic interchange; 2) The Roman army on the frontier participated in and benefited from this interchange. The guiding principle of frontier locations was linked to supply not to defense, although the supply was internal as well as external to the Empire; 3) The deep zone of the frontier, therefore, reflected in part, at least, a natural limit which had little to do with ethnic or cultural divisions -- a fact which further encouraged interchange of men and supplies. 4) Such exchanges, and the wealth created by them, were inevitably unevenly distributed. The effect was unequal vertical development across the frontiers rather then uniform horizontal acculturation. The Romanized elite among the external *gentes*, i.e. the people living outside the borders of the Empire, as contrasted with a little- or non-Romanized populace, accelerated social divisions within the emergent frontier states. This process may, paradoxically, have eased the process of integration between Romans and Barbarians when the invasions of the later empire took place.[104] The similarities between

---

Lombardo, 5 ser., 46 (1919): 1-2; White jr., "Technology and Invention," 12-13, 15.

[99] V. La Salvia, *Archaeometallurgy*, 7; R. Francovich, "Introduzione," in R. Francovich ed., *Archeologia delle Attività Estrattive*, 9.

[100] S. Mazzarino, *L'Impero Romano*, vol. 3 (Rome-Bari: Laterza, 1980), 803-15; Id., *Stilicone* (Milan: Rizzoli, 1990), 237; O. Bertolini, *I Germani. Migrazioni e Regni nell'Occidente già Romano*, part 1. vol. 3 of *Storia Universale. Nuove Formazioni Politiche nel Mondo Mediterraneo Medievale* (Milan: F. Vallardi, 1965), 21-22. With regard to the Late Empire see A. Cameron, *The Late Roman Empire* (London: Fontana Press, 1993); E. Demougeot, *De l'Unitè á la division de l'Empire Romain (395-410). Essai sue le gouvernament imperial* (Paris: Maisonneuve, 1951); W. Goffart, *Barbarians and Romans*; A.H.M. Jones, *The Later Roman Empire: A Social, Economic and Administrative Survey*, 3 vols. (Oxford: Blackwell, 1964), 264-602; F. Lot et al., *Les Destinees de l'Empire en Occident (395-888)* (Paris: P.U.F., 1940-41).

[101] A. Cameron, *The Mediterranean world*, 36, 44; S. Mazzarino, *L'Impero*, 778, 787; O. Bertolini, *I Germani*, 13-17; J. Chapman and H. Hamerow , "Introduction," 6.

[102] A. Dopsch, "Istituzioni agrarie dei regni germanici dal V al IX secolo," in *Storia Economica Cambridge. L'agricoltura e la società rurale nel Medioevo*, vol. 1 (Turin: Einaudi, 1976), 223-24; T.L. Markey, "Germanic in the Mediterranean: Lombards, Vandals, and Visigoths," in F.M. Clover and R.S. Humphreys eds., *Tradition and innovation in Late Antiquity* (Madison (USA)-London: The University of Wisconsin Press, 1989), 52.

[103] A. Melucco Vaccaro, *I Longobardi*, 48; A. Tagliaferri, "Note sull'economia longobarda dagli stanziamenti nordici al primo Ducato italiano," *Economia e Storia* 4 (1972): 425-43.

[104] D. Whittaker, "The Economy of Roman Frontier," in E. Aerts, J. Andreau and P. Ørsted, eds., *Models of Regional economies in Antiquity and the Middle Ages to the 11th. Century. Proceedings of the 10th International Economic History Congress. Leuven, August 1990* (Leuven: University Press, 1990), 62.

populations on both sides of the borders eased the process of mutual integration.[105] The Rhine front appears to have been restored and to have remained intact through most of the AD 5th century. The process of cultural integration was similar in the Middle Danubian territories as well. The problem of the western part of the Empire did not lie in the decay of the western frontiers but in the collapse of a coherent, western, political administration. Thus political stability, related to the inability of the province to collect provincial taxes, lost the loyalty of the provincial and frontiers elites in the centre. In such a fluid situation, the new federated warlords and the old provincial landlords made common cause to avoid state control of their revenues. [106] However, the AD 5th century marked a qualitative change within the framework of contacts between Romans and Barbarians as well.[107]

Both archaeological evidence and written sources indicate that there had been continuous commercial exchange between the Roman Empire and the regions located east of the Rhine and north of the Danube River. [108] Nonetheless, it is still difficult to fully understand when and what form this trade took. However, many classical authors, starting from *Caesar*, already mention trading activities along the *limes* from the first century BC[109] Later accounts, from *Tacitus*[110] and *Cassius Dio*,[111] also suggest continuity in commerce well into the AD 2nd century. The movement of goods along the *limes* is clearly documented by two interesting finds: the first, a wooden tablet found at *Tolsum* in *Frisia* and dated between AD 29 and 116, contains evidence for a cattle sale between a Roman purchaser and a Germanic seller. Moreover, the price is given in currency, *nummis CXV*.[112] The other is an epigraphic inscription dated to the second half of the AD 1st century from near the village of *Boldog* in present-day Slovakia. [113] The inscription refers to *Q(uintus) Atilius Primus*, possibly a tradesman working within the *Barbaricum*.[114]

Between the Empire and the Barbarians, thus, there existed a certain level of exchange of goods. Pottery, weapons, brooches, and golden coins came out of the Empire and were spread within *Barbaricum*. This trade was, however, just one aspect of the complex relationship between Rome and Barbarian Europe. Thus, it is extremely difficult to reduce Romano-Barbarian relationships to a single explanatory model. The military confrontation, consistently emphasized in the written sources, was dominant in some given periods, however, Romans and Barbarians had moments of peaceful contact as well. These must have constituted the relevant portion of the general framework within which Rome and Barbarian Europe found their way to a mutual coexistence.[115]

However, the temptation to overestimate the importance of trade beyond the frontiers simply because it is highly visible from an archaeological point of view must be resisted. On the one hand, there is no precise information about the Roman need for metals or about the supply of metals or grain in any quantities from beyond the frontier for this period. The trade in slaves is mentioned by some literary sources but again its importance is hard to grasp. On the other hand, the import of luxury goods of Mediterranean origin among Barbarian elites may not have been as important as believed for the maintenance or changes in the power structure within *Barbaricum*, just as simple luxury goods or spices of Indo-Chinese origin were not crucial for the Roman Emperors in preserving their power.[116] In any case, a certain degree of penetration of Roman goods into *Barbaricum* is clearly evident in the archaeological record.

---

[105] As already noted in classical Soviet anthropology, repeated and constant economic relationships between different people facilitate cultural integration. Concerning this argument see, J.V. Bromlej, *Etnos e Etnografia*, 183: "... lo sviluppo delle relazioni economiche in un determinato territorio non soltanto esige l'integrazione della cultura (e in primo luogo del linguaggio), ma nello stesso tempo la rende possibile. Infatti, al rafforzarsi dei contatti tra la popolazione che si verifica in tali circostanze si accompagna inevitabilmente l'incremento dell'informazione culturale sincronica."

[106] C.R. Whittaker, "What happens when frontiers come to an end?," in P. Brun, S. van der Leeuw, C.R. Whittaker, *Frontieres d'empire*, 135-37 and 139; B. Ward-Perkins, *The fall of Rome*, 41.

[107] M. Richter, *The oral tradition in the early middle ages*, in *Typologie des sources du Moyen Age occidental*, 71 (Turnhout: Brepols, 1994), 22-23.

[108] O. Brogan, "Trade between the Roman Empire and the free Germans" *Journal of Roman Studies* 26 (1936): 195-222; H.J. Heggers, *Der römische Import im freien Germanien* (Hamburg: Hamburgisches Museum für Völkerkunde und Vorgeschichte, 1951); M. Wheeler, *La civiltà romana oltre i confini dell'impero* (Turin: Einaudi, 1963), 11-101; J.P. Callu, "I commerci oltre i confini dell'Impero," in *Storia di Roma 3, L'età Tardoantica I, Crisi e trasformazioni* (Turin: Einaudi, 1992), 487-524; R. Wolagiewicz, "Der Zufluß römischer Importe in das Gebiet nordlich der mittleren Donau in der altern Kaiserzeit," *Zeitschrift für Archäologie* 4 (1970): 222-49; K. Godłowski, "Der römische Handel in die *Germania Libera* Aufgrund der archäologischen Quellen," in K Düwell, H, Jankhun, H. Siems and D. Timpe eds., *Untersuchungen zu Handel und Verker der vor- und frühgeschichtlichen Zeit in Mittel-und Nordeuropa* (Gottingen: Vandenhoeck & Ruprecht, 1985), 337-66; M.G. Fulford, "Roman material in barbarian society c. 200 B.C.- c. A.D. 400," in T.C. Champion and J.V.S. Magaw eds., *Settlement and Society: aspects of West European prehistory in the first millenium B.C.* (Leicester: Leicester University Press, 1985), 91-108.

[109] Caesar, Gaius Iulius *The Gallic war*, H. J. Edwards, trans (London/ Cambridge, Mass.: Heinemann/ Harvard University Press, 1958), 4, 2, 1 and 1, 39, 1.

[110] Tacite, *La Germanie*, ed. J. Perret, (Paris: Les Belles Lettres, 1949), 41, 1; Id. *Annales*, P. Wuilleumier, trans. and ed., (Paris: Les Belles Lettres, 1974), 2, 62; Id. *Histoires*, Henri Le Bonniec, trans. and ed., (Paris: Les Belles Lettres, 1989), 4, 65.

[111] Cassius Dio mentions trading activities in connection with the peace treaties from the years AD 171 and 172 between the Emperors *Marcus Aurelius* and *Commodus* with the tribes of Marcomanni and Quadi. See Cassius Dio, *Dio's Roman history*, E. Cary, trans., (London/ New York: Heinemann/Macmillan, 1914-1927), 71, 11, 2-4; 71, 15, 1; 71, 18, 1.

[112] *Fontes Iuri Romani Anteiustiniani* (Florence, 1940-43), III, 137.

[113] *L'Anèe Épigraphique* 1978, 635, *Q(uintus) Atilius / Sp(uri) f(ilius) Vot(uria tribu) Pri/mus, inter(p)rex / leg(ionis) XV, | (= centurio) / negotiator, an(norum) / LXXX, / h(ic) s(itus) e(st), / Q(uintus) Atilius Cog(i)ta/tus, Atilia Q(uinti) l(iberta) Fau/sta, Privatus et / Martialis hered(es) / p(osuerunt)*. For a closer discussion on this source see, J. Kunow, *Negotiator et vectura. Händler und Transport im freien Germanien* (Marburg: Philipps-Universität Marburg, 1980), 17 note 83.

[114] T.M. Lucchelli, *La Moneta*, 166 and ff.

[115] T.M. Lucchelli, *La Moneta*, 2; J. Kolendo, "Les influences de Rome sur les peuples de l'Europe central habitant loin des frontieres de l'Empire," *Klio* 63 (1981):453-72.

[116] G. Woolf, "European social development and roman iperialism;" in P. Brun, S. van der Leeuw and C.R. Whittaker, *Frontieres d'empire*, 15, 18-19.

In addition to this influence on their material culture, the 'Barbarians' were also deeply involved in Roman politics. Already in the period of the Emperor *Honorius*, the Goth's king *Ataulf*, married *Galla Placidia* and presented himself as the *Romanae restitutionis auctor*. Following his example, another king of the Goths, *Theodoric II*, taking part in the policy of the Gallic-Roman aristocracy against Byzantium and the Italian-Roman nobility, became one of the most important personalities in the western part of what was still the Roman Empire. Other examples where the heritage of ancient world was encountered by the 'Barbarians' can be recognized in the use of Latin to record the old customs of the Germanic tribes[117] as well as the fact that the cultivated class within the western Roman-Barbarian kingdoms still comprised elements of the former Roman leading class such as *Avitus*, bishop of Vienne, *Sidonius Apollinares*, *Boethius*, *Simmacus*, and *Cassiodorus*.[118] Therefore, the fact that some of them, i.e. Barbarians, were already able, in the 5th century, to force their will and intentions on Roman society and to rationally cope with its surviving structures does not simply and simplistically pinpoint the level of Roman influence in shaping their social structures and social desires but helps us to assess the degree to which these *gentes* were *already* socially and politically self-organized. As a matter of fact, the archaeological evidence from their North European homelands also indicates that during the centuries of our concern these peoples were not "erratically raiding, undisciplined, socially amorphous hordes, but societies with socio-political and administrative structures."[119]

The result of this integration between the Roman and Germanic worlds was a sort of *transformation of the Roman world*, a gradual process which is apparently a more adequate term than the traditional label of the fall of the Roman Empire.[120] However, as correctly questioned by J. BINTLIFF and H. HAMEROW, this idea of a transitional period, this transformation, is just as historically and conceptually unsatisfactory as the notion of continuity and/or gap/disruption.[121] Actually, on the one hand, it is rather striking that so important a concept as 'the Roman World/Civilization' stands unquestioned and unchallenged as a historiographic monolith, in the face of the deconstructed and still being deconstructed notion of 'Barbarian Civilization.' On the other hand, it might be said that no period has existed in human history that was not a transition/transformation, a passage from one phase to another.[122] Certainly, some institutions of the Roman world self-evidently survived into the post-Roman world, particularly in regions such as Italy, southern Gaul and Byzantine territories. Their nature was, nonetheless, deeply altered. Therefore, this period of sweeping changes cries out for a more sophisticated approach which does not seek to "iron out" this immensely complex series of transformations in Europe by viewing them as a single transition or within a single explanatory model.[123]

In fact, in contrast to the general influence of Roman customs on the Germanic ones,[124] the evaluation of the grave goods found in the Germanic necropolises indicates that the weapons and war equipment of the intruding Barbarians were not at all influenced by Roman traditional forms of warfare. On the contrary, with the Migration Period, Roman armaments completely disappeared and the swords which developed in this period had no relationship to those produced and used during the Roman time.[125]

[117] I use here the term tribe based on Gasparri's definition. See S. Gasparri, "I Germani immaginari," 28.

[118] Bertolini, *I Germani*, 48, 63-65, 144, 323, 327-28.

[119] H. Schutz, *Tools*, 3.

[120] M. Richter, *The oral tradition*, 23.

[121] J. Bintliff and H. Hamerow, "Europe between Late Antiquity and the Middle Ages. Recent archaeological and historical research in Western and Southern Europe," in J. Bintliff and H. Hamerow eds., *Europe between Late Antiquity and the Middle Ages. Recent archaeological and historical research in Western and Southern Europe* (Oxford: BAR, I.S. 617, 1995), 6; on the same argument see, A. Cameron, *The Mediterranean world*, 198: "But societies do not exist in a vacuum. The world itself is a constant state of change ... myriads of small and large scale changes happened within the vast territories of the Empire and outside its borders."

[122] This attitude toward this question is, in my opinion, not accidentaly expressed in a few lines in W. Pohl, "Introduzione" in W. Pohl, *Le origini etniche dell'Europa*, 3: "L'impero romano aveva dato relativamente poco spazio alle identità etniche; l'appartenenza allo stato alla *civitas*, e il diritto dei cittadini romani erano molto più importanti." However, especially for Late Antiquity, terms as Roman and Byzantine should be questioned as much as Germanic or Barbarian. See, R. Bianchi Bandinelli, *Roma. L'arte nel centro del potere. Dalle origine al II secolo D.C.* (Milan: RCS Libri, 2005), 369-70: "Le province dell'Impero divengono protagoniste della storia, sia dal punto di vista politico sia dal punto di vista dell'organizzazione amministrativa e finanziaria, e quindi anche culturale e artistica. Roma inizia da ora un suo lungo ruolo di passività culturale, di quasi parassitario centro di assorbimento e di incontri. ... Ma da quelle difficili e drammatiche situazioni, quelle che erano culture artistiche periferiche di paesi quasi coloniali, andranno poi costituendo le tradizioni formali che saranno attive nelle età successive."; C. Azzara, *Le invasioni barbariche*, 18: "Studi recenti hanno messo in luce come le province del tardo impero fossero realtà tutt'altro che pacificate e omogenee sul piano etnico e culturale; in esse persistevano, infatti, specificità locali ben vive e coscienti, soprattutto negli strati inferiori della popolazione, di matrice eterogenea rispetto alla romanità ... "

[123] On this argument see, J. Bintliff and H. Hamerow, "Europe between Late Antiquity and the Middle Ages," 6; A. Cameron, *The Mediterranean world*, 56: "But simple explanations are always inadequate for complex historical change."; C. Azzara, *Le invasioni*, 70: "Peraltro, riconoscere i precisi tratti di continuità che regioni quali l'Italia, la Gallia, la penisola iberica mantennero nel passaggio dall'ordinamento imperiale a quello dei regni, rispettivamente, ostrogoto ..., franco, e visigoto (in parte diverso fu il caso dell'Africa vandala, in cui la rottura ebbe aspetti più traumatici), se per un verso corregge l'errata impressione di troppo drastiche cesure, per un altro rischia di svilire in modo eccessivo i fenomeni di trasformazione che pure vi furono, e che, tra l'altro, distinguono lo sviluppo delle province occidentali da quello dell'oriente bizantino, in cui l'eredità di Roma appare essersi trasmessa in forma senz'altro più lineare. È impossibile, allo stesso tempo, pretendere di ridurre a tipologie generali le dinamiche che si verificarono nei singoli regni e che furono invece molto diverse, tanto che ogni vicenda va studiata nella sua specificità, come caso a sé, senza la pretesa di astrarre da un insieme tanto eterogeneo categorie interpretative che semplifichino arbitrariamente ciò che fu articolato e diseguale."

[124] This influence was indeed mutual. The laws of the *Codex Theodosianus* forbidding people to adopt Barbarian fashions in their clothing and the way they dressed their hair in the city of Rome are a good example; on the subject see, T. Momsen and P.M. Meyer eds. *Codex Theodosianus* (Berlin: 1905), XIV, 10, 2-4; B. Ward-Perkins, *The fall of Rome*, 41.

[125] R. Forrer, *Die Schwerter und Schwertknaufe der Sammlung Carl von Schwerzenbach-Bregenz* (Leipzig: Hiersemann, 1905), 9 and 12; E. Salin, *La Civilisation merovingienne*, vol. 3, 6, 81-82; V. La Salvia, *Archaeometallurgy*, 20-22.

This situation is obviously related to the fact that the Roman army should not automatically be considered superior to that of the Barbarians. In fact, in the period between the battle of *Hadrianopolis* (AD 378) and the death of the Emperor Gratian (AD 383) the Imperial army faced a difficult crisis which finally exploded in the first half of the AD 5th century.[126]
The barbarization of the Roman army also has to be considered as being closely tied to the technical superiority of the Barbarian methods of weapon production[127] and not only to difficulties in recruiting soldiers within the borders of the Empire.

Thus, it seems that the "intruding" Barbaric populations not only had a simple, simplistic sponge-like approach to Roman civilization, that is to say, were undergoing an acculturation process which led to their formal, uncontested and "desired" peaceful assimilation. There are certainly two sides to the picture. Moreover, due to the nature of the evidence, there is a tendency to overrate the Roman element which is so much more obvious and also archaeologically much better attested.[128] However, according to the French anthropologist A. LEROI-GOURHAN, contacts between populations and different cultural entities, as for example between the Roman and Germanic worlds, do not imply any sudden, complete and mutual substitution and/or transplant of cultural traditions. On the contrary, these contacts point towards a general trend of cultural osmosis. The latter is largely due to the pressure of the *external environment* which is significantly altered for both actors, that is to say, for the old residents and for the newcomers. Moreover, as far as intercultural loans are concerned, they can become long-lasting only if the material objects loaned meet the following conditions: they either satisfy a pre-existing need, or satisfy it better or create a new one that is compatible with the present life of the group. In other words, the loaned objects encounter a *favorable environment.* This means that a loan occurs more easily between cultures at similar technical/social levels, confirming the old axiom that things tend to be borrowed that one is already equipped to invent. Once assimilated, this object satisfies two other conditions: it acquires the "personal touch" of the borrowing group, a so-called local *facies*, and, eventually fits itself to the requirements of available raw materials and the technical know-how of the new habitat. In fact, in ethnographic literature there are no examples of loans that do not show signs of pressure from the new internal environment: a loan becomes a proper object within the new ethnic context only through becoming an integral part of the *internal environment* of the given human group.[129]

Therefore, the presence of Roman goods within *Barbaricum* has a deeper and more complex economic and social meaning (even from an ethnogenetic point of view) than the simplistic assumption of foreign models regarded as "signs/markers" of new social conditions within a more primitive society. The presence of such goods should, on the contrary, help archaeologists to explore the internal dynamics of the societies living within *Barbaricum.* It should be possible to trace development processes in the *Barbaricum* and loans from abroad, rather than imposing on archaeological data paradigms developed thorough historical sociology, mainly based on written sources.

As noted by J. BINTLIFF and H. HAMEROW, in recent studies the crucial question of the developing lines of "Barbarian" societies living outside the borders of the Roman Empire during the first centuries of our era has finally come to the attention of scholars. Through massive excavation campaigns in northern and central Europe, it has been clarified that Barbarian societies did not wait for the decline of Rome as a passive, unchanging and relatively primitive form of society, like the traditional 'Germanic mode of production,' 'Military Democracy,' 'Tribalism' or the like. In this way, any of the observations of outsiders such as *Caesar* or *Tacitus* might apply to any phase of the Iron Age. On the contrary, these societies are now recognized as having undergone complex transformations in parallel to the rise and fall of the Roman Imperial power. Various models have been used to explain these changes: purely indigenous development, core-periphery relationship involving contacts with the Empire, and, finally, a combination of the two models. Whichever approach is taken, however, the result has been to encourage us to see the Migration Period as potentially the result of growing convergence between Barbarian communities *increasing* in their complexity and a Late Roman society that in many or most provinces is *deconstructing* its complexity. The archaeological evidence shows that by the Late Roman Iron Age (from the AD 2nd -4th centuries) many settlements from Scandinavia to the Low Countries displayed a complex, controlled and orderly layout. A disruption and break-down of settlements is apparent in the AD 5th century in many, though certainly not all, areas. This break-up of the highly structured layout resulted in a decrease in the average length of the longhouses and an increase in the number of houses without byres, in a shift in emphasis from arable farming to craft production, reflected archaeologically in fewer stalls for cattle, dwindling numbers of granaries and increased numbers of sunken huts-cum-workshops. According to some German scholars this period marked a change from fenced, ancestral *Mehrbetriebgehöfte* to

[126] A. Cameron, *The Late Empire*, 134, 136-37, 141; Ead., *The Mediterranean world*, 49-56; *Maurice's Strategikon*, trans. and ed. G. T. Dennis, (Philadelphia: University of Pennsylvania Press), viii: "Warfare as practiced in the ancient world underwent a series of transformations ... The tough, disciplined legionary who, covered by his shield, used his short sword to cut his way through all opposition, ... no longer corresponded to the needs of the time. Instead of fighting on foot, the legions mounted horses and became archers and lancers. As with their armament, the size of the units, tactics, also underwent significant changes. ... In the course of the third and fourth centuries the legions came to be reduced in size, and large numbers of new units were created. More emphasis was placed on mobility, and thus on cavalry, who could move more rapidly ... The enemies of the Romans also depended more and more on horses in their attacks. ... It was not long before the victors at Adrianople were being hired to serve in the Roman army."; see also V. La Salvia, *Archaeometallurgy*, 21-22; and E. Salin, *La Civilisation merovingienne*, 90, 114 ff..

[127] Salin, *La Civilisation merovingienne*, 90

[128] M. Richter, *The oral tradition*, 22.

[129] A. Leroi-Gourhan, *Milieu et Techniques*, 351-73.

unenclosed *Großgehöfte* in which several family groups lived together. Nonetheless, the explanations put forward by archaeologists are inevitably tentative and frequently couched in terms of climatic change and demographic pressures, culminating in an agricultural "crisis". Any such hypothesis always requires considerable critical revision. The investigation of settlement layouts is very important since through its investigation it is possible to infer economic and social changes within any given society. These changes are linked to the way in which a community arranges its living space. Anthropologists have 'demonstrated' that these process are only partially related to technical or formal considerations and that shared attitudes and social relations play a major role in determining the layout of settlements. Consequently, spatial order in a settlement reflects both social order and helps to regulate social relations, providing, literally, a framework for living. It is possible to correlate increased economic complexity and increased complexity and regularity in settlement structure and, thus, it is also possible to relate the evidence of increasingly complex settlement organization and social stratification to economic and ecological factors, although exogenous influences might be solely or partially responsible. Contemporary with changing settlement structures, other changes in Barbarian societies included developments in crop growing, stock breeding, and internal responses to natural transformations in the world around them (including climatic deterioration). Moreover, these societies were affected by the collapse of the Western Empire and the formation of "successor societies" to differing degrees and responded in different ways.[130]

From what has been said above, it should be clear that the complexity of various societies that existed beyond the *limes* cannot be considered only as the result of their continuous relationship with areas controlled by the Roman Empire or as simple creations of the political genius of Rome,[131] just as today we cannot assume that the development of contemporary China or India can be ascribed to their historical and important relationships with the modern industrialized West. Today, just as previously, the task of a historian and/or an archaeologist is not to offer a single explanatory model but to delve into complexities.

From economic and also social points of view, the *Barbaricum* and the various populations that comprised its demographic structure must have had a well-defined ethnic identity. The existence of such an identity is proven by the fact that each of these groups possessed technical objects and production strategies that are "absolutely" distinct from those of other groups (including the 'Romans'). These, objects are totally shaped by the pressure of the *internal environment* of a given group. Indeed, this identity does not correspond objectively either to what a group may call itself which, as stressed above, is only a conventional definition useful for research and even less to a biological entity such as race, as defined in physical anthropology. Nevertheless, it corresponds to a process of ethnogenesis which is based on the formation of individual ethno-cultural identities, the exact opposite of a melting-pot. In this sense, the development of human communities is similar to the general trend of biological evolution which does not lead to the development of standardized hybrids (without specific characteristics, melting-pot like entities) but to the "creation" of individual forms with a positive fitness relative to their habitat.[132]

Moreover, while for other aspects of human tradition, such as the definition of a religious or social/status, the tradition and respect for it are highly conscious, active and verbal (oral), the preservation of craftsman's technology is almost never theoretical or spoken and never corresponds to a fully conscious practice. Technological traditions form an unstable basis that assures the possibility of material reproduction of the majority of objects pertinent to a given human group for successive generations. The ultimate role of these traditions is, thus, that of guaranteeing the continuous survival of technical know-how through generations and promoting social solidarity.[133]

The *routine*' the custom, the repetition of the *chaine opératoire* of appropriate technical gestures through time is the core of this system of accumulating technical

[130] J. Bintliff and H. Hamerow, "Europe between Late Antiquity and the Middle Ages," 1; H. Hamerow, "Shaping settlements: Early Medieval communities in Northwest Europe," in J. Bintliff and H. Hamerow eds., *Europe between Late Antiquity and the Middle Ages*, 8, 14, 16.

[131] As in P. Geary, *Before France and Germany* (New York: Oxford University Press, 1988), vi.

[132] On the issue of preserving ethnic/group identity in human communities see the fundamental studies of I. Eibl-Eibesfeldt, *Etologia Umana. Le basi biologiche e culturali del comportamento* (Turin: Bollati Boringhieri, 1993), 210-215: "... il comportamento degli altri componenti del gruppo è in complesso ben prevedibile; il partner insomma non uscirà dal ruolo che gli è stato assegnato culturalmente in base all'età, al sesso e allo status. La norma del gruppo si manifesta nel linguaggio, nelle usanze, nell'abbigliamento, negli ornamenti personali e in molte altre azioni quotidiane e persino tradizioni comportamentali e spirituali sono improntate secondo questa norma. La cultura si dimostra in ciò capace di plasmare e di condizionare come una "seconda natura," in quanto il patrimonio delle usanze trasmesse culturalmente non lascia troppa libertà di scelta ... Allo stesso tempo, il mantenimento della norma specifica di gruppo è un mezzo per distinguersi dagli altri ... La tendenza all'accentuazione del contrasto, basata sulla volontà di evidenziare le proprie caratteristiche, ha condotto a un rapido differenziamento culturale (pseudospeciazione) che ha permesso agli uomini di adattarsi rapidamente a spazi vitali assai diversi. La molteplicità delle culture, le quali utilizzano strategie di sopravvivenza molto differenti, è un'espressione di questa tendenza, che a sua volta, produce diversità e certamente ha effetto creativo. La norma di gruppo viene difesa ed esiste un'aggressività in difesa della norma rivolta contro quei membri del gruppo che deviano in modo evidente da essa."; R.A. Hinde. *Le basi biologiche del comportamento umano* (Bologna: Zanichelli, 1979), 284; A.F.C. Wallace. *Culture and Personality* (New York: Random House, 1967), 120-63; see also B. Ward-Perkins, *The fall of Rome*, 77: "There is no reason to believe, as people once did, that ethnic behaviour and identity are genetically transmitted and, therefore, immutable. But experience suggests that a great deal of an individual's identity is acquired during childhood and early youth, from parents, the wider family, and companions, and this identity is not easily forgotten. This being so, individuals have never been entirely free to choose what they wish to be."

[133] M. Godelier, *L'Idéel et le Matériel*, 109: "Ainsi l'appartenance à un group garantit à un individu l'accès aux ressources et les rapports des groupes entre eux multiplient et distribuent le nombre des possibilités offertes à l'individu."

knowledge. At the same time, this technical tradition seems to be the only way to maintain the style of objects that suits the tastes of a given group. Indeed, while remaining attached to one's own style, at the same time, craftsmen were also bound to preserve their traditional technical know-how.[134] As G. STABILE has poignantly marked, it is not a projection of an individual language into a visible [object], but a material expression of a technical slang assimilated through memory and tradition and, therefore, the contribution of single craftsmen must conform to this common language.[135]

Based on this belief, archaeological artifacts must be "interpreted" first as a way of understanding how efficient objects were produced according to different means and strategies of production by various workforces who had diverging technical backgrounds and, thus, were related to a different organization of production and, therefore, to different societies and related social values.[136]

[134]A. Leroi-Gourhan, *Milieu et Techniques*, 342-43, 409, 421-23, 425-26.

[135]G. Stabile, "La torre di Babele," 276; on the same topic, see as well A. Leroi-Gourhan, *Il gesto*, 322 and ff., 356.

[136]J.V. Bromlej, *Etnos e Etnografia*, 182: "In quanto si accompagna di solito a un progresso tecnico, la riproduzione dei beni materiali trascina inevitabilmente con sé modificazioni culturali ... Ma l'influenza esercitata dalla produzione sulla modificazione della cultura materiale non si riduce, s'intende, alla semplice incorporazione in seno ad essa di oggetti prodotti con materiali nuovi; a una forma importante di rinnovamento della cultura materiale si perviene anche per altra via: la creazione di oggetti di tipi nuovi partendo da materiali di impiego tradizionale. Nell'uno e nell'altro caso tali innovazioni si riferiscono non soltanto agli oggetti di consumo, ma alle stesse forze produttive. Costituendo, infatti, insieme con i rapporti di produzione, il modo di produzione, cioè la base economico-sociale, le forze produttive sono componenti della cultura contemporaneamente in qualità di risultati (strumenti di lavoro) e di modo (conoscenze ed esperienze) dell'attività delle persone."

Chapter 3
Lombard Iron-making within Northern and Central Europe.
The building up of a craftsmanship tradition.

3.0 Introduction

The Suevians were an important west Germanic group originating around the Baltic Coast, Scandinavia and Northern Jutland. They were part of a great population movement toward Celtic territories which had been underway since the 4th century B.C. and ended in a geographical area stretching between the Rhine and the Danube until the *Elbe* in the east. This was the zone occupied by western Germanic civilization. According to *Strabo*, around the year AD 20, this area was fully populated by the Suevians, among whom he also includes the Ermodures and Lombards although a part of the latter also inhabited areas beyond the *Elbe*.[137] This was a part of a larger movement of Germanic peoples that also spread east, far beyond the *Elbe*, all the way to the flatlands of Western Russia giving rise to the formation of the so-called eastern Germanic cultures.

In these regions, particularly the northern parts of present-day Germany, to the west as well to the east, and southern Denmark, the Netherlands, north-western Slovakia and Poland (especially the area of the Holy Cross mountains home of the important *Přezworsk* culture), there is wide-spread evidence of iron production centers. Iron-making was the dominant economic activity in these centers, developing into traditional activities. This shift in emphasis from elite, specialist to everyday life implies a long time-frame from the Early Germanic Iron Age [138] till the Migration Period (AD 5th-8th centuries), and, in many cases, till the high Middle Ages. (**see figure 2**). Recently, numerous archaeological excavations have brought to light important archaeometallurgical evidence in these territories.

Denmark

A farmstead, a "smithing farm," was discovered in Denmark, in *Snourp*. Here, iron production probably had already gotten underway by the 5th –4th century BC. From later time periods 4000 to 5000 slag-pits furnaces distributed over 35 hectares were found and excavated, apparently indicating an intense period of production of two hundred years, from the AD 4th to the 6th century. Bog ore was used, and there is evidence of blacksmithing activities beside the general smelting process. Moreover, from the evaluation of the slag-pits, it was possible to establish that about 280 kg of roasted iron was necessary to produce 200 kg of slag, which hypothetically suggests 60 kg of unworked iron. The stratigraphy of the furnaces, of those distributed linearly as well as of those grouped in clusters, is not chronologically homogeneous: the furnaces were thus, not operating at the same time. Most likely, each farm ran 10 furnaces per season for a total of less than 20 active furnaces per season. The enormous number of furnaces must have been the result of continuous smelting year after year throughout the site's existence.

Another, important site is that of *Drengsted* where 224 slag pit furnaces were discovered. Iron-making was concentrated here around the three farms on the site, representing the entire life span of the settlement, dating from the Early pre-Roman Iron Age to the Early Germanic Iron Age. Moreover, the furnaces' locations, all around a farmyard, suggest that the farm units were surrounded by a fence which was not discovered during excavations. In this case as well, surface limonitic iron ores constituted the raw material.[139]

Northern Germany

Moreover, the excavations carried out in the area between the *Elbe* and the *Weser* rivers have identified 59 sites with traces of organized production using slag-pit furnaces. Iron production was continuous in this region for an extremely long time-period between 1stcentury BC till AD 14th century. The vast majority of the archaeometallurgical finds have been subjected to archaeometric analyses. Bog ore was used as raw material in this region as well. During prehistoric times, this mineral was abundant in the wetlands beside lakes and along the rivers close to the northern sea coast. The continuous exploitation and intensification of agricultural finally brought the exploitation of this natural resource to an end. Charcoal was made from oaks and the smelting furnaces were of the slag pit type. Iron-making activities took place only in areas with more abundant supplies of ores and fuel (wood).[140]

---

[137] See, M. Menke, "Archeologia Longobarda," 47; L. Schmiedt, *Geschichte der Germanische Völker* (Munich/Berlin: Oldenbourg, 1909), 23; and O. Bertolini, *I Germani*, 17.

[138] This system of dating, commonly used for the archaeology of northern continental Europe and Scandinavia, does not imply any racial, biological, or even cultural affiliation. It is a normal and simple conventional terminus technicus for indicating specific periods of time within the Iron Age (beginning around 500 BC): Early pre-Roman Iron Age (500-300 BC); pre-Roman Iron Age (300-100 BC); Roman Iron Age (AD 1-200); Late Roman or Early Germanic Iron Age (AD 200-600); Late Germanic Iron Age (AD 7th century).

[139] On the Scandinavian situation see, O. Voss, "Eisenproduktion und Versorgung mit Eisen in Skandinavien vor der Vikingerzeit," *Early Medieval Studies* 3 (1971): 22-30; on the site of Snourp see, O. Voss, "Snourp: An Iron Producing Settlement in west Jutland, 1st-7th century AD," in G. Magnusson ed., *The Importance of Ironmaking*, 133-34, and 137-38; Id., "Snourp – An Iron Age settlement with iron production in the 2nd to the 7th Century AD," *Early Iron* 1 (1996): 19-28; about Danish situation see also, L. Norbach, "Iron Smelting in Denmark from 100 BC to c. 400 AD," *Early Iron* 1 (1996): 9-18; Id., "Organising iron production and settlement in Northwestern Europe during the Iron Age," in C. Fabech and J. Ringtved eds., *Settlement and Landscape. Proceedings of a Conference in Århus, Denmark. May 4-7 1998* (Århus: University Press, 1999), 238-42.

[140] As far as these zones are concerned see P. de Rijk, "Eiseverhüttung und Eisenverarbeitung im nordwestlichen Elbe-Weser Raum," *Ethnographisch-Archäologische Zeitschrift* 37 (1996): 325-33; Id., "Iron production and manufacturing in the Elbe-Weser-area," in G. Magnusson ed., *The Importance of Ironmaking*, 81-86; Id., "Examinations on iron slags in the north-western German coastal region," *Early Iron* 1 (1996): 49-52.

Further to the east, in the area between the rivers *Elbe* and *Oder* since the AD 1st century iron exploitation and smelting was taking place on a relatively large scale and continued in this manner until the Migration Period. Within these densely populated zones, there is evidence of iron-making activities in every second settlement. Bog ore, that is to say, limonitic iron ores were used as raw materials. The richness of these ores in metal content varies consistently, ranging between 30% and 50%. Eleven different zones can be distinguished as important iron production centers: the area between *Wismar* bay and the mouth of the river *Oder*; the region between *Greifswald* and *Wolgast*; southwestern *Mecklenburg* between *Boize* and *Löckniz*; the eastern and central *Mecklenburg*; southern *Vorpommern* and the central *Oder* area; the *Havelland* and the *Spree-Havel* zone; the region of *Lausitz* till the mouth of the *Elbe* river; the mouth of the *Elbe*; central and northern *Altmark*; the zone between *Jeetze* and *Oste*.

Among these areas the most significant ones, in terms of levels and fineness of production, are southwestern *Mecklenburg*, where the largest bog ore deposits in northeastern Germany are located, the area around *Greifwald*, and Lower *Lusatia*, where remarkable batteries of slag pit furnaces from Late Antiquity could be found. Moreover, the shape of the slag pit furnaces (such as the *Tulaty* types) demonstrates direct Celtic influence from Bohemia on iron technology as early as AD 1st century.[141]

The Netherlands

There is also evidence of iron production in the Netherlands. Bog iron ore and *Pierre d'Aigle* were mined in open-cast pits. In the *Vecht* area over a period which extends from the AD 2nd to 4th century, iron production took place on a non-organized basis primarily in order to supply local needs. Nevertheless, during the first half of the AD 4th century, apparently within a generation, production assumed a strictly organized face. In the *Veluwe* area, land not suitable for agriculture but inhabited continuously from the Paleolithic until the Late Middle Ages, large-scale iron production was on-going between the AD 7th and the 10th century.[142]

The settlement of *Heeten* in the same region (province of *Overijssel*), excavated in 1994, is extremely interesting. The site dates to the first half of the AD 4th century and is located on the plains in front of the Germanic *limes*. More precisely, it lies almost 50 km from *Arnheim-Meinerswijk* where a Late Roman *castrum* was placed. *Heeten* is a fortified settlement as well, surrounded by a fence enclosing an area of 110 x 77 m. A large quantity of iron slag was found inside pits and sunken-floored houses. The smithing activity was apparently located inside a rectangular house and in its close proximity. At the site 1000 smelting furnaces were also unearthed together with around 45/50 tons of slag (some still within the foundation line of the fence) together with 15 tons of iron bloom. Iron production must have been significantly high (more then 100 kg per year) since the site itself is relatively short-lived. *Heeten*'s production was, therefore, quite important and possibly related to the manufacture of weapons and other military equipment. Whether the production was destined for the Roman or the Germanic markets is still difficult to say. As a matter of fact, the quantity of Roman goods discovered at *Heeten* does not exceed that which would normally be expected at any other site in the nearby region. Thus, there does not seem to have been any special connection with the Roman world. However, it is also clear that a Germanic elite was involved in the organization of large-scale iron production as a main source of income during the first half of the AD 4th century.[143]

The Holy Cross Mountains (Poland)

The metallurgical centre in the Holy Cross Mountains in present-day Poland, among its long-lasting production activities from the *La Téne* period to the Migration period (five hundred years of production), was able to produce 3800-5400 tons of iron. It therefore, must represent the main production center for the *Barbaricum*. Here, traces of a settlement and a cemetery of the *Přezworsk* culture, pertaining to the main area of production, were also discovered. The largest smelting site here, *Witomarz*, consisted of two organized fields with a total of 231 furnaces. In this area, hematite iron ores were exploited. This marks a slight difference with the rest of the Germanic production sites where surface limonitic iron ores (Bog ore) were the main source of raw material: this resource, abundantly available in prehistoric times in the low-lying areas adjacent to streams and lakes in northern Europe, has been totally exploited or destroyed by later intensive agricultural works over the centuries.

Dating to the Roman Iron Age and flourishing between the 3rd-4th century AD, the iron-production centre in the Holy Cross Mountains displays an organized production structure with many parallel lines of furnaces indicating

[141]See, J. Schneeweiss, "Die ur- und frügeschichtliche Eisenverhüttung und –verarbeitung im westlichen Odergebiet," in *Beiträge des Symposium, Eisengewinnug und –verarbeitung in der östlichen Germania Magna.' Humboldt Universität Berlin 26-27 April 1996*, *Ethnographisch-Archäologische Zeitschrift* 37.3 (1996): 335-63; A. Leube, "Die Eisengewinnung und –verarbeitung während der römischen Kaiser- und Völkerwanderungszeit im Gebiet zwischen Elbe und Oder," *Ethnographisch-Archäologische Zeitschrift* 33.4 (1992): 471-98; Id., "Studien zur Wirtschaft und Siedlung bei den germanischen Stämmen im nördlichen Mitteleuropa während des 1.-5./6. Jh. u.Z.," *Ethnographisch-Archäologische Zeitschrift* 33 (1992): 132-36; I. Spazier, "The Germanic Iron Smelting Complex at Wolkenberg in Lower Lusacia, Southern Brandenburg," in L. Nørbach ed., *Prehistoric and Medieval Iron Smelting in Scandinavia and Europe* (Åarhus: University press, 2003), 37-42.

[142]See, M. van Nie, "Three Iron Production Areas in the Netherlands: Contrasts and Similarities," in G. Magnusson ed., *The Importance of Ironmaking*, 101-02.

[143]For this argument see, A. Dirk Verlinde and M. Erdich, "Eine germanische Siedlung der späten Kaiserzeit mit umwehrter Anlage und umfangreicher Eisenindustrie in Heeten, Provinz Overijssel," *Germania* 76 (1998): 693, 698, 699, 700, especially 704-08 for the catalogue of the metal finds, 709-11 for iron production, 714, 715, 716, 718; see also, I. Joosten, M. van Nie, "Introducing the early iron production in the Netherlands," *Early Iron* 1 (1996): 29-42.

intensive exploitation of natural resources. The organization of such large-scale iron production certainly necessitated some kind of structured and centralized power, able to provide fuel and a specialized work force, and presupposes the existence of a market. Moreover, the increasing, continuous and exclusive use of the slag-pit furnace would lead Germanic iron production in the entire *Germania Libera* until the Migration Period. This type of furnace was used in little settlements for small-scale production as well. It is still a question whether all this worked iron was meant, as believed by Polish scholars, to fulfill the needs of all the Germanic groups belonging to the *Přzeworsk* Culture, or traded to the Danubian provinces of the Roman Empire.[144]

From the above it is evident that there were specialized iron-makers among the Germanic peoples. This is confirmed by the discovery of 33 metal-workers graves in *Germania Libera* from the pre-Roman Iron Age and Roman Iron Age, as well as another 28 such graves from the Migration Period. Among the latter group, graves from *Brno* (Moravia in the Czech Republic) and *Poysdorf* (Austria), are attributed to the period of Lombard occupation of these areas.[145] Thus, from the perspective of iron production, it appears that the *limes* truly functioned as a technological barrier, revealing that an essentially uniform and original *Barbaricum* existed actually from the technological point of view, a real technological *koinè* as far as iron production and particularly as reduction technology was concerned. The two key-stone features of the latter process were: the use of a particular type of a furnace – *the slag pit furnace* (**see figure 3**) – and an extensive and decentralized production organization which was aimed at satisfying regional demands. Moreover, another important factor is that with the exception of the area of the Holy Cross Mountains in Poland where hematite was used, bog ore was the only raw material. Within this latter Polish region, there is evidence that iron production was organized differently as well as being characterized by a more intensive activity.[146]

Despite the great number of furnaces, total productivity did not attain the production level of Roman mining enterprises, amounting only 10% of it. But this was not due to supposedly inadequate technology, but to a completely different technological style which being extensive and diffused can be sharply distinguished from the intensive and centralized Roman-Mediterranean system. The production of greater quantities was assured due to multiple and successive operations of reduction in the same furnaces which were, usually, loaded only once per use. Such organization of production was found also in territories closer to the border with the Roman Empire, such as in southern Slovakia, Moravia and along the left bank of Danube. The only exception seems to have been the area of the *Puchov* culture, between *Slovakia* and *Silesia* (Poland), which some scholars would link to the Cotins, recorded by *Tacitus*, Celts subjected to western Germanic peoples who settled in this area (*Quadi* and *Marcomanni*) and to whom they paid their tribute in iron.[147]

---

[144] On the Polish situation see, K. Bielenin, "Frühgeschichtliche Eisenverhüttung im Heiligenkreuz-Gebirge (Góry Świętokrzysjkie). Allgemeine Bemerkungen," in *Beiträge des Symposium 'Eisengewinnug und –verarbeitung in der östlichen Germania Magna,'* 293-308; K. Bielenin, "Frhügeschichtliche Eisenverhütung auf den Gebiet Polens," In J. Gomori ed., *Traditions and Innovations in the early Medieval Iron Production.* Sopron/Somogyfajsz: Dunaferr-Somogy Archaeometallurgical Foundation/ Soproni Muzeum, 1999, 57-62. R. Pleiner, "Extensive Eisenverhüttungsgebiete im freien Germanien," in *Symposium. Ausklang der Laténe-Zivilisation und Anfänge der germanischen Besiedlung im mittleren Donaugebiet* (Bratislava: Slovak Academy of Sciences, 1977), 298-99; Id., "Der Handel mit Eisen im östlichen Mitteleuropa im 4. bis 9. Jahrhundert," *Early Medieval Studies* 3 (1971): 16; Id., "The Technology of Iron in the Bloomery Period: A Brief Survey of the Archaeological Evidence" in R. Francovich ed., *Archeologia delle Attività Estrattive*, 540 and 553; M. Muller-Wille, "Settlement and non-agrarian production from the high mountain region to the shoreline" in C. Fabech and J. Ringtved eds., *Settlement and Landscape*, 210.

[145] As reported in R. Pleiner, *Iron in Archaeology*, 275: "In any case, it would seem that all of the iron-producing communities must have included trained metallurgists who had gradually become established and who promoted the technology." On the same topic see also, J. Henning, "Schmiedegräber nördlich der Alpen. Germanisches Handwerk zwischen keltischer Tradition und römischem Einfluß," *Saalburg Jahrbuch* 46 (1991): table 1, 71; A. Leube, "Studien zur Wirtshaft," 136.

[146] On the difference in iron working between the Roman and the Germanic world see R. Pleiner, *Iron in Archaeology*, 45: "... decentralized Germanic iron metallurgy began in some cases, to concentrate in several regions within '*Germania Libera*,' presumably in an effort to increase production and was achieved simply by deploying ever-larger assemblages of traditional slag-pit furnaces ... Such a way of working may be described as 'extensive,' as distinct from the 'intensive' work of the Roman world which was based on large, long-lived shaft furnaces."; 272: "Even close to the *Limes*, in southern Slovakia and Moravia and the lands on the left bank of the Danube in Austria, the remains of barbarian iron smelting explicitly indicate small-scale production and do not show any Roman influence."; and 274: "The metallurgy of the non-Roman territories was dominated by a particular smelting device – the slag-pit furnace – whose origin are unknown, but which was typical for Germanic tribes who used it in the extensive smelting works ... In principle it was a shaft furnace, allowing a sufficiently long exposure of the ore-and-fuel charge to the process, and was adapted for the removal of slag by gravity down to the sunken hearth ... Slag-pit furnaces were possibly in use from pre-La Tène times. The eastern Celts definitely used it as did non-Celtic populations from Volynia and Ukraine in the east, to England in the west and from Austria and Moravia up to central Norway in the north."

[147] With respect to Germanic iron technology again see, R. Pleiner, *Iron in Archaeology*, 45, 47, 272, 274. On the question concerning the existence of a Germanic cultural and technological *koinè* see V. La Salvia, *Archaeometallurgy*, 68; V. La Salvia and F. Zagari, "Cultura materiale e tradizione tecnica," 952; N. Christie, *The Lombards* (Oxford: Blackwell, 1995), 4, 12; Id., "Longobard weaponry and warfare, AD 1-800," *Journal of Roman Military Equipment Studies* 2 (1991): 2-3; M. Todd, *Northern Barbarians, 100 BC- AD 300*, (London: Hutchison, 1985), 64-72; L. Paroli, "La cultura materiale," 258. On the *Pùchov* culture see R. Pleiner, *Iron in Archaeology*, 45: "They (Cotini) did not escape the attention of ancient writers such as Tacitus ... Ptolemy of Alexandria in his Geography (2.11) drew particular attention to iron mines (*siderucheia*) which he located somewhere in the lands of the Quadi ... In relation to this, a relatively small area, still only partially investigated, at Varín, east of Žilina in northern Slovakia should be noted. The smelting of hematite ores resulted in the accumulation of seven heaps of slag and furnace debris heaps measuring up to 10- 20m in diameter. In addition, a number of banks of slag on forested slopes in this area have been identified ... numerous furnace fragments and tap slags permitted the reconstruction of large shaft furnaces, built out of brick-like blocks standing to well over 1m in height. These are somewhat unusual among the Romano-Barbarian contexts, differing from the Barbarian slag-pit furnaces which appear everywhere else in the surrounding territories. Pottery finds belong to the Pùchov culture of

Moreover, the iron production in *Germania Libera* seems to have experienced a significant increase in volume between the AD 2nd and 3rd centuries. In the same time period and in the same geographical area the development of intensive agricultural practices can be observed which may very well have been related to the availability of more iron for the production of agricultural tools. Such transformations in economic organization also led to changes in the social structure of the Germanic peoples, something that can be reconstructed based on archaeological evidence, since changes in settlement structures and production models evolving towards a more stratified system have been observed. Thus, the social differentiation within Germanic tribes did not, in fact, begin only as a result of their contacts with Roman military organization, that is to say, from their "Romanization," nor was it related to any inherent predilection for a hierarchical-military organizational system. Thus, while written classical sources such as *Tacitus* insist on the existence of a hierarchical society among the Germans as the result of a warriors-chieftains organization, the archaeology of Northern Europe reveals a more complex situation. It, actually, seems that a long lasting process of development in agricultural technology and surpluses in production and, consequently, in land ownership, took place over the period stretching from the Late pre-Roman Iron Age well into the Roman Iron Age. This radical change in the strategy of obtaining means of subsistence (from an extensive farming to a more intensive mode of production) may have affected the social structure of the Germanic populations as much as their contacts with the Roman culture. It is also striking that the earliest stratified societies of northern Europe emerged far from the frontier (further north). This is particularly important since the possible acculturation of the military elites does not at all imply the acculturation of the rest of the populace.[148]

Thus, the connection between the agricultural boom, the wide-spread diffusion of iron working tools and the "consequent" formation of a stratified society took place within a complex framework in which different socio-economic dynamics, as revealed through archaeological excavations, each played a significant role. Technological innovations, access to natural resources, topography and distribution of settlements, system of land and tools ownership, economic, technological and cultural

the 1st – 2nd centuries AD, which was much affected by Celtic influences, or even immigrants."

[148] On the agricultural boom see A. Leube, "Die Eisengewinnung," 471, and Id., "Studien zur Wirtschaft," 130-31. On the enhancement of iron production in the *Barbaricum* see R. Pleiner, *Iron in Archaeology*, 274; on the general economic situation beyond the *limes* e the interaction between Roman culture and endemic socio-economic dynamics, see L. Hedeager, "Empire, frontier and the barbarian hinterland: Rome and Northern Europe from AD 1-400," in M. Rowlands, M. Larsen and K. Kristiansen eds., *Centre and Periphery in the Ancient world* (Cambridge: University Press, 1990), 127, 130, 133 ("The changes observed in political, social and military organization during the period AD 1-400 were accompanied by a profound change in the pattern of settlement and production throughout northern Europe"), 134 (traditional farming methods were abandoned between the first and the second centuries AD "... in favor of more intensive agriculture ... infield system became established over the same period, and this represents the earliest integration of arable farming and stock breeding ... transition from extensive agriculture to intensive kind of farming ... Throughout the Roman Iron Age, this is precisely what characterized the cultural landscape of northern Europe."), 137 ("This development, however, was not exclusively determined by the influence of the Roman Empire, because already before the expansion of the Empire it is evident the Germanic tribes of northern Europe were structurally peripheral in relation to the archaic Celtic state system. The completition of Roman expansion in Europe took place in accordance with their pre-existing framework. But the Germanic periphery was also characterized by regional variation in social structure as reflected by differences in the distribution of Roman prestige goods in burials."), 138, 139 ("Thus, Germanic societies in the period AD 1-400 experienced population growth, expanding settlement, and increasing demand for surplus production in order to feed their population and as payments of tribute, which were channeled further into alliances and long-distance trade in Roman prestige goods. To meet these pressures intensive agriculture was developed through the infield/outfield system, the redistribution of land took place, as well as the replanning of the villages ... It was the changes in the organization of production which gave the system a new basis for expansion. ... Therefore, the transformation in the organization of production after AD 200 was of paramount importance for the evolution of a class-based society and the state."); M. Parker Pearson, "Beyond the pale: Barbarian social dynamics in western Europe," in J.C. Barrett, A.P. Fitzpatrick and L. Macinnes eds., *Barbarians and Romans in North-west Europe from the later Republic to Late Antiquity* (Oxford: BAR, I.S. 471, 1989), 200-03, 206, 208, 210, especially 221: "Contrary to expectation hierarchical societies emerged further north of the frontier than other chiefly societies and without any evidence for a productive basis... other than agriculture." Moreover, on the concept of Romanization based on the analysis of the Roman finds within *Barbaricum* see, F. Curta, *The making of the Slavs* (Cambridge: University Press, 2001), 25: "To claim that acquisition, imitation, and use of Roman silver plate reflects the degree to which barbarians were Romanized is simply to ignore that the symbolic system changed with the changing context in which imported objects were employed;" thus, the simple presence of several objects imported from Roman territories within *Barbaricum* does not imply either the beginning of a simplistic one way acculturation process nor its eventual homogeneous distribution through all the different social strata. In fact, as noted in P.S. Wells, *Culture contact and Culture change: Early Iron Age Central Europe and the Mediterranean World* (Cambridge: University Press, 1980), 7-8: "when relations are established with outside groups ... the chief or king generally controls the circumstances of the interactions ... often the rest of the society, aside from the chief's or king's agents, have little to do with the interactions with the outsiders ... A flourishing intersocial trade can exist between elite members of societies without the general populace taking any significant part in the interaction or their results ... In any social context, different goods circulate through different mechanisms depending on the nature of the goods, the social subgroups involved in giving and receiving them, and the distances over which they are carried;" and in J.H.F. Bloemers, "Acculturation in the Rhine\Meuse basin in the Roman period: some demographical considerations," in J.C. Barrett, A.P. Fitzpatrick and L. Macinnes eds., *Barbarians and Romans in North-west Europe*, 178: "When two culture systems meet, acculturation can occur. By this I mean the resulting change ... The term often used to denote process of acculturation during the Roman period in northwestern Europe is Romanization. However, this suggests a unilateral and complete absorption of Roman culture by the indigenous population: Gallic and Germanic cultures in their turn also influenced the Roman culture, albeit on specific levels and at specific times." and 179: "The degree of acculturation depends on the level at which it takes place, be it state or civil law, administration and organization. It may concern matters of spatial organization, the economy or religion. On all these levels a distinction must be made between form and content." To sum up, a contextual approach to the question of Romanization is apparently the most useful one, as suggested by S. Jones, *The Archaeology of Ethnicity*, 128-44 and especially 133 where it is emphasized that care should be exercised concerning the idea of Romanization as an inevitable and uniform process of acculturation and the associated categories of culture and identity, such as "Roman" and "Native." Therefore, any simplistic correlation of Roman-style material culture with Roman identity should be rejected, and the existence of cultural and ethnic entities such as "Roman" and "Native" questioned.

relationships with the Celtic as well as the Roman populations, were all important factors within this process.

In summary, there was a considerable increase in iron production, starting from the late Early Roman Iron Age which coincides with the introduction of the slag pit furnace. This fact, led to a substantial modification in agriculture, to changes in the structure of Iron Age farmsteads, and to a general reorganization of settlements which become evident especially between the AD 2nd and 3rd century. On the basis of archaeological evidence, it can be concluded that, even within sites with large-scale iron production through slag pit furnaces, not every population unit took part in the iron-making process. In contrast to this, iron-making was controlled by an elite, or simply by a particular sector of society possessing the economic capacity to maintain continuity of production over years. The layout of the slag-pit furnaces location and distribution within settlements and in farms suggest that access to raw material was not evenly distributed among the populace and, thus, that some kind of regulation in raw material ownership must have existed as well. Therefore, iron production became a specialized craft within the hamlet.[149]

3.1 'Lombard' Iron-making in Northern Europe. A survey of the evidence.

As far as 1st century BC to 4th-5th century AD Lombard settlements along the *Elbe* river are concerned, there is clear evidence of iron production took place within them. The iron-making structures within the Lombard area perfectly match the general features of the Northern European iron production system (**see figures 4, 5, 6**).

In the northern part of the *Altmark* region, at the site of *Zethlingen*, eleven iron smelting furnaces, dated between AD 3rd and 4th centuries, were unearthed within a 180 square meters area. Bog ore was used in this site as well for iron making. The iron producing workshop is located not far from a contemporary settlement and a necropolis. In this case as well, the furnaces typology fits in the general horizon of the iron smelting devices used during this period in northern Germany and Scandinavia (i.e. slag pit furnaces). Moreover, through the archaeometallurgical analyses of the slags, it was possible to underline a slight difference between the outcomes in iron of the different furnaces. In addition to that, the diverging Manganese oxides (MnO) content in the slags revealed also the highly varying quality of the ore used in iron production at this site.[150]

In the area of *Harburg*, more precisely, near *Scharmbeck*, an interesting site was discovered containing archaeometallurgical evidence datable to a period from the 1st century BC to the AD 2nd century. Archaeological traces of iron production, large amount of slag and just as many remains of slag-pit furnaces, have come to light at this site. Despite the fact that neither *tuyers* or remnants of bellows pipes have been found, their use for ventilation purposes is revealed through the traces left by these same instruments in the sections of furnace walls that have been preserved. It is rather plausible that the surface limonitic iron ores of this area were used to extract the ore. In this case as well, then, the region demonstrates a strong proclivity towards iron metallurgy, especially taking into consideration archaeological data which supports the existence of such production from the Iron Age till the high Middle Ages.

Another site of archaeometallurgical interest in the *Elbe* basin has been identified and studied near *Göhlen*, in south-west *Mecklenburg*. In this case as well, the demand for primary materials was satisfied mainly through the exploitation of the iron-rich sands of the low marshlands that reach the *Elbe* from the south, between *Sude* and *Rögnitz* rivers. Already in 1972, archaeological surveys indicated that this site was of archaeometallurgical interest. Between 1989-1990 regular excavations took place there as well. A total of 88 slag-pit furnaces were found during the 1990 excavations; the slag pits, almost always found full of discarded material, were generally 40-50 cm in depth and 40-60 cm in diameter. A new excavation expedition in 1992 provided many other furnaces, slightly smaller than those discovered before. These furnaces have parallels in other areas inhabited by Germanic peoples, such those discovered in *Schleswig-Holstein* and Poland. The total number of productive structures seems to have comprised 275 slag-pit furnaces that seem to have produced around 40-100 kg of slag. Within the same site, a furnace, different from all the others, was brought to light as well. Its dimensions and the associated materials were rather peculiar: a stone anvil, various charcoal fragments, pieces of ore, remains of blooms, smithy slags, with their typical bowl shape,

[149] See, L. Nørbach, "Organising iron," 237, 244-46; again on slag-pit furnace cluster diffusion as the main device for iron making within *Barbaricum* see, R. Pleiner, *Iron in Archaeology*, 70: "Most of the larger ironworks of this type are found to the east and north of the Roman military *limes*, in the barbarian hinterland ... They date in some cases to an early phase around the beginning of the Christian era, but most commenced operations in the middle or late Romano-Barbarian period. These furnace fields are generally irregular in layout, having developed haphazardly as new smelting units were installed;" and 74: "The emergence of smelting installations of this kind in non-Roman Europe in the first four centuries AD indicates a response to a perceived need to increase the production of iron within the capacities of the various socio-economic systems."

[150] On the iron production site of *Zethlingen* see, R. Leineweber, "Langobardenwerkstatt Zethlingen –lebendiges Museum mit archäologischen Experimenten," *Archäologische Mitteilungen aus Nordwestdeutschland* 6 (1991): 119-29; Ead., "Römerzeitliche Eisenverhüttung in der Altmark. Archäologischer Befund und Rekonstruktion," in A. Espelund ed., *Bloomery Ironmaking during 2000 Years* (Trondheim: Metallurgisk Institutt. Universitet i Trondheim, 1993), 41-50; Ead., "Ein spätrömerzeitlicher Verhüttungsplatz Bereich einer zeitgleichen Brandgräberfeldes von Zethlingen, Kr. Salzwedel," *Jahresschrift für mitteldeutsche Vorgeschichte* 72 (1989): 97-120; Ead., "Ein germanischer Siedlings- und Bestattungsplatze bei Zethlingen in der Altmark," *Magdeburger Blätter* (1990): 39-48; R. Leineweber and F. Kirsch, "Ein römerzeitlicher Werksattkomplex zur örtlichen Eisengewinnung von Zethlingen, Kr. Salzwedel," *Ausgrabungen und Funde* 34 (1989): 180-86; M. Fennert, "Metallurgiche Aspekte zur Eisengewinnung im Rennofen unter direkter Bezugnahme auf eine spätrömerzeitliche Verhüttungsstelle bei Zethlingen, Kr. Salzwedel," *Archäologische Informationen aus der Altmark* 3 (1992): 36-40.

and a noteworthy quantity of hammer scales. Therefore, the furnace area was recognized as a smithy. In addition to this, on the basis of the archaeometallurgical analyses performed on both slags and pieces of ores found at the site, it was possible to hypothesize the use of fluxes during iron production. The iron content of the raw material could, actually, vary a lot: ranging from as high as 70% and as low as 60/40% or even less. The iron-making site of *Göhlen*, presents features of a complex production structure within which all the different phases of iron production were carried out, from ore exploitation and reduction to the smithing of semi-products, such as blooms as well as objects. Moreover, since the excavations have failed to provide any evidence concerning residential structures, the site apparently must be a settlement specialized in iron production and smithing. The general feature of iron-making at *Göhlen* and within southwestern *Mecklenburg* in general, suggests that this activity represented one of the main sources of income for the local population and, thus, formed the basis for prosperity in the region, from the AD 2nd century onwards, as demonstrated by the quality of the finds in contemporary *Fürstengräbern* in the zone. The dating of the site, between the AD 4th and 5th centuries suggests possible attribution to a Lombard-like group occupying the area at that time.[151]

In addition to this, it is important to note that this archaeometallurgical evidence is found in the core of what has been called the Lombard territory within the lower *Elbe* basin. Southwestern *Mecklenburg* and northern *Lüneburg* till the *Elbe* river should, actually, be regarded as the main zones of the so-called Lombard settlement in northern Germany, a group generally characterized by a western Germanic culture, pertaining specifically to the Elbian Circle cultural milieu.[152]

3.2.0 The situation in Central Europe: continuity and transformations.

The 'Lombard' settlement in Central Europe, particularly its later phase in *Pannonia*, raises several historical-archaeological questions such as the location and topography of settlements and their relationship to the locations of necropolis and the distribution of ores in the territory. These problems are, in fact, extremely important for understanding the different strategies used by the various Germanic peoples concerning settlement, managing and controlling territories and maintaining access to the natural resources in those territories.[153] Moreover, it was within this Central European area that Germanic iron technology, developing from the late Early Germanic Iron Age, started to approach the organization of production of Roman provincial origin. This meeting of Germanic peoples and Romans, far from resulting in a simple melting pot, was a complex process. It actually marked the beginning of a confrontation between two different material culture traditions, each with its own structure and form. It was not, therefore, a one way process but on the contrary its development grew along complex lines rather than along simple and determined ones.

3.2.1 Beyond the limes: traces of 'Lombard' Iron-making within the territories of Bohemia and Moravia.

The 'Lombard' evidence regarding metalworkers and iron-making workshops is often connected with regions known to be mining regions such as the Moravian-Bohemian basin. This is, in fact, the case of both the tomb found in *Brno* in Moravia and grave # 6 in *Poysdorf* near *Mistelbach* in the northern part of present-day Lower Austria (**see figure 7**). The tomb of the smith in *Brno* (dating to the beginning of the AD 6th century) was discovered in 1937 during the construction of a building on *Kotlarska* Street in *Brno* (Czech Republic). The grave yielded the following evidence: an iron anvil, a pair of tongs, two iron hammers, a bronze bowl, a small bronze scale, four stone weights, one horn or probably antler comb, one iron axe, one iron rod, a large sheet of bronze foil, one iron disk, several bronze fragments, and a sandstone file. Grave # 6 in *Poysdorf* (dated between the end of the AD 5th century and the first half of the AD 6th century) yielded very similar tools: a yellow sandstone file, one iron anvil, two hammers, a curved iron knife, an iron stud, an iron padlock, two bronze moulds for fibulae making, an iron file, a fragment of a pair of iron blacksmithing pincers, a pair of iron tongs, flint stones, an iron clamp, a pair of bronze tweezers, one iron buckle, an iron knife, one horn/antler comb, and the iron umbo from a shield. The bronze casting moulds from grave # 6 in *Poysdorf* were used in the production of characteristic Lombard fibulae. To prepare the casting mould, the smith impressed a design in the desired shape into the clay which was then fired.

[151] On the sites of Scharmbeck and Göhlen see, W. Wegewitz, "Ein Rennfeuerofen aus einer Siedlung der älteren Römerzeit in Scharmbeck (Kreis Harburg)," *Nachricthen aus Niedersachsens Urgeschichte* 26 (1957): 3-25; U. Yalçin and B. Wollschläger, "Die Eisenverhüttung bei den Elb-Germanen in Südwestmecklenburg," *Das Altertum* 40, (1994): 5-12; A. Leube, "Die Eisengewinnung," 479-80; M. Muller-Wille, "Settlement and non-agrarian production," 209; H. Jons, "Iron production in Northern Germany during the Iron Age," in C. Fabech and J. Ringtved, *Settlement and Landscape*, 251-55; see also R. Pleiner, *Iron in Archaeology*, 70. On the iron production boom see, A. Leube, "Studien zur Wirtschaft," 135: "Das beträchtliche Produktionsvolumen der Eisengewinnung im südwestlichen Mecklenburg führte zu einer Blüte der materiellen Kultur, die in den zahlreichen Grabbeigaben- im Vergleich zu den Nachbarräumen- ihren archäologischen Niederschlag fand."

[152] M. Menke, "Archeologia Longobarda," 36, 46, 77.

[153] On this topic see the remarks in J. Driehaus, "Fürstergräber und Eisenerze zwischen Mittelrhein, Mosel und Saar," *Germania* 43 (1965): 33 and 44-45; and L. Nørbach, "Organizing iron," 242: "In general the relation between settlements with iron production from the Late Roman and Early Germanic Iron Age and the post-glacial sediments of freshwater peat and freshwater turf is evident, since these wetland areas with their potential for hay-harvesting, are located here. But also the fact that bog iron ore is precipitated in the slopes near these areas must have governed the location of the settlements." On the topic see as well, K.R. Dark, *Theoretical Archaeology*, 143: "The location of an archaeological site in its contemporary landscape has long been recognized as a source of evidence for its economy. Access to land, water-supply, costal resources, and minerals are clearly factors in the selection of sites."

It is important to stress that the partial set of metalworking tools, from the grave of the smith unearthed in 1826 in *Grupignano* in *Cividale del Friuli* (Italy), indicates that there was a strong continuity in the use of such tools. The grave yielded one iron anvil, one iron tool, possibly a rivet, and a silver buckle. [154] However, at the same time, the Friulian grave indicates a certain degree of Roman influence suggested by the shape of the anvil which resembles types of Mediterranean origin. In contrast to this, both the Central European graves present a structure and a tool types characteristic for the late Early Germanic Iron Age as it developed in the area along the river *Elbe*. The conservatism of grave # 6 in *Poysdorf* and the partial lack of influence with regard to Roman provincial culture are striking, as the grave lies behind the *limes*, that is to say, already on former imperial territory. Its peculiar character can be, however, explained by its attribution to the immigrant generation, to those who first stepped into former Roman soil.[155]

Excavations over an area of 7 hectares at *Březno* (Bohemia) are also relevant to the presence of the 'Lombards' in Central Europe and the traces of their iron-making activities. There, a slag-pit furnace for reducing iron ore was found among a group of 21 sunken-floored houses from the Migration Period (around the year AD 500). It was apparently used only once. The furnace is similar to other slag-pit furnaces in this period, recalling a type which originated in the La Tène period and changed shape around the Early Middle Ages (**see figure 8**). Since the presence of a lone furnace cannot even be regarded as small-scale production, it has to be considered an occasional production structure. This suggests that the single furnace at *Březno* represents the work of an itinerant blacksmith, invited to the village during a period in which the settlement suffered a shortage of iron. The production capacity of this furnace would have been of about 20-25 kg of unworked iron, that is, 3-4 kg of forged iron. This quantity could have been used to produce a more than sufficient number of tools and arms for a small community.[156]

3.2.1.1 Metallographic analyses on three iron knifes from the iron-making workshop of Březno

Thanks to a study by Professor R. PLEINER, there are metallographic analyses on three iron knifes from the same period as the iron-making workshop of *Březno*. These are sample no. 618, from house 8, and nos. 619 and 620 from house 43. All the objects come from the foundation levels of the houses and, thus, it is extremely unlikely that they had suffered any other accidental heating.[157]

As far as sample no. 618 is concerned, the entire surface of the blade was cut through a cross-section (12 mm. wide). The metallographic analysis reveals the metal to be quite free from slag. As a matter of fact, the mass of slag inclusions barely reaches Level 2 (according to *Järkontoret scale*). The inclusions are generally grey and glassy. Among them, there is an oblique line of inclusions running toward the cutting edge that is worth mentioning: under the microscope it appears to be a

[154]As far as Lombard smiths' graves are concerned see, specifically for *Brno*, J. Tejral, *Grundzüge der Völkerwanderungszeit in Mähren* (Prague: Studie Archeologické no ústavo Cekoslovenske Ved v Brne, vol. 4.2, 1976), 81, pics. 9-11; Id., "Die Langobarden nordlich der mittleren Donau," in R. Busch ed., *Die Langobarden von der Unterelbe nach Italien, Veröffentlichung des Hamburger Museums für Archäologie und Geschichte Harburgs (Helms-Museum)*, 54 (Neumünster: Wachholtz, 1988), 230, n. 69; for Poysdorf see, E. Beninger, *Die Germanenzeit in Niederösterreich von Marbod bis zu den Babenbergern* (Vienna: Dr. E. Stepan, 1934), 111 and ff., pic. 54; E. Beninger and H. Mitscha-Märheim, "Der Langobardenfriedhof von Poysdorf, Niederösterreich," *Archaeologia Austriaca* 40 (1967): 167 and ff.; J. Werner, "Zur Verbreitung frühgeschichtlicher Metallarbeiten," *Antikvarist Arkiv* 38 - *Early Medieval Studies* 1 (1970): 67-71; G.C. Menis ed., *I Longobardi* (Milan: Electa, 1990), 20-21, 32-33, 372; M. Brozzi, "Strumenti di orafo longobardo," *Quaderni ticinesi*1 *(NAC)* (1972): 167-74. For other early medieval graves of metalworkers see J. Decaens "Un nouveau cimetière du haut moye age en Normandie, Hérouvillette (Calvados)," *Archéologie Medievale* 1 (1971): 1-126; particularly, 12-17, for tool inventories; 65-66, for the comments on AD 6th century graves; 83-90, for the remarks on the smith's grave. See the bibliography quoted in the article for the graves of Neuwied in the Rhine valley, Beckum in Westfalia, and Jutas in Hungary; in addition to this see, D.A. Hinton, "A smith's hoard from Tattershall Thorpe, Linconshire: a synopsis," *Anglo Saxon England*, 22 (1993): 147-66.

[155]On the tools of the Germanic smith and their relationship with the Roman technical tradition see, J. Henning, "Schmiedegräber," 71-74 and, for a rich catalogue of finds and illustrations of tools see, H. Ohlhaver, *Der germanische Schmied und sein Werkzeug* (Raumburg (Saale): Lippert & C., 1939). On the graves of Lombard smiths see the remarks in V. La Salvia, *Archaeometallurgy*, 68-69. To support the hypothesis that grave 6 of *Poysdorf* pertains to the earliest generation who had immigrated beyond the Roman *limes* around AD 510-512, see M. Rotili, "Le necropoli di tradizione germanica," 164 reporting the thesis of I. Bóna who believed that the Lombards crossed the Danube in three successive phases, the first being, between AD 510-512. This initial movement brought them into the area between *Vindobona* and *Aquincum*; the second phase, around AD 526 and 527 involved occupation of the zone south of present day Budapest, and the last phase in the years AD 546-547 when the Lombards reached *Pannonia Secunda*, *Savia* and *Noricum Mediterraneum*.

[156]With respect to finds and excavations and their archaeometallurgical evaluation see, R. Pleiner, "Small-scale rural Ironmaking in Migration Period Bohemia," in G. Magnusson ed., *The Importance of Ironmaking*, 116-18; R. Pleiner, *Iron in Archaeology*, 48 and I. Pleinerová, "Germanische und slawische Komponenten in der altslawische Siedlung Březno bei Louny," *Germania* 43 (1965), 136 who compares the situation of the Czech site with that of Röderau (in the region of Reis) dated between AD 450 and 525; moreover, in the same article see 121-38 and particularly, relative to some forms of pottery decorations that are found in the Pannonian phase of Lombard settlement as well, see 134-135: "Er (J. Werner) hält dieses Motiv für ein zeitlich begrenztes Element, wenn wir seine alten Wurzeln im Gebiet der oberen Elbe außer acht lasen. Sein Hauptvorkommen in Böhmen fällt in der Ende des 5. und uns 6. Jahrhundert. Nur ausnahmweise kommt es noch in der Pannonische Phase der langobardischen Keramik vor, und zwar in Grab 5 aus Várpalota, das zur ältsten Stufe der Pannonische Phase zählt, nach Werner in die Zeit zwischen 530 bis 550 n.C.". Moreover, with respect to Lombard archaeology in Bohemia, see also B. Svoboda, "Zur Frage der Langobarden in Böhmen," in A. Tagliaferri ed., *Problemi della civiltà e dell'economia longobarda* (Milan: Giuffrè, 1964), 55-64; Id., *Cechy v dobe stehovani narodu* [*Bohemia during the Migration Period*] (Prague: Academia, 1965), 309, 310, 335, 357-58; N. Christie, *The Lombards*, 19; Id. "Longobard weaponry," 4: "The sparse material recovered from Bohemian cemeteries of the second half of the 5th century compares closely with finds of early 6th century data in lower Austria and Moravia, zones which are historically attested as Longobard by that time."

[157] R. Pleiner, "Spätkaiserzeitliche und völkerwanderungszeitliche Stahlklinge,“ 606-11 for presentation and discussion of the metallographic analyses.

welding line which marks the border between the cutting edge and another section of the blade characterized by a *pearlitic* structure. Moreover, along the width of the metallographic cross-section there are other two welding lines that do not join materials with different metallographic structures. The knife is, thus, formed by three different iron sheets welded together. It is basically composed of a fine grain *Ferrite* structure with small amounts of *Pearlite* at its borders. The Phosphorus and Nickel content is moderately high (0,218% P, 0,178% Ni). The hardness tests indicate a level of 200mH_V for the *ferritic* structure and around 250mH_V for the *pearlitic* one although, in some places near the welding line, it reaches 350mH_V. The general quality and structure of the knife makes it clear that the smith had brilliantly mastered an advanced but common production process. Using this method it was possible to weld a steel bar within the framework of a three strata blade. However, given that the bar did not consist of very hard steel, the final product was not really first rate.

Sample no. 619 comes from the 12mm cross-section of the blade. It presented a metallographic surface often segmented by slag inclusions (Levels 2-3 *Järkontoret scale* in its average content). Among them, the glassy ones were partially broken in the smithy through repeated thermal shocks. As for the other sample, an oblique chain of slag inclusions accumulated along the welding line. This knife as well has a composite structure made of different iron sheets welded together. It consists of a *pearlitic-ferritic* steel with *Widmanstätten* structures. There is also a highly carburized zone on one side which runs into a *sorbitic* structure. The same situation is found in material of the welded iron sheet where a fine grained *pearlitic* structure assumes a clear *sorbitic* character. Only the corner portion near the ridge is *ferritic* and it was possibly welded to the carburized iron sheet. The hardness tests presented the highest values near the edge (400-400 mH_V) while they gradually become lower (between 300-400 and 200-250 H_V) toward the ridge. The content of Copper is 0,247%, of Phosphorus 0,297% and of Nickel 0,068%. To conclude, the knife consisted of three different iron bars with diverging carbon content and, therefore, different mechanical characteristics. The ridge was made of a simple iron piece whereas the core of the knife and the edge were made of carburized iron, that is to say, steel (a binary alloy between Carbon and Iron). These parts underwent short thermochemical treatments. The knife was, thus, masterfully produced.

Sample no. 620, covers the 16mm cross-section of a knife blade. The quantity of slaggy inclusions is relatively high. Their distribution is, however, interesting and useful for understanding the metal structure: while within the core and edge they form oblique lines, near the ridge they have spiral shapes. This patterning is better understood after etching with the *Oberhoffer* reagent, through Phosphorous segregation. The edge was surely welded to the core and consisted of fine *Martensite* in a fine-grained *pearlitic* matrix. The core of the knife is *ferritic-pearlitic* and presents curved stripes of slaggy inclusions layered one over the other. Moreover, in the middle of the ridge, there is an oval *martensitic* spot which is possibly the result of flattening a rolled steel band, through hammering. The *martensitic* structure is extremely hard (980mH_V 30g), while the *pearlitic* or *sorbitic* portions registered hardness values of 400mHV. The content in Phosphorus is 0.262%, that of Nickel 0.076%.
The production procedure can be reconstructed as follows: a steel band, possibly made of different sheets, was welded, through heat treatment, to another hard steel band. This had then been rolled together along its width with another steel sheet. This composite structure was finally welded and hardened together. The quality of the knife was, thus, extremely high.

Moreover, the methods of production of this set of objects from *Březno* were compared with those used to make the artifacts from the nearby site of *Opočno*, dated to the Late Roman period (AD 3rd-4th century). The resulting data showed that the iron-making technology diverged strikingly for the two different periods especially concerning steel-making and iron hardening techniques. The knives from *Březno* were produced throughout the composition of their bodies by hammering together bars of iron with different carbon contents. This method, according to R. PLEINER, is the more progressive one and, actually, became a standard technique for blade-making throughout the Middle Ages.[158]

As concerns the Moravian-Bohemian mineral basin, it is worthwhile pointing out that the mineral resources of the region were always exploited by western Germanic peoples, particularly the Quads and Marcomans, who inhabited the area from the Late Roman Iron Age till the Migration Period. There exists copious archaeometallurgical evidence in the region, characterized by a typically Germanic production system, linked to the functioning of small workshops – with the exception of the *Sudice* site – which exploited the resources of the area for an essentially local market. This type of economic organization seems to have taken shape during the first centuries of the Christian era in numerous regions outside the Imperial borders and was distinguished by the vast diffusion of iron production. Nevertheless, despite the great number of furnaces, the total production level did not attain that of the Roman mining companies. This difference was the result of a completely different work style: not intensive and centralized, but rather extensive. Thus, the production of bigger quantities of iron were secured through multiple and successive reductions in the same furnace and usually aimed at satisfying regional needs on a smaller scale. The quantities produced in such a system came to between 20-40 and 30-60 kg of wrought iron per smelting furnace, which corresponded to around 15-30 kg of soft iron ready to be smithed.[159]

---

[158] R. Pleiner, "Spätkaiserzeitliche und Völkerwanderungszeitliche Stahlklinge," 610-11.

[159]On the iron-making tradition of these regions see, K. Motykovà and R. Pleiner, "Die römerzeitliche Siedlung mit Eisenhütten in Orech bei Prag," *Památky Archeologické* 78 (1987): 372, 374, 400-02; R. Pleiner, *Iron in Archaeology*, 36, 40. In addition see V. Souchopová, *Pocátky západoslovanského hutnictivi zeleza ve svetle prameni z Moravy* [*The*

After his analysis of the typological evolution of furnaces in this area between the Late Roman Iron Age till the beginning of the Migration Period, R. PLEINER hypothesized a direct influence of Celtic iron-making traditions on the Germanic one. He even suggested that some groups of Celtic metallurgists were assimilated by the Germanic population. [160] That such a close relationship, or rather overlap of Celtic and Germanic iron metallurgy existed is quite plausible especially if one remembers a famous passage in *Tacitus*: ... *Cotinos Gallica, Osos Pannonica lingua coarguit non esse Germanos, et quod tributa patiuntur. Partem tributorum Sarmatae, partem Quadi ut alieginis imponunt: Cotini, quo magis pudeat, et ferrum effodiunt.*[161] The Celtic tribe of Cotins possibly lived in Moravian territory or in northern Slovakia, in close contact with western Germanic populations, more precisely the Quads and the Marcomans, who were also wide spread in Bohemia.[162]

Beginnings of the Metallurgy of Iron among Western Slavs in the Light of Sources from Moravia] (Brno: Archeologicky ústav Akademie ved Ceské Republiky, 1995), where the author underlines the continuity of the iron-making tradition from Celtic times to the Middle Ages when metallurgy became one of the most significant features of the economy in the area. On the diverging iron working traditions of the Romans and the Germanic populations see also R. Pleiner, "Small-scale iron-making," 115. The main contribution of the Roman world to iron-making was not strictly technological. It was, in fact, related to the better organization of production: in this way the Roman economy was able to systematically exploit mineral resources from over the whole of the great Imperial territory and to continuously and consistently respond to market demands, on this topic see, R. Pleiner, *Iron in Archaeology*, 41-45; R.F. Tylecote, *A history of Metallurgy* (London: The Metals Soc, 1976), 53; A. Leger, *Les travaux public, les mines et la metallurgie aux temps des romains* (Paris: 1875), 689.

[160] See, R. Pleiner,"Extensive Eisenverhüttungsgebiete," 297-305, see especially 297: "Dank der archäologischen Forschung wissen wir jedoch dass die germanischen Völkerschaften das Eisen verhütteten, und zwar völlig auf ostkeltische Weise und - möglicherweise- mit Hilfe keltischer Schmelzer, die den ansässigen Schichten der älteren Bevölkerung enstammten." R. Pleiner, *Iron in Archaeology*, 272: "The Celtic population of the non-Roman areas managed to transmit many of the achievements of its material culture to Germanic newcomers, but succeeding generations were assimilated relatively quickly. In certain areas such as Bohemia, the smelting of iron profited from Celtic traditions. At the beginning of the Christian era, sunken-floored ironworks equipped with small slag-pit furnaces were not only identical in general lay-out but also in shape and detail."; In addition to this, on the relationship between the Celtic and Germanic civilizations see P.S. Wells, "Identities, material culture and change: Celts and Germans in late Iron Age Europe," *Journal of European Archaeology* 3.2 (1995): 169-85; N. Roymans, "Romanisation and the transformation of a martial elite-ideology in a frontier province," in P. Brun, S. van der Leeuw and C.R. Whittaker eds., *Frontieres d'empire*, particularly 33-34: "The Celto-Germanic societies were related cultural groups in a linguistic and cultural-historical sense, and had many meaningful similarities and differences. The focus themes within these societies were clientship, gift-exchange, martiality;" and A. Woolf, "Romancing the Celts. A segmentary approach to acculturation," in R. Laurence and J. Berry eds, *Cultural Identity in the Roman Empire* (London & New York: Routledge, 1998), 116: "... medieval Celtic and Germanic societies which not only shared a wealth of cultural traditions with the late pre-Roman Iron Age populations but also had access to similar technological resources and exchange mechanisms." On the assimilation of Celtic tribes by the Germanic population in Central Europe see as well, R.Wenskus, *Stammesbildung und Verfassung* (Köln-Graz: Bönlau Verlag, 1961), 397.

[161] Tacite, *La Germanie*, chap. 43, 96.

[162] On the possible location of Quadi and Marcomanni see L. Schmiedt, *Geschichte der Germanische Völker*, 21, 23, 36-37, 172; Id., *Geschichte der deutschen Stämme* (Munich: C.H. Beck'sche Verlagbuchhandlung, 1938), 153-94; on the location of the Cotins see, J. Piaskowski, "Studia

Moreover, it is rather possible that when *Tacitus* was writing these famous lines, the relationship between the two cultures was reduced to a simple subordination or complete assimilation of the Celtic inheritance by the Germanic populations. At the time Celtic tribes had to pay tribute to the Quads in iron. Probably the origins of the slag-pit furnace can be traced to exactly this area of criss-crossing and integration of diverse craftsmanship know-how, given that this furnace was of remote Celtic origins. It dominated iron metallurgy for almost five centuries in areas that were never directly subject to the influence of the Romans. This is exactly the type of furnace that can be considered progressive: in fact, the slag-pit furnace, having developed into a classic shaft furnace, can be considered, from a technical-architectonical, that is to say, structural, point of view, the predecessor of the blast furnace.[163] Therefore, it is likely that such furnaces diffused through former Roman Europe with Germanic populations. Iron production fits into the general cultural continuity of populations and culture between the lower basin of *Elbe* and the Bohemian-Moravian area, confirmed by paleoethnoarchaeological data. The production of iron has rather similar characteristics in these two geographical areas both with regard to the extensive organization and the diffusion of production centers in the territory, and the typology of the furnaces. Taking into consideration this particular aspect of the material culture, the Lombards can be considered as an integral part of the Western Germanic cultural *koinè*, especially taking into account the fact that the Lombards together with the Obis, according to written sources, were present in the area already in the years AD 166/167-180 AD. They were the ones who incited the so-called Marcomannic wars, exercising pressure along the Danube frontier, especially in the territory between *Arrabona* (*Győr*, Hungary) and *Brigetio* (*Szőny*, Hungary).[164]

nad lokalizacja starożytnych Kotynów [Study on the localization of the ancient Cotins tribe]" *Acta Archaeologica Carpathica* 3/1-2 (1981): 77-102.

[163] See, Pleiner, "The Technology of Iron in the Bloomery Period," 543 and 545.

[164] Archaeological excavations carried out in southern Moravia (Krepice, Pollau, Vicemilice) apparently identified the area where the 6000 Lombards and Obi, mentioned by written sources, were located. On this discussion and on the close cultural relationships between Moravia, Bohemia and the region of the lower Elbe see, M. Menke, "Archeologia longobarda," 58, 75, 80, 86, 91, 93, 94, 98; R. Hachmann, *Les Germains* (Geneve: Les Éditions Nagel, 1971), 87: "Les fouilles ont permis de délimiter les regions suivantes: la culture 'des Germains de l'Elbe,' s'étendant du bassin inferieur de l'Elbe, au nord, en passant par l'Almark, le Brandenburg, la Thuringie et la Saxe jusqu'a la Bohemie, la Moravie et la Slovaquie occidentale, au sud ..."; S. Gasparri, "Tra antichità e Medioevo, 324: "Importante è poi l'area dei Germani dell'Elba – caratterizzata da ricchi corredi tombali e dal rituale dell'inumazione – della quale fecero parte i Longobardi, un'area che ebbe legami con Austria, Slovacchia, Moravia. Era abitata da genti Sveve, ... In stretto contatto con la zona dell'Elba era la Boemia, già centro importante di cultura celtica ..."; K. Godłowski,"The Chronology of the Late Roman and Early Migration Periods in Central Europe," *Universitas Jagellonica Acta Scientiarum Litterarumque CCXVII Schedae Archaeologicae*, fasc. 11 (1970): 59, 77, 87, see particularly 82 : " This area was the starting point of the famous 'amber route,' which played such an important a part in contacts between the ancient world and Central European Barbarians. Contacts with the Sarmatians should also be mentioned here which were very active at

3.2.2.0 Crossing the borders. Lombard Iron-making on former Roman soil.

The passage across the Danube presents further problems for the evaluation of the state of Lombard metallurgical development. It is during their incursions into Roman provinces along the Danube – *Pannonia*, *Savia* and *Noricum Mediterraneum* – that the Lombards came into a closer contact with Roman provincial culture, something attested also by necropolises of the Lombard period. I. BÓNA has indicated that not only objects but also individuals of Roman provincial origin were buried there.[165] This is especially evident in those regions that could boast of a long Celtic mining tradition that had developed during the Roman occupation such as the territory around *Scarbantia*, today *Sopron* in Hungary, *Norikòn Polis* or *Poetovio* (*Ptuj,* Slovenia), and the *Urbs Pannoniae*, that is, *Siscia* (*Sisak*, Croatia), all areas given to the Lombards by the Byzantines after their conflict with Theodoric the Great (around AD 540) and which they held for at least twenty years. In fact, for these areas the written sources indicate the existence of a tradition of plant cultivation and processing of metals that lasted from the Celtic period until the time of Theodoric the Great and Justinian.[166]

As various studies have brought to light independently from each other, first by J. HENNIG and more recently by F. CURTA, the late Antique productive structures survived at least till the AD 6th century. This continuity is shown by the uniformity of both the distribution and the quality and shape of some working tools produced in workshops in the region. However, these were no longer productive structures centered in the cities and linked to long-distance trade, but to the needs of great estates and to the *castra* along the *limes*. The strong defense of the *limes* and of the great estates of the Danubian provinces promoted by Justinian and recorded by *Procopius*, seems to have effected its longevity and directly affected some mining areas, such as those located along the border between *Dacia Ripensis* and *Moesia inferior* and the valley of *Morava*, where many forts were built at this time. Such an intensive program of territorial defense was not simply aimed at opposing "barbarian invasions" but rather at attaining full control of the territory, including its economic and strategic resources.

Archaeological evidence such as the continuous production and diffusion of the axes without *Schaftlochlappen* and with *schiedformige Nacken* supports this hypothesis. An example of this kind of axe was found in grave 34 at *Hegykő*. This occurrence not only demonstrates that it was being manufactured in the AD 6th century but also attests to the direct contacts that the Lombards maintained with such iron production organizations. These workshops were possibly still located in areas that during Late Antiquity already had well established iron-making traditions.[167]

A similar continuity in iron production is also attested in *Noricum* and *Carinthia.* Suffice it to say, finds of iron tools in present-day Slovenia such as those from *Nove Antične najdbe Povirju* (second half of the AD 4th century); *Limberk nad Veliko Racino* (circa AD 400); *Zidani gaber nad Mihovim* (AD 5th-7th centuries); *Ajdovski gradec nad Vranjem* (AD 5th-6th centuries); *Tinje nad Loko pri Zusmu* (end of the AD 6th century), particularly important because of the "typical" Lombard pitcher found there; and *Ljubicna nad Zebelovsko Goro* (AD 6th-7th centuries). This distribution of finds demonstrated an 'uncontested' continuity of production between Late Antiquity and the Early Middle Ages. Moreover, these finds not only underline iron-making continuity but also the permanence of the system of craftsmanship in the area since they are all working tools **(see figure 9)**.[168]

this time; and last but not least, with the sub-stratum of Celtic traditions, which played such important a role in the formation of the European culture milieu in the Roman period. One of the most prosperous active centers lay on the middle Danube. Thus, in the densely populated lowlands of Slovakia and Moravia all the necessary conditions existed for the creation of a cultural milieu which would adapt and transform in a unique way influences coming directly from the civilization of the ancient world, ... "; see also V. Salač, "Die Bedeutung der Elbe für die böhmisch-sächsischen Kontakte in der Laténezeit," *Germania* 76 (1998): 601, 604.

[165] I. Bóna, "I Longobardi e la Pannonia," in *Atti del Convegno Internazionale sul Tema: La Civiltà dei Longobardi in Europa. Roma-Cividale 1971* (Rome: Accademia Nazionale dei Lincei, 1974), 252-53; A. Tagliaferri, "Aspetti e Limiti dei Traffici Internazionali Longobardi (489-568 d.C.)," *Economia e Storia* 4 (1963), 540; A. Melucco Vaccaro, *I Longobardi*, 43-46; V. La Salvia, *Archaeometallurgy*, 9-10.

[166] I. Bóna, "I Longobardi in Pannonia," in G.C. Menis ed., *I Longobardi*, 14; V. La Salvia, "L'artigianato metallurgico," 15; R. Koščević and R. Makjanić, *Siscia. Pannonia Superior* (Oxford: BAR, I.S. 621, 1995), 2, 23.

[167] J. Henning, "Eisenverarbeitungswerkstätten im unteren Donaugebiet zwischen Spätantike und Frühmittelalter," *Zeitschrift für Archäologie* 21 (1987): 60, 61, 64, 68, 70, 71; J. Henning, "Untersuchungen zur Entwicklung der Landwirtschaft in Südeosteuropa im Übergang von der Spätantike zum frühen Mittelalter," *Etnographisch-Archäologische Zeitschrift* 25 (1894): 129-30; and F. Curta, *The making*, 145-46, 149, 153, 163, 184, 189. Moreover, see J.A.S. Evans, *The Age of Justinian* (London/New York: Routledge, 1996), 79: "Yet in a law of 535, he (Justinian) could boast that both banks of the Danube were full of our cities, though he named only three north of the river (cod. Theod. 16.8.15). However, from the archaeological evidence it seems clear that imperial control persisted well past Justinian's accession on a strip of territory north of the Danube;" 218, and, for the Balkans, 223: "the picture that Procopius conveys is not inaccurate;" certainly, as stated on page 225, "Cities as the classical world had known them were undergoing change, and by 600, the amenities of urban life which still existed a century before had disappeared in most of them," and 231: "A new type of urbanism was emerging, one that cared little for the ordered space and the amenities of the classical city. Urban settlements responded to the changing pressures that beset them by developing new forms. The pressures varied from region to region." A different opinion is expressed in J.K. Knight, *The End of Antiquity. Archaeology, Society and Religion in early medieval Western Europe AD 235-700* (Strond: Tempus, 1999), 166: "The Justinian revival was brief, as it was in the Balkans, where Justinian work of urban renewal was lost to the Slavs within twenty years of his death in 565." In addition to this, see J. Tejral, "Les fédérés de l'Empire et la formation des royaumes barbares dans la règion di Danube moyen à la lumière des données archéologiques," *Antiquités Nationales* 29 (1997): 137-66 for a detailed and sound discussion on the destiny of the region of the middle Danube during Late Antiquity.

[168] On the Slovenian material and the continuity of iron production see, P. Bitenc and T. Knific eds., *Od Rimiljanov do Slovanov* (Lubiana: Narodni muzej Slovenije, 2001), 32 n° 87; 49 n° 140 e 141; 52 n° 146; 57 n° 167; 58 n°168 and as well the bibliography mentioned in the volume; S. Ciglenečki, "Die Eisenwerkzeug aus den befestigten

A similar situation can be inferred for *Transylvania*. The Gepids, while occupying the region, most likely continued to exploit the Transylvanian ores since the tomb of a smith found in this very region, namely, in *Mezőbánd* (*Band*, present day Romania), yielded evidence of several smithing tools, among which there was a pair of large tongs (45 cm), possibly used to hold the heated iron in the smithy forge and, later, on the iron anvil. However, the typology of these tools clearly indicates the influence of provincial Roman culture. The grave goods placed in the *Mezőbánd* burial clearly reflect two coexisting but differing metal-working traditions: the Germanic one, as shown by the rite of burying tools as grave goods, and the provincial Roman influence suggested by the form of tools and the geographical location of the grave in a mining basin formerly controlled by the Romans. The location of the ores and Roman provincial know-how were, thus, welded together with the Germanic metallurgical tradition into a unitary technological framework, as reflected in the composite character of the smith's grave from *Mezőbánd*.[169]
Moreover, for the same region there is another indirect indication for continuity in exploitation of mining resources during the Migration Period. A gold necklace was discovered amongst the treasures from *Szilágysomlyó*, with miniature tools from various crafts hanging on it, including those of miners and smiths. This piece of jewelry, in my opinion, is a symbolic representation of the control of the owner over all the crafts embodied by each tool and, therefore, emphasizes the importance of these specializations within metallurgy.[170]

Nevertheless, there are no such iron-ores sources in *Pannonia*, since its geological situation is very different from *Transylvania*. Moreover, there are no direct data for the exploitation of local iron ores in the Lombard period. However, it is difficult to imagine that iron was consistently imported from nearby *Noricum*, which was still under the jurisdiction of Theodorician Italy, at least until the first half the AD 6th century. At that time, the relationship between Lombard period *Pannonia* and Italy were quite troubled, the first being firmly under Byzantine political and military influence. In order to produce their iron implements, largely present from the end of the AD 5th and the first half of the AD 6th century, groups like the Lombards needed to discover and exploit at least small quantities of surface-iron ore.[171]

However, the simultaneous presence at *Várpalota* of both a necropolis of the Lombard period and rather important iron ore deposits, may possibly indicate that groups like the Lombards were still exploiting these types of natural resources. However, archaeologically attested mining of iron ores in the region is convincingly dated only for the 10th century.[172] As a matter of fact, the use of surface iron ores, even if very limited should not be regarded as unusual. With respect to the given period, the processing of iron ore, geologically a widely diffused mineral, was generally characterized by dispersion of the workshops in

---

Höhensiedlungen Slowenien aus der Völkerwanderungszeit," *Balcanoslavica* 10 (1983): 45-54; Id., "Romani e Longobardi in Slovenia nel VI secolo," in *Paolo Diacono e il Friuli alto medievale*, 179-200; P. Petru, "Zgodn jesredn jeveška naselbina na Polhograjski Gori nad Polhovin Gradcem" [Early Medieval settlement of Polhograjska Gora ob Polhovo Gradec], *Arheoloski Vestnik* 18 (1967): 455-60; N. Osmuk, "Nove Antične najdbe Povir" [New Roman finds from Povir], *Goriški Letnik* 3 (1976): 71-87; Ž. Trkman, "Sv. Pavel nad Vrtovinom" [St. Paul above Vrtovin], *Varstvo Spomenikov* 31 (1989): 243; A. Valič, "Ajdovska Luknja," *Varstvo Spomenikov* 27 (1985): 272-74; A. Gaspari, M. Guštin, I. Lazar, Ž. Trkman, "Late Roman tool finds from Celje Gradišče at Zbelovska gora nad Sv. Pavel above Vrtovin," in M. Feugère, M. Guštin eds., *Iron, Blacksmith and Tools. Ancient European craft. Acts of the Instrumentum 12 Conference at Podjreda (Slovenia). April 1999* (Montagnac: Éditions Monique Mergoil, 2000), 187 and ff.; M. Sagadin, "Late Antique wood working tools from Grdavov hrib near Kamnik (Slovenia)," in M. Feugère, M. Guštin eds., *Iron, Blacksmith and Tools*, 205-08; R. Krempuš, "Krvavica beu Vransko in Slowenien, Höhensiedlung des 3. bis 6. Jahrhunderts," M. Feugère, M. Guštin eds., *Iron, Blacksmith and Tools*, 209-31; R. Pleiner, *Iron in Archaeology*, 47; R. Sprandel, "Bergbau und Verhüttung im frühmittelalterlichen Europa," in *Artigianato e tecnica*, 600; L. Aitchinson, *A History of Metals*, 2 vols (London: Macdonald & Evans, 1960), 245. Moreover, on Late Antiquity and the Lombard settlement in present day Slovenia see, S. Ciglenečki, "Poznorimski depo z Rudne pri Rudnici" [Late Roman finds from Rudne near Rudnici], *Arheoloski Vestnik* 42 (1991): 225-32; T. Knific and D. Svoljšak, "Grobovi Langobardskih vojščacov iz Solkana (Nove Gorice)"[Graves of Lombard warriors from Solkan (Nova Gorica)], *Arheoloski Vestnik* 35 (1984): 277-90; T. Knific, "Vojščaki iz mesta Karnija" [Warriors from Karnij], *Kranjski Zbornik* (1995): 23-40.

169 The custom of depositing tools as grave goods in the burial pit is, apparently, deeply-rooted in Germanic traditions as is evident from analyses of northern sagas such as the *Egill Saga*. The latter narrates historical and legendary early medieval events which took place between AD 874 and 1000 mainly within Norway, Iceland and Finland; it was possibly standardized in written form around the years 30/40 in the AD 13th century. This legend has, in my opinion, a broader meaning of memory in the sense of the social heritage of an ethnic group, in this case the Icelanders. With regard to the death (around the year AD 946) and the burial of *Skalla-Grimr*, warrior, smith and peasant, the story goes as follows: "Egill ordered a tumulus to be constructed ... there was placed *Skalla-Grimr*, together with his horse, his weapons and his smithing tools as grave goods. It has not been handed down to posterity whether precious objects were also placed beside his body" (English translation by the author).On this subject see, "La Saga di Egil," in M. Meli ed., *Antiche saghe nordiche*, vol. II (Milan: Mondadori, 1997), 613. See also V. La Salvia, "L'artigianato metallurgico," 19.

170 About the treasure of Szilágysomlyó and its interpretation see, J. Hampel, *Alterthümer des frühen Mittelalters in Ungarn*, 3 vols. (Braunschweig: F. Vieweg und Sohn, 1905), 70-71 and ff.; Gy. László, *The Art of the Migration Period* (Miami: University Press, 1974), 19-21, 24; T. Capelle, "Bemerkungen zu einer Untersuchung der goldenen Miniaturenkette von Szilágysomlyó," *Communicationes Archaeologicae Hungariae* (1992): 75-79; M. Martin "Die goldene Kette von Szilágysomlyó und das frühmerowingische Amuletettgehänge der westgermanischen Frauentracht," in U. von Freeden and A. Wieczorek eds., *Perlen. Archäologie, Techniken, Analysen. Akten des Internationalen Perlensymposiums in Mannheim vom 11. bis 14. November 1994* (Bonn: Dr. Rudolf Habelt GmbH, 1997), 349-72; V. La Salvia, "L'artigianato metallurgico," 13.

171 I. Bóna, *The Dawn of the Dark Ages*, 46.

172 The hypothesis for the possible exploitation of the iron ores around Várpalota was first offered by I. Bóna, *The Dawn of the Dark Ages*, 48. On distribution of ores in Hungary and their historical exploitation, see J. Gömöri and P. Kisházi, "Iron ore utilization in the Carpathian Basin up to the Middle Ages with special regard to bloomeries in western Transdanubia," in M. Kretzoi and M. Pécsi eds., *Problems of the Neogene and Quaternary in the Carpathian Basin. Geological and geomorphological studies. Contributions to the 8th Congress of the Regional Committee on Mediterranean Neogene Stratigraphy* (Budapest: Hungarian Academy of Sciences, 1985), 323-55; again on the natural resources of the region see as well, H. Schmid, "Die montangeologischen Voraussetzungen des Ur- und Frühgeschichtlichen Eisenhüttenwesens im Gebiet des mittleren Burgenlands (Becken v. O. Pullendorf)," *Wissenschaftliche Arbeiten Burgenland* 59 (1977): 11-23.

the territories. Production was based on the exploitation of the superficial outcrops of various metal ores.[173] For groups like the Lombards the utilization of such superficial limonitic iron ore was not at all a novelty, since along the banks of the *Elbe* river this kind of mineral extraction constituted the main natural resource (**see figure 10**).

I. BÓNA's evaluation of the archaeological evidence suggests that Lombards in *Pannonia* during the AD 6th century relied on imports of certain goods from Frankish regions, such as western damascene swords, bronze vessels, and glass objects. Nevertheless, also craftsmen of *Pannonia* during the Lombard period were an integral part of the skilful vanguard of a common Germanic patrimony of technical knowledge with respect to blacksmithing, goldsmithing, and pottery-making.[174]

3.2.2.1 Metallographic analyses of swords from the necropolises of Hegykő (Hungary) and Benevento (Italy)

The level of iron metallurgy within the Lombard milieu can be evaluated through the metallographic analyses carried out on two iron swords from the Hungarian necropolis of *Hegykő*, dated to the Lombard period.[175]

The following results indicate on the one hand that the technological level was perfectly in line with contemporary standards in Merovingian Europe and, on the other hand, within local 'Lombard' society. This craftsmen's tradition, with its own conventions in the manufacturing processes, would continue to function and evolve in Italy as well. With regard to this particular technical aspect, the data presented in 1977 by M. ROTILI on sword production in Lombard *Benevento* (Campania, Italy), as we shall see in detail in the next chapter, should serve as excellent grounds for comparison. The insistence on studying weapon production is not accidental since this class of artifacts represents a vital clue to an understanding of the technical knowledge of Lombard blacksmiths, especially in light of the words of Paul the Deacon: "*Arma quoque precipua sub eo* - king *Alboin* -

---

[173] See R. Farinelli and R. Francovich, "Potere e attività mineraria," 446.

[174] I. Bóna, "I Longobardi e la Pannonia," 252: V. La Salvia, "L'artigianato metallurgico," 14; Id., *Archaeometallurgy*, 68.

[175] The first finds of Migration period weapons in the region of *Sopron* were found near the suburb of *Hegykő* in 1955. These were afterwards stored in the Sopron museum and published by *Gy. Nováki* in the *Soproni Szemle* in 1957. Based on this scant evidence, I. Bóna began archaeological investigations in the region and, between 1959 and 1962, completely excavated the came to light necropolis. During the excavations, 81 graves were discovered. Generally, all graves were orientated west to east and placed one next to the other at depths of between 1.50-2.00 m. The skeletons were laid in a sandy bed and were in good condition; only three of the deceased had been buried in wooden coffins; these may possibly have been the burials of higher ranking people. Anthropological and chemical analyses performed on skeletal remains yielded evidence about gender, the average life expectancy of individuals and blood groups. It was possible to divide the graves into two different groups which were divided by an empty strip of land some 6 – 8 m wide. There were 49 graves in group I, and 32 graves in group II. It was not possible to recognize any relevant difference between the two groups with respect to the chronology of interment. However, the older tombs and the richest grave goods were mainly found in the graves of group I. The ethnic features of the necropolis are difficult to evaluate. It seems likely that, between AD 500-568, members of a Germanic tribe were buried there. From the first half of the sixth century, the same region was occupied by another Germanic group coming from the banks of the river Elbe, that is, a group termed Lombards in the literature. After the excavation of more than 300 'Lombard' graves in western Hungary, it was stated by Bóna that the Lombards in Pannonia used a different burial system (still heavily influenced by the traditions of dress characteristic for the western Germanic cultures of the lower Elbe basin) compared to the graves found at the Hegykő cemetery. In fact, the general evaluation of the Hegykő necropolis yielded evidence of a completely different range of burial patterns compared with those coming from northern Pannonia and yet also assumed to be Lombard finds. The common features between them are tied to the relevant and general influence of the eastern Merovingian culture dominating throughout the whole area. The archaeological investigations within the area yielded much evidence relevant to establishing a comparison with the Hegykő-type of necropolis: the cemetery of Nikitsch with its 29 graves discovered in Burgenland and only 15-17 km. distant from Hegykő is clearly connected with the latter; westwards there are the necropolises of Mannersdorf am Leithagebirge and Sommerein; eastwards there is the cemetery of Fertőszentmiklós which was, unfortunately, badly disturbed. Through these data, it is possible to separate out a Germanic population group from the pre-Lombard period which still maintained its 'ethnic' autonomy after the 'Lombards' had entered the region. The archaeological evidence from the Hegykő necropolis indicates the existence of a culturally and ethnically mixed population. The typology of the finds underlines that people buried here belong to a Germanic group living in the Hegykő region, the northern part of *Transdanubia*, who dressed differently from other groups occupying Pannonia in the Lombard period. This may suggest that 'Lombards' attracted parts of other Germanic groups by assimilating them within their *exercitus* while always maintaining certain ethnic distinctions within a general horizon typical of the eastern Merovingian culture. Anthropologically, the vast majority of buried people consist of the vigorous northern proto-European type; concerning the blood group, the predominant component is A, characteristic of contemporary Germanic populations. Another 20-25% of the people buried there, mainly women, consist of graves without funerary goods; their blood groups clearly suggest a Mediterranean or mixed-Mediterranean configuration. These people were interred with their arms brought together and crossed over their chests, a burial custom that may indicate an adherence to the Christian faith. This burial custom has been found at many other cemeteries in the Late Antique period, such as those at Lorch-Lauriacum and Dunaújváros-Intercisa. Thus, it can be also be inferred that this type of archaeological feature in Hegykő provides evidence for the presence of remnants the Christian provincial populace. The remaining graves have to be placed between these two main groups; in fact, these tombs clearly represent a transitional period: both the anthropological features and the burial rituals are consistently mixed, a compound between the eastern Merovingian culture and that of provincial origin. Remarkable funerary goods were found in the warriors' graves, including swords, lances, axes, arrowheads, and the umbo of a shield; these same graves also yielded silver and bronze belt buckles and belt ornaments. The Hegykő necropolis produced crucial evidence for the process of cultural assimilation between Germanic and local populations. Moreover, the finding in *Cividale* of some fibuale which had their prototype in those of Hegykő, may suggest that the people who used this Hungarian cemetery did eventually migrate, together with the 'Lombards', into Italy around the year AD 568. Concerning these data, see, I. Bóna, "Das langobardische Gräberfeld von Hegykő, Komitat Gyor-Sopron," in P. Anreiter, L. Bartosiewicz, E. Jerem and W. Meid eds., *Man and the Animal World. Studies in Archaeozoology, Archaeology, Anthropology, and Palaeolinguistic in memoriam Sándor Bökönyi* (Budapest: Archaeolingua Foundation, 1998), 109-20; I. Bóna, "Hatodik századi germán temető Hegykőn" [Sixth-century German Cemetery at Hegykő], *Soproni Szemle* 14 (1960): 233-41, 15 (1961): 131-40, 17 (1963): 136-44; I. Bóna, "I Longobardi e la Pannonia," 244-45; Bóna, *The Dawn of the Dark Ages*, 31, 41; V. La Salvia, *Archaeometallurgy*, 43-49; for a deep and detailed analysis and revision of the finds of this necropolis see, I. Barbiera, *Changing Lands in Changing Memories. Migration and Identity during the Lombard Invasions* (Florence: All'Insegna del Giglio, 2005), ch. 1.

fabricata fuisse ..."[176]

The metallographic analyses were carried out on two samples taken from swords: Inv. # 65.59.1 and # Inv. 65.43.1.[177]

The sample taken from sword Inv. # 65.59.1 (sample no. 1) was cut from near the tang and did not include the whole cross-section. The metallographic surface to be analyzed, which extended from the edge to the core of the sword, was prepared by grinding and polishing. The grinding was made using metallographic grinding papers and the polishing with diamond polishing pastes in grain sizes 3 and 1 m. The metallographic surface was first observed under a metallographic optical microscope. Afterward, the surface of the sample was etched in *nital* (3% concentrated nitric acid in alcohol) in order to visualize the structure of the metal. After evaluation of the iron structure, the sample from sword # 65.59.1 was re-ground with fine grinding paper (600 mesh) and re-polished using the diamond pastes. The surface of the sample was then etched in a *Oberhoffer* reagent.[178] This reagent makes it possible to distinguish structures through the segregation of phosphorus or some other alloying elements (such structures are not etched by the reagent). This reagent is therefore conventionally used to detect the various iron materials used in the pattern welding. In fact, phosphorus has the property to move inside the iron structure and to become segregated on the surface of *ferritic* grains or in the grains themselves. Eventually, phosphorus can replace carbon in iron during decarburization.

The observation of non-metallic inclusions, under a metallographic optical microscope, clearly indicates the presence of cracks, pores, and other discontinuities in the iron structure. The presence of bands of non-metallic inclusions probably indicates welding lines. These non-metallic inclusions are of two different sorts: those tied to the reduction conditions inside the furnace, which are the furnace's slag types, and those connected to activities in the smithy. The presence of the first class of non-metallic inclusions stresses the fact that the iron production to be used for sword manufacturing took place in small reduction furnaces. With regard to the smithy inclusions, they have to be considered the results of the oxidic scale reaction with silica from silica sand, which is ferrous silicate. Silica sand was used by blacksmiths in order to dissolve the oxidic scales which formed on the surface of the heated iron semi-products or objects. The band of smithy inclusions propagates in direction of the welding line.

By etching the surface with *nital*, it was possible to evaluate the general distribution of the iron structures. Two different areas with various structures were recognized. One of them, characterized by pattern welding which consisted of alternative sheets of coarse grained *ferritic* iron and fine grained *ferritic-pearlitic* iron, has to be identified as the core. Two different types of material were used to manufacture the pattern welding. With respect to the other area, identified with the edge, it has a structure of fine grained *Ferrite* with both soft and tough properties which are not generally functional with regard to this particular part of the sword. Moreover, in the very same area near the edge there is a medium-size *ferritic-pearlitic* structure covering most of the zone up to the pattern welding. A thin strip of carburized *pearlitic* and *pearlitic-ferritic* iron was recognized in a zone near the surface of the blade. By using the *Oberhoffer* reagent, it was possible to recognize distinct sheets composed of two iron materials differing in their carbon and phosphorus content in the core of the sword, the area where pattern welding took place. The etching of the edge indicates a structure which is mostly *ferritic-pearlitic*, and the distribution of phosphorus attests to a structure mildly and not homogeneously carburized (for the mentioned metallographic structures **see figure 11, A**).

The sword # 65.34.1 (sample no. 2) was sampled somewhat closer to the centre of the blade. Again, the sample did not cover the whole cross-section but only the part from the core to the edge. The method used to perform the metallographic analyses was exactly the same as the one previously described for the sample taken from sword # 65.59.1.

Under the metallographic optical microscope, bands of coarse smithy inclusions of ferrous silicate were observed on the metallographic surface.

After etching the sample in *nital*, a very homogeneous structure was found over nearly the whole surface. Only in one small section of the surface there were bent bands with fine-grained *ferritic* structure surrounded by fine-grained *ferritic-pearlitic* structure. This precise structure must be considered the remnants, that is the last part, of pattern welding. The structure of the remaining surface is characterized by decomposed and globularized *Pearlite* which is a *sorbitic* structure. This structure, which is extremely homogeneous, is the result of annealing the sword's blade. After this evaluation of the iron structure, the sample was re-ground and re-polished and then etched in *Oberhoffer* reagent. This treatment, on the one hand, confirmed the presence of pattern welding; on the other, the *Oberhoffer* etching of the *sorbitic* structure indicated its very high level of homogeneity (for the mentioned metallographic structures **see figure 11, B**).

[176] *Pauli Historia Langobardorum*, L.C. Bethmann and G. Waitz eds., *MGH, SRL*, bk. 1, ch. 27. On this topic, N. Christie remarks in "Longobard weaponry," are of interest 7: "The short sax may also be a Merovingian introduction and marking the later stages of Longobard rule in Pannonia." The conquest of *Sirmio*, which also occurred in the last period of the Lombard occupation of Pannonia, might also have been important within this context since an important Roman weapon making industry was located there, *Sirmiensis scutorum, Scordiscorum et Armorum* as reported in *Notitia Dignitatum, Notitia Occidentis, VIII, Insigna viri illustris magistri officiorum* (Bonn: Böcking, 1839-1853), 2, 43-44.

[177] The following analyses were carried out by Prof. Lubomir Mihok of the Faculty of Metallurgy of the Technical University of Kosice (Republic of Slovakia), and in the laboratory of that University. Moreover, all the stages of the analyses were documented with photographs. The results have been published in V. La Salvia, *Archaeometallurgy*, 33-66; I. Bóna, J. Gömöri, L. Mihok and V. La Salvia, "Metallographic Analyses of Lombard Swords," *Východoslovenský Pravek* (special issue, 1999):180-86.

[178] The *Oberhoffer* reagent is composed of 0.5 gr. $SnCl_2$, 1 gr. $CuCl_2$, 30 gr. $FeCl_3$, 42 ml. concentrated HCl; 500 ml. alcohol.

The sample from sword # 65.59.1 yielded evidence that the blade was constructed in at least two parts: a central part, the core of the blade, consisting of pattern welding which was made by welding thin sheets to two different kinds of iron materials, differing in their carbon contents, that is, alternating sheets of coarse-grained *ferritic* iron and fine-grained *ferritic-pearlitic* iron. In fact, it was made by hammering distorted 'faggots' of these sheets of metals together. To this central part, a side element, the edge, was always attached by hammering. The edge was mildly and unhomogeneously carburized. Actually, no deep carburization was found at the edge although this would have enhanced the hardness of this part which consisted of fine-grained *Ferrite*. Possibly, since the sample was not cut far from the handle and considering the way the sword was used, the properties of hardness in that part of the blade were not so important.

With regard to sword # 65.43.1, the metallographic analyses indicate the composition of the sword's body, made by hammering bars of iron with different carbon contents, and the presence of pattern welding. The central part of the blade, identified with the core where the pattern welding was found and which contained bent bands with a fine-grained *ferritic* structure surrounded by a fine-grained *ferritic-pearlitic* structure, ends near the edge at the place where the sample was taken. It was not possible to recognize whether or not the edge was made from one or more pieces of iron semi-products. However, the iron structure of the edge reveals a deeply carburized area. Finally, the blade was annealed, producing quite a homogeneous *sorbitic* structure. In fact, annealing removed the heterogeneity of the *pearlitic* structure resulting from imperfections which occurred during the carburization process and relieved strains in the iron, conferring a very homogeneous structure on the material of the sword. This homogeneous *sorbitic* structure had a level of hardness suitable for the good functional properties of the blade. Moreover, the production process was exemplary, and, therefore, the utility value of the given sword was very high. The metallographic analyses of both samples did not reveal *quenching* and *tempering*.

3.2.2.2 The analyses of iron objects from the necropolis of Kajdacs-Homokbánya and Tamási-Csikólegelő (Hungary)

Moreover, another set of metallographic analyses, carried out using a *Scanning Electron Microscope* (SEM) by Dr. ÁRPÁD KOVÁCS and Dr. BÉLA TÖRÖK, from the Department of Physical Metallurgy of the University of *Miskolc* (Hungary), on 'Lombard' finds generously donated by the *Wosinsky Mór County Museum* (*Szekszárd*, Hungary), confirm the ability of these smiths and help to clarify their metallurgical traditional know-how. All the iron tools come from the necropolis of *Kajdacs-Homokbánya* and *Tamási-Csikólegelő*. Both necropolises date to the last phase of the 'Lombard' occupation of *Pannonia*. They are, in fact, located within the territories donated to *Audoin* king of the Lombards by Justinian I around the year AD 546. These zones were former Ostrogothic possessions.[179]

The analyzed samples were taken from an iron short sword of the type known in the literature as *Webschwert*, 1) "weaving sword" (# 73.148.1); 2) an iron knife (# 75.6.4); 3) an iron drill (#75.18.4) and 4) a fragmentary iron handle of a shield (# 75.22.2).

1) Iron short sword of the type known in the literature as *Webschwert*, "weaving sword" (# 73.148.1). This artifact comes from grave 2 from the necropolis of *Kajdacs-Homokbánya* in southwest Hungary. This "weaving sword" is all together 45.7 cms. long and 3.8 - 2.3 cm (towards the tip) wide. Its tang is 6 cm long. Together with the "weaving sword," other grave goods included an undecorated bronze hairpin (23 cm long), 59 orange and

---

[179]According to I. Bóna these burial sites can be classified within the Vörs-Kajdacs type. These cemeteries were around 80x80 m^2 similarly to the Szentendre type of necropolises. These kinds of cemeteries were opened and used by groups of 'Lombards' and comprise ca. 80-120 individuals and contained, at the end of their use, 80-90 graves. In contrast to this, the Vörs-Kajdacs types had only half the number of graves, that is to say, only around 40-50 burials. In Bóna's opinion, this situation suggests a shorter period of utilization for the necropolises. Actually, Kajdacs-Homokbánya contains 48 graves and Tamási-Csikólegelő 53 graves. In addition to this, both necropolises share other important characteristics. First, in both cemeteries, together with inhumation graves there is still evidence for the cremation rite. As far as Tamási-Csikólegelő is concerned, graves 27 and, possibly, 44 contain cinerary urns. The same situation occurred at Kajdacs-Homokbánya where cinerary urns were unearthed in graves 1, 19, 30 and possibly also 22, 25 and 43. Moreover, in both necropolises, the costumes of the deceased, as understood through the investigation of the archaeological features, are quite peculiar. Concerning the Tamási-Csikólegelő cemetery, the general feature of the grave goods, together with the form of weapons and coffins (inside which half of the cemetery population was buried), reveals a strong influence from the Thuringian cultural milieu. Nevertheless, this is not surprising since groups of people from Thuringian territory may well have escaped into Lombard controlled regions after the fall of their independent kingdom in AD 533. The dynastic relationships between the Lombard kings Wacho and Audoin and the Thuringian royal family certainly would have favoured this movement of people. Therefore, it seems that different groups of Germanic people were buried together inside this necropolis. Some of them display features influenced by Thuringian costume traditions while the others, some kind of western Germanic people, possibly migrated into the region even before the Lombards. These newcomers were to be heavily influenced by local provincial customs. With regard to the Kajdacs-Homokbánya necropolis, the general character of the grave goods as well as the style of depositions appear to be extremely similar to those of 'Lombard' settlements in Bohemia, Moravia and Austria, in areas north of the right bend of the Danube, thus pertaining to the Elbian cultural milieu. This archaeological situation confirms the late migration into Pannonia of a part of the Lombards who occupied those regions north of the Danube. Thus, the Lombard migration was a medium- to long-term process of relocation of people and not a sudden and definitive movement of populace. As a matter of fact, the same framework is indicated by the presence of some forms of pottery decorations that are found in the Pannonian phase of the 'Lombard' settlement as well as in the earlier phase, as is shown by finds from Várpalota and Březno. Neither of these Hungarian necropolises are fully published and the data are therefore scattered, however, to gain some general information see I. Bóna "I Longobardi e la Pannonia," 243-44 and fig 2; Id., "Langobarden in Ungarn," *Arheološki Vestnik* 20-21 (1970-1971): 49-50; I. Kiszely, *The Anthropology of the Lombards* (Oxford: BAR, I.S., 1979), 107, 113; for a deep and detailed analysis and revision of the finds of this necropolis see, I. Barbiera, *Changing Lands in Changing Memories*, ch. 3. I owe the information on the grave goods found inside the graves together with the analyzed items to Dr. Irene Barbiera. For her kind help, her suggestions and cooperation I am grateful and sincerely thankful.

green beads from a necklace, an S-shaped fibula, a disc brooch, a large decorated bead, a loom weight, a buckle, two arched brooches, an iron knife and, several unidentified copper alloy fragments.
A similar 'weaving sword' was found at the same necropolis in grave 18. In this case, the artifact was 49 cm long and 4.4 - 4.1 cm wide (the tang measured 13.5 cm). The 'weaving sword' was unearthed together with a spindle whorl, 49 beads of various form from a necklace, including a bead carved from a rock crystal and several glass beads with floral patterns, an iron buckle, and two hairpins. This grave was badly disturbed.

Three different samples were taken from this object (**see figure 12**). The metallographic structure of this artifact is quite fascinating. On the basis of the evidence from sample no.1, it is possible to conclude that this "weaving sword" was made of two different iron components. The inner part consisted of a soft iron bar which seems to be almost pure *Ferrite* with very low carbon content. There is a different iron element attached to its external sides. This element displays a typical pattern welding structure which consists of successive bands of *Ferrite* and *Ferrite-Pearlite*. Within these areas, the carbon content is higher. Moreover, the thick layer of ferrous-oxides which occurred between the two different materials indicates an intensive smithing process. It appears that a certain period of time passed before the pattern welding structure was attached to the soft iron part, a fact which allowed the formation of the thick oxidation layer and the accumulation of dust. It is, therefore, clear that the two diverse portions of the object were separately worked as different semi-products and only later joined together though hammering in the smithy.

In **figure 13**, Picture 1 can be seen a) on the bottom, the soft iron core containing pure *Ferrite* and b) on the opposite side, the pattern welding structure; here the *ferritic-pearlitic* areas are thicker with coarse grains; c) between them the welding line is clearly visible, characterized by a big accumulation of dust and ferrous-oxides.

Figure 13, Picture 2 shows the same zone as Picture 1 but under higher magnification. Here, in the upper part, the *Ferrite* can be examined together with some inclusions while the *ferritic-pearlitic* structure is located toward the bottom. More precisely, the *Pearlite* appears as a lighter color. In **figure 13**, Picture 3, the pattern welding structure is analyzed under higher magnification. Thus, traces of a long smithing activity become evident. A complex structure consisting of different iron layers and inclusions can be observed. In fact, through a careful investigation it is possible to observe a woven structure and a peculiar distribution of slag inclusions. This is clearly the result of numerous and successive hammerings on this part of the object during the smithing process.

Sample 2 was taken out from the edge corner, not far from the tang (**see figure 12**). This sample is, therefore, still a portion of the pattern welding structure. It has, in fact, a *ferritic-pearlitic* structure. The carbon content is quite high (about 0.6%) which corresponds, in modern terms, to a C60-type steel. This means that the iron was ready to be tempered. In **figure 13**, Picture 4, the *Ferrite* is dark and within the lighter zones, a huge amount of *Pearlite* can be detected. As shown in **figure 13**, Picture 5, the *Pearlite* has a foliated shape. Moreover, a slight decrease in the general amount of *Pearlite* is evident showing that general decarburization of the area took place. The smithing of the artifact and the consequent long exposure to high temperatures are responsible for the formation of such a structure. However, there is no direct proof of intentional heat-treatments. The corrosion of the edge is, nevertheless, too deep to risk any definite conclusion about heat-treatments. Again in **figure 13**, Picture 6, the traces and direction of the smithing process appear clearly delineated as lined structures.

Sample 3 was cut from the tip of the tang (**see figure 12**). It consists of pure *Ferrite*. It is composed therefore of the same material that composes the entire inner core of the instrument. This is absolutely a different sort of iron, basically a low carburized soft iron, compared to the iron forming the edges (pattern welding structure).

In **figure 14**, Picture 7 and 8 the coarse grains of *Ferrite* are visible. The quality of the iron of the tang, therefore, confirms that this 'weaving sword' was manufactured by joining through hammering on two diverging iron materials in the smithy: one, the inner core, consisted of a soft iron bar and the other had a pattern welding structure comprising bands of *ferritic* and *ferritic-pearlitic* iron. Within these zones the carbon content became higher.
To sum up, the structure of this tool is rather odd. The two diverging iron semi-products (the inner core and the edges) were separately produced and presented different mechanical characteristics. However, what is striking is that the pattern welding structures are usually placed into and used for the core whereas in the case of "weaving sword" # 73.148.1 it is used for the edges. As is clear from evaluation of the metallographic analyses of the knives from *Březno* and the swords from *Hegykő*, pattern welding was used in order to provide the inner part of the blade with hardness and tensile strength simultaneously so that it would be able to resist hits without warping. Through this method the core, composed of metal sheets differing in their carbon content and, thus, possessing diverging mechanical properties, attained a harmonious compromise between the various properties of the different iron bars: a compromise between elasticity and strength. At the same time, it ensured that each part of the sword -- such as the edge, core or body, and handle - maintained its own functional peculiarities.[180]
In contrast to this, "weaving sword" # 73.148.1 has a completely different structure. The core consists of almost pure *ferritic* iron while the edges are made with pattern welding. Moreover, the quality of welding which unites the two different materials, that is, the soft iron inner core and the pattern welding portion, indicates that

[180]On the mechanical characteristics of pattern welding see, C. Panseri, *Ricerche metallografiche*, 10, 24; and A. Leroi-Gourhan, "Notes," 36.

this tool may not have been particularly efficient: Picture 1 shows a huge accumulation of dust and Fe-oxides all along the welding line. Such a working procedure produced an extremely composite structure because it requires too many welding junctures. Thus, in addition to the main weld between the core and the pattern welding edge, there was also a seamed structure in this latter part where low carbon iron bands were joined together with more carburized ones through hammering. Not only did such a structure not improve the efficiency of the instrument, but it also required extensive application of one of the most difficult crafts in pre-industrial times: welding. Problems during the welding process, such as the oxidization which occurred between the different parts, as well as the difficulties of controlling the exact distribution of slag inclusions within the metal surface, could have severely and profoundly affected the functionality of the juncture. Therefore, even the positioning of the pattern welding within the edges would not have been a technically beneficial choice. As a matter of fact, the welded portions of a tool would only have been able to resist a lower degree of mechanical stress.

To draw definite conclusions on the basis of a single analyzed object is scientifically hazardous. Nevertheless, it is still possible to put forward some hypotheses, further substantiated by the finds from grave 71 from *Unterthürhein*, and graves 36, 22 and 579 from *Schretzheim*. The results of macroscopic analyses (i.e. a non-archaeometallurgical investigation) indicate once again that such tools have a composite structure. A central band differing from the rest of the surface is clearly evident within the body of these "swords".

This type of object is widely distributed only in the eastern Merovingian area and in the Danubian zone. To the west of the river Rhine these "weaving swords" are, in fact, rare and their presence is commonly interpreted as the direct influence of eastern Merovingian burial traditions. Within the archaeological record, in association with specific pottery types, they are specifically regarded as indicators of a Thuringian population or direct Thuringian influence. Their appearance in the grave goods is dated between the mid- and the second half of the AD 6th century. Moreover, their shape and length, which is extremely similar to real swords, have led to the suggestion that they were manufactured from old weapons. Traditionally, scholars had thought that these types of tools were instruments used in weaving. It has been usually assumed that these tools served to warp fabrics. It was also believed that existed similar wooden instruments, considered a sort of prototype, which do not appear in the archaeological record because wood does not preserve. For a long time archaeologists considered them to be typically female grave goods. However, the evidence for such a conclusion is rather scant. Actually, their presence in woman's graves, during the AD 6th century, is neither significant nor homogeneous. Moreover, even if these tools were mainly deposited in female graves, they can also be found in male tombs as well. Such is the situation with the grave goods placed in the burial at *Roztoky* near Prague, where a bone or antler comb, the umbo of a shield, two spear heads were unearthed. In addition to this, recently a similar item came to light in Italy in the Lombard necropolis of *Collegno* (Province of *Turin*, Piedmont) again in the tomb of a man (grave 60) dated after AD 660 (**see figure 15**). It has the same dimensions as a small knife: 13.8 cm long and 1.8 cm wide with a tang measuring 3.5 cm. Nevertheless, its shape is exactly the same as a bigger "weaving sword," even if the object is miniaturized. The object was unearthed together with a scramasax; a small strip of iron, possibly placed at the base of the wooden handle of the scramasax; six pieces of belt decorated with *agemina*; two copper alloy rivets and a stone spindle whorl. Moreover, it was positioned very close to the scramasax, possibly inserted in a special pocket of its sheath. That a shorter form of these tools, not exactly resembling real swords, existed is also confirmed by the finds from the necropolis of *Schretzheim*, from graves 311 and 26, related to *Stufe 3* (AD 565-590/600), that measure around 35 - 36 cm.

According to the data presented above, the identification of the object # 73.148.1, and consequently of all the other similar artifacts, as weaving tools is apparently not at all certain. First, they may not only be a female accessory and, given the Piedmontese knife, their shape may have varied a lot. In addition, there are many technical problems that suggest that these tools were not particularly efficient. In fact, on the one hand, archaeometallurgical analyses yielded evidence of a composite structure which requires too many welding junctures. On the other, their ergonomic features are apparently not suitable for warping fabric simply because a heavy iron tool, with too short a tang, would have been rather difficult to use with precision while standing in front of a loom plus moving your arms from top to the bottom and vice versa.

A much more plausible explanation seems to be that which ascribes a certain symbolic status to these tools, something already proposed by J. WERNER. This theory can be tested against the set of these instruments found in the Italian necropolis of *Nocera Umbra* (Province of *Perugia*, Italy) from the Lombard period (AD 568-774). Here, eleven of such items came to light in eleven different tombs: 148, 64, 100, 60, 23, 160, 144, 107, 85, 95, 69 (in chronological order). These graves can be further subdivided into different categories according to their chronology and to the richness of their funerary equipment. Graves 148, 100, 60, 23 and 69 are among the richest depositions in the entire necropolis. Grave 148 is related to the *Zeitstufe 1* (AD 572-590) and to quality group 1a. This tomb is the richest among all the others of the same period. It was the tomb of a young lady and, together with many pieces of gold jewelry, there was an extraordinary ivory casket, probably a reliquary. Graves 100 and 60 belong to *Zeitstufe 2* (AD 590-610) and, beside precious and superb jewelry, each contain a folding iron chair with *agemina* decorations. These items clearly indicate the high status of the buried persons. Therefore, they can be ascribed to quality group 1. A similar situation is indicated by the typology of the grave goods from tomb 23. Here, along with gold jewelry an ivory *pyx* was again found. However, since the folding iron chair is missing, this grave was placed within a different quality group, that is, 1b. Also in *Zeitstufe 2* (AD 590-610), tomb 160 contained a rich set of grave goods. Among the other things, the golden trims from stockings that resemble those from grave 100 are worth mentioning here. Grave 95 can be placed in *Zeitstufe 2* and contained objects made from gold, including a golden cross and golden beads. Grave 69 came from the last phase of the use of the necropolis and is, thus, dated between *Zeitstufe 2* and *3* (AD 610-620/30). The presence of a small ivory box together with gold jewelry is noteworthy here too. The rest of the tombs, that is to say, graves 64 (*Zeitstufe 1*, quality group 3), 144 (*Zeitstufe 2*, quality group 2), 107 (*Zeitstufe2*, quality group 2) and 85 (*Zeitstufe2*, quality group 2), can be considered "normal" and their grave goods are not particularly rich. Within these tombs, therefore, there are no signs emphasizing the rank of the buried person.[181]

The general framework of graves at *Nocera Umbra* suggests that *Webschwerter* as indicators of the rank of the engraved person is too simplistic a theory as well. The distribution of these objects does not seem to consistently follow the distribution of the richest grave goods. However, there is still the possibility that iron *per se* should be considered an important indicator of the richness of a given set of grave goods and, not only when connected with other precious items. During the Migration Period iron itself was regarded as a precious material. If we consider the *Webschwerter* not as objects produced to fill a given function but as a form of accumulation/tresaurization of wrought iron, 'weaving swords' would become as important as weapons and jewelry for indicating the quality of a particular set of grave goods as well as the rank of the deceased. In fact, only the richest people could afford to remove such great an amount of raw material ready to be worked from the world of the living. It may not be simple chance that within the testaments and charters of the AD 8th century and soon after the interruption of grave goods depositions, iron tools, *ferramenta*, together with weapons and jewelry are often recorded as well.

However, in order to resolve the question of the 'weaving swords' more securely, more archaeometric analyses on such artifacts would be required. Metallographic examinations or less invasive X-ray investigations may well shed new light on this problem, helping to evaluate the structure and the efficiency of this category of objects. Without this type of research, it remains impossible to decide whether or not all the "weaving swords" were produced according to a single standard method, thus, sharing the same technical and mechanical characteristics.

2) Iron knife (# 75.6.4). This object is 13.9 cm long and 2.1 cm wide. It is broken into two pieces and was found in grave 8 of the necropolis of *Tamási-Csikólegelő* (**see figure 16**). A belt element, a loom weight, two S-shaped fibulae, 20 necklace beads, a fragmented iron knife, a copper alloy key, an iron object which may be a circular buckle with a diameter of 6. 4 cm and a 7 cm long tongue

[181] On the question of *Webschwerter* see, B. Schmidt. *Die späte Völkerwanderungszeit in Mitteldeutschland* (Halle (Saale): Max Niemeyer Verlag, 1961), 144; J. Werner, *Die Langobarden in Pannonien. Abhandlungen, Bayerische Akademie der Wissenschaften, philosophisch-historische Klasse*, n. F. H. 55 A/B (Munich: Verlag der Bayerische Akademie der Wissenschaften, 1962), 34 and ff.; M. Hoffman, "The warp-weighted loom" *Studies in the History of Thechnology. Studia Norvegica*, 14 (1964): 39 and ff.; J. Bank-Burgess, "An Websthul und Webrahmen," in *Die Alamannen* (Stuttgard: Theiss, 1997), 371-78 especially 373-74: "Im ostfränkischen un alamannische Raum lassen sich eiserne Webschwerten häufig dort beobachten ... Abgesehen vom ostfränkischen Gebiet ist die Beigabe von Webschwerten bei den Franken nicht üblich. Sie finden sich bis auf wenigen ausnahmen in reiche Frauenbestattungen des 6.Jh;" P. Zenetti and H. Zeiß,"Alamannische Webschwerter im Museum Dillingen a.d. Donau," *Germania* 16 (1932): 307-08; W. Veek, *Die Alamennen in Würtemberg*, 2 vols. (Berlin-Lepzig: 1931), 5, 87; G. Behrens, "Eiserne Webschwerter der Merowinngerzeit," *Mainzer Zeitschrift* 41-43 (1946-1948): 138-43; Ch. Grünewald, "Das alamannische Gräberfeld von Unterthuerhein, Bayerisch-Schwaben," *Materialhefte der Bayerische Vorgeschichtsblätter* 59 (Munich: Michael Lassleben Kallmuenz, 1988): 128-30; R. Christlein, *Die Alamannen. Archäologie eines lebendigen Volkes* (Stuttgart: Theiss, 1991), 103 where these objects are identified with tool for weaving; B. Svoboda, *Cechy v dobe stehovani narodu*, 351; U. Koch, "Alamannische Gräber der ersten Hälfte des 6. Jahrhunderts in Südbayern," *Bayerische Vorgeschichtsblätter* 34 (1969): 162-93; see especially 187-89, table 2 where the early 6th century female graves from the Frankish, Alamannic and Thuringian zones are noted; U. Koch, *Das fränkische Gräberfeld von Klepsau im Hohenloherkreis*, in *Forschungen und Berichte zur Vor- und Frühgeschichte in Baden-Würtemberg* Band 38 (Stuttgart: Theiss, 1990), 171:"Die Beigabe eines Webschwertes beschränkte sich in der Mitte und zweite Hälfte des 6. Jh. nicht nur auf reiche Frauengräber, sondern stand ganz offensichtlich unter starkem östlich-merowingischen Einfluss. Angesehen von Nordgermanen und Anglesaschsen kannten Thüringen und Langobarden die Beigabe eiserner Webschwerter; bei den Franken westlich des Rheins waren sie nicht üblich. In ostfränkischen und alamannischen Raum lassen sich eiserne Webschwerter häufig dort beobachten ...;" U. Koch, *Das Reihegräberfeld bei Schretzheim*, in *Germanische Denkmäler der Völkerwanderungszeit*, 2 vols., ser. A, Band 13, (Berlin: Gebr. Mann Verlag, 1977), especially 93-94, 93: "Von dem Exemplar aus Grab 311 (*Taf. 83*) ist noch die Länge von 36,5 cm bekannt. Es entspricht damit den ebenfalls kurzen Exemplaren aus Grab 26 (*Taf. 11, 17*). Beide Gräber gehören in die Stufe 3 (565-590/600). Das Webschwert aus Grab 26 besitz eine gewölbte, in einem Stück geschmiedete Klinge. Vergleichbar ist das großere, nämlich 43 cm lange, Exemplar aus Grab 126 (*Taf. 31, 12*). Angesetzte Schneiden besitzt das ebenfalls fast 43 cm lange Webschwert aus Grab 36 (*Taf. 13, 8*); die Mittelbahn ist jedoch nicht damasziert, sondern zeigt durchgehende streifige Struktur. Dieses Exemplar ist in Stufe 1 (525/30-545/50) datiert. Angesetzte Schneiden und zwei damaszierte Bahnen in der Mitte haben die Webschwerter aus Grab 22 (*Taf. 9, 16*) und Grab 579 (*Taf. 151, 9*). Das Schwert aus Grab 579 ist nur fragmentarisch erhalten, dasjenige aus Grab 22 ist 41,2 cm lang. Damaszierte Webschwerter waeren zweifellos häufiger, wenn entsprechende Untersuchungen angestellt wurde. Bekannt sind daher nur eines von Finglesham in Kent und eines von Viel-Aître (Dép. Meurthe-et-Moselle) Webschwerter sind bisher nicht in Gräbern der sozial führenden, wohl alldingen Bevölkerungsschicht nachgeweisen, obgleich sie offensichtlich Privileg reicher, angesehner Frauen waren; vermutlich war das Webschwert für die Frau Zeichen ihrer Würde als Hausfrau und Herrin eines Hofes. In fränkischen Gräberfeldern sind Webschwerter selten anzutreffen; dort beschränken sie sich überwiegend auf ostfränkisches Gebiet. In alamannischen, thüringischen und langobardischen Gräbern kommen sie viel häufiger vor." S.E. Chadwick, "The Anglo-Saxon Cemetery at Finglesham/Kent," *Medieval Archaeology* 2 (1958): 1 ff.; for the Piedmontese find see, C. Giostra, "Catalogo," in L. Pejrani Baricco, *Presenze Longobarde. Collegno nell'Alto medioevo* (Turin: Soprintendenza per i Beni Archeologici del Piemonte, 2004), 131-32 and note 186; on the finds of *Nocera Umbra* see, C. Rupp, "La necropoli longobarda di Nocera Umbra (loc. Il Portone): l'analisi archeologica," in *Umbria Longobarda* (Rome: Edizioni De Luca, 1997), 27-29, 33-34, 38.

came to light in the same grave. Under a scanning electron microscope (SEM), it was not possible to clearly identify any definite metallographic structure. The sample appears to be oxidized. **Figure 17**, Pictures 9 and 10 shows the inner portion which is less corroded. Here, under higher magnifications (**see figure 17**, Picture 11), a sponge-like heterogeneous structure (traces of bloomery iron) becomes evident. This situation indicates that there was not much smithy working of the artifact and that the production processes involved were very simple. The quality of the knife is, thus, not high though it was surely a functional tool.

3) Iron drill (#75.18.4). This tool came from grave 20 at the necropolis of *Tamási-Csikólegelő* (**see figure 18**). Within the same burial, there were 10 arrow heads, an iron circle (4.4 cm diameter), a 16 cms long undecorated bone comb, a ceramic pot, three iron knives, (one 32 cm, the other 14.4 cm and 10.7 cm long) 4 flint stones, a small iron knife, a drift (6.8 cm across), an iron nail, a possible flint-lock, two iron buckles, a fragment of glass, a Roman gem, a glass-paste bead and a Roman coin. The archaeometallurgical analysis under the *SEM* indicates that again there is a high degree of oxidation and corrosion. As shown in **figure 19**, Pictures 12-15, the metallic iron (Fe) is marked by the light areas whereas the darker zone denotes the oxides (FeO, Fe_2O_3, Fe_3O_4). The temperature the metal was exposed to was hot enough to allow some local smelting. However, as it is evident from **figure 19**, Pictures 16 and 17, production through direct iron-making is confirmed by the sponge-like structure. Thanks to these pictures, traces of burnt charcoal grains also become visible.

4) Fragmentary iron handle from a shield (# 75.22.2). This artifact comes from grave 24 (**see figure 20**). It was discovered together with a shield, a leaf-shaped spear head broken in two pieces (the tang is 16 cm and the tip 10 cm long), an undecorated bone or antler comb, a sword, a fragmentary oval iron buckle, copper alloy rivets from a belt, an iron knife (11.5 cm long), badly damaged iron tweezers, two iron buckles. The sample taken from the shield handle also yielded evidence of iron with an extremely heterogeneous structure. Only *Pearlite* can be identified within the lighter areas in **figure 21**, Pictures 18 and 19. The chemical composition was also analyzed using a spectrometric analysis. Three different spots (marked in **figure 21**, Picture 19 by the numbers 1, 2, and 3) were then selected and analyzed. Non-ferrous components occurred in the black area (where spot 1 is located). Iron (Fe), together with a significant increase in the amounts silica (Si) comprise the grey zone (where spot 2 is placed). The area inside the white area is almost pure concentrated iron (Fe) (where spot 3 is positioned). This area was specifically analyzed using a chemical analysis. The result was as follows:
C: 1.00%; Mn: 0.20%; Si 1.14%; P: 0,44%; S: 0.00%; Cu: 0.01%; Cr: 0.01%; Ni: 0.38%
No Sulphur content could be detected. In the period when the artifact was produced, there were no technical means for complete desulphurization since the temperatures attained in the furnaces through direct iron reduction did not permit such a chemical process. Therefore, the quality of the raw material must have been excellent, that is to say, already with no trace of sulphur (see the spectrum in **figure 22**). This artifact was not subjected to intense smithing. In fact, for such an object, a simple iron handle, not much work would have been required to achieve sufficient efficiency.

3.3 Conclusions

From the data presented above, it is clear the level of iron-making technology had already reached a high standard in the 'Lombard' milieu during their settlement in northern and central Europe. Iron reduction was carried out in slag-pit furnaces based on a general procedure used throughout the entire Germanic world. Smithing was masterfully handled as well. A technical comparison between the analyzed artifacts from *Březno*, *Hegykő* and *Kajdacs-Homokbánya* and *Tamási-Csikólegelő*, provides enough evidence for demonstrating that there was an established craftsman's tradition with its own production methods: the mastering of pattern welding[182] and steel-making. Moreover, the smith was apparently well aware of the amount of work necessary for the production of a well made object.

Certainly, once the Lombards had crossed the borders of the former Roman Empire, they encountered the living remnants of Roman provincial culture with its own technical background. However, it is also clear that the Lombards came to this meeting fully equipped with their own patrimony of technical know-how. In this context, the information provided by Paul the Deacon regarding the numerous external/outsiders ethnic elements that joined the Lombards during their invasion of Italy, including the *Norici* and *Pannoni* who were traditionally linked to iron metallurgy, acquires a fundamental importance. In fact, on the one hand, it has to be emphasized that the geological features of the iron ore of *Noricum* and Lombardy are similar. On the other, from the point of view of history of technology, evaluation of Lombard iron-making traditions, as developed in North and Central Europe, permit us to reflect upon the various contributions to metal technology brought to Italy through the mediation of the Lombards and that played a role in the general reorganization of metal production in the Peninsula.[183] Actually, as stated by Paul the Deacon, the Lombard migration involved the entire population: *Igitur Langobardi, relicta Pannonia cum uxoribus et natis* ***omnique supellectili,*** *Italiam* ...[184] Therefore, it is evident that the 'Lombard' craftsmen brought with them

[182] R. Tylecote, *A History of Metallurgy*, 64: "It is not surprising, therefore, to find that in the Migration Period there was a fairly widespread knowledge of pattern welding in sword making throughout Western Europe. ... The techniques used were those known in the Roman period ... They were brought to full fruition in the early phases of this period."

[183] *Pauli Historia Langobardorum*, bk. 2, ch. 26: "Certum est autem tunc Alboin multos secum ex diversis ... gentibus ad Italiam aduxisse ... Gepidos, Vulgares, Sarmatos, Pannonicos, ... Noricos, sive aliis." Moreover, see the remarks reported by V. La Salvia, *Archaeometallurgy*, 69 and Id., "L'artigianato metallurgico," 15.

[184] *Pauli Historia Langobardorum*, bk. 2, ch. 7.

their working traditions and all the necessary tools.[185]

[185]N. Åberg, *Die Goten und Langobarden in Italien* (Uppsala: Almqvist & Wiksell, 1923), 147. Here it is stated that the Lombards in Italy "brachten...verschiedene Altertümer mitteleuropischer Typen und eine Ornamentik germanischen Charakters mit;" M.Cl. Bournand, "Les artisans dans le royaume lombard," *Arte Lombarda* 4 (1959): 13; U. Monneret de Villard, "L'organizzazione industriale," 10; G. Fasoli, "Aspetti di vita economica e sociale nell'Italia del secolo VII," in *Caratteri del secolo VII in Occidente. Atti della Settimana di Studio del Centro Italiano di Studi sull'Alto Medioevo, 5. (Spoleto 23-29-IV-1957)* (Spoleto: CISAM, 1958), 148.

Chapter 4
The Lombards in Italy.
The interaction of material cultures particularly referring to Iron making.

4.0 Introduction

The evaluation of both the archaeological and archaeometallurgical evidence from the Lombards in Northern and Central Europe has shown a strong "inclination" to iron metallurgy among them. The connection between Lombards and iron making is strengthened in light of the special attitude towards metalworking that was a common feature of Western Germanic culture. Moreover, it is important to stress here that this metallurgical activity took place in areas known to be important and traditional mining basins.[186] In addition to this, the methods of metalworking, as determined from the evaluation of metal objects from Lombard necropolises and settlements in northern and central Europe, demonstrate that the Lombard craftsmen were an integral part of the skilful vanguard of the common Germanic patrimony of technical knowledge.[187] The homogeneity and the high quality of the metal artifacts produced clearly indicate, one the one hand, the consistent utilization of standardized production processes and, on the other, the presence in this society of a firmly established group of specialized craftsmen as is shown by the finds from the actual graves of these metalworkers (*Brno*, *Poysdorf*, *Cividale*) and the nature of the workshops (*Scharmbeck*, *Göhlen*, *Březno*). Therefore, by the time of their migration to Italy (AD 568), the Lombards had fully mastered iron making technology.[188] In this respect, among the ***omnique supellectili***[189] that the Lombard brought with them to Italy, as stated by Paul the Deacon, one has to include their metal manufacturing traditions, and all the tools necessary to that tradition including, of course, craftsmen.[190] Thus, the Lombard migration may have involved not only the army but also relatively large portions of "civilian" population. Moreover, it is also crucial to consider the impact of the people of *Noricum* on the Lombards at the time of their migration to Italy, granted that there survived a fairly strong iron making tradition also during the Migration Period. The similarities of features of the iron ore between *Noricum* and Lombardy could have helped the re-establishing of ore exploitation in northern Italy as well.[191]

Finally, the evaluation of pre-Italian metallurgy, especially in light of Pannonian archaeological evidence, reveals that when the Lombards arrived in Italy the organization and technical level of their metal workshops was complex and on a very high level. Moreover, various funerary goods that have been discovered are defined and uniform enough to allow us to conclude that there was a diffusion of workshops which were homogenous in terms of their technical-cultural style.[192] Therefore, Lombard impact on Italian metal-working practices and the Mediterranean metallurgical heritage is more than simply one of acculturation, that is a one-way relationship, but rather represents a reciprocal exchange of ideas, ways of life, habits and customs. Certainly, it was a non-linear process, profoundly influenced by complex and varied sub-regional workshop contexts.[193]

With regard to metallurgical craftsmanship which was, as shown above, particularly specialized in the Lombard milieu the following questions should be raised:
a) The modes, time-frame and problems of integrating the Romano-Byzantine technical-artistic heritage with the Lombard heritage;
b) The eventual transmission of metal tools previously "unknown" in the Graeco-Roman world through Lombard mediation.
c) The results of the Lombard impact on those Italian regions that were traditionally mining basins (Piedmont, Lombardy, Tuscany). Their arrival may have exerted influence on the organizational structures of production but also possibly resulted in improvements in metallurgical production. That is, it would be interesting to see whether there was a revitalization of the mining industry on the peninsula in this period, especially considering that fact that since Late Antiquity at least mining had undergone a period of crisis in parallel with the whole classical production system.[194]

4.1.0. The material framework of integration

As to the first question, it is important to note that Lombards had a doubly positive effect on the total volume of production of metallic manufactures in Italy, especially the non-ferrous metal objects. First of all, having become the only large land-owners and therefore

---

[186] In my opinion the best example may be found in the Bohemian-Moravian mining district. See note 159 in Ch. 3 of the present study.
[187] I. Bóna "I Longobardi e la Pannonia," 252.
[188] The presence of specialized craftsmen underlines the existence of a stratified society and of a system of technically and culturally homogeneous workshops; in this respect see, P. De Vingo, "Metal handicraft among Western Germanic population in the 6th-7th centuries: working, manufacturing and decoration techniques," in *Proceeding of the International Conference Archaeometallurgy in Europe (Milan 24-26 September 2003)*, vol.1 (Milan: AIM, 2003), 607-16; moreover, see the classic studies of J. de Baye, *Étude Archaelogique. Epoque des Invasions Barbares. Industrie Langobarde* (Paris: Librarie Nisson, 1883), 111; I. Pieri, "Da Totila ad Autari," *Atti dell'Accademia di Scienze, lettere ed arte di Palermo* s. 4, 12 (1951-52): 68-69; A. Tagiaferri, "Aspetti e Limiti," 522, 526, 535-36, 540; Id., "Note sull'economia," 432; A. Melucco Vaccaro, *I Longobardi*, 37, 38, 42-46; I. Bóna "I Longobardi," 246-47, 250, 252, 253.
[189] *Pauli Historia Langobardorum*, bk. 2, ch. 7.
[190] N. Åberg, *Die Goten und Langobarden*, 147; here it is stated that the Lombards in Italy "brachten...verschiedene Altertümer mittel-europäischer Typen und eine Ornamentik germanischen Charakters mit;" M.Cl. Bournand, "Les artisans," 13; U. Monneret de Villard, "L'organizzazione industriale," 10; and G. Fasoli, "Aspetti di vita economica," 148.
[191] G.B. Brocchi, *Trattato mineralogico e chimico sulle miniere di ferro del dipartimento del Mella*, 2 vols. (Brescia: 1808), 34-36.
[192] See note 188.
[193] C. Citter "I corredi funebri nella Toscana," 270-71, I am intentionally widening the original concept of the author here to include broader geographical and chronological aspects.
[194] As far as the crisis of the mining industry at the time of Theodoric the Great, see C. Citter, "I corredi nella Tuscia," 180-81: V. La Salvia, "L'artigianato metallurgico," 16.

those who benefited most from land profits, they also became the only group that could afford to encourage the production of craftsmen.[195] Thus, they recreated – also from a strictly demographical point of view –a social group of potential commissioners whose ethnic-cultural structures were a mixture of classical and Germanic elements. This encouraged also, to a certain degree, the revitalization of the market through an innovation in styles and esthetical cannons applied to the production of luxury goods and their consumption. Moreover, the infusion of Lombard artists/artisans – bearers of their own well-defined traditions of craftsmanship — opened the way for the emergence of a "new craftsmanship" in Italy, created by the encounter of Lombard technical experience with that of the local native type of Roman origin.[196] This process reached its culmination in the 8th century when the ethnic affiliation of craftsmen became completely devoid of significance when they were recruited.[197]

Certainly, the assimilation of the two traditions did not occur immediately. The finds from Lombard necropolises in Italy, attributed to the newly immigrated generation, contain luxury objects mainly reflecting the Germanic tradition, most likely still produced in *Pannonia* or by the same craftsmen who crossed the Friulian passes into Italy along with the army.[198] Those few funerary goods from the same graves that seem not to belong to the traditional artistic culture of the Lombards, besides being sporadic, are very difficult to attribute to any other definite production milieu.[199] Thus, the *Kunsthandwerk* of the Lombards on the eve of their entry to Italy was typically Germanic and, as far as ornamentation is concerned, can be assigned to the animalistic style I. The latter was unknown in Italy prior to the Lombards.[200] Since there is a close relationship between artistic culture and the people producing it, i.e. the artist himself, the analysis of ornamental forms can help us trace the evolution of the artistic culture in Lombard Italy. Clearly, this evolution reflects the technology of craftsmanship. The artist, by which we mean the author of an aesthetic motif, and the artisan, the material executor of that motif, are not two different persons, but two aspects of the manual work of the same person.[201] Therefore, since for each artistic "style" there is a corresponding "technique" of crafting, the meeting of Germanic technique with the Roman-Byzantine "*intreccio*" motif, which gave rise to the so-called style II, presupposes continuity in way craftsmanship was organized. This means a stable workshop organization which allows continuous up-dating of technical procedures necessary to modify products according to changing trends in commissioners' tastes. Indeed, the typological evolution of the so-called *Bügelfibeln* provides an important example of this sort of metallurgical craftsmanship.[202]

Even if Lombard craftsmanship had developed original models within its own patrimony of traditional technology – something that is particularly evident in metalworking – it was operating in a context where Late Antique and eastern influences were always present. The inclusion of Roman-Byzantine models and their use in a wider artistic context which, nevertheless, must have taken into consideration the technical-productive capacities of Lombard craftsmen, had to go through production as organized in workshop.[203] The diffusion and substantial homogeneity of many of the products common in the Lombard period, which have been analyzed according by archaeological context, appear to have been an outcome of repeated craftsmanship production. Certainly, next to workshops focused on large-scale production, there continued to operate small ateliers and craftsmen who supplied their individual production to smaller markets. Therefore, the traditional metallurgy of craftsmen, represented especially by goldsmiths who cast gold-plated silver *fibula* such as the one buried in *Poysdorf*, continued to function within Lombard communities for a long time. Based on this, one may even hypothesize that having come into contact with the Roman-Mediterranean technical heritage, Barbarian craftsmen in some cases used the pre-existing structures of workshop organization typical of the Roman tradition, for their own purposes. Thus, the technical production of the Lombard period continued within the structure of workshops of Roman origin which remained at the heart of the manufacturing production.[204] The process of

[195]See A. Tagliaferri, "Industria artistica nell'Italia longobarda nel VII secolo," *Economia e storia* 3 (1964): 475; Id., *Strutture sociali e sistemi economici precapitalistici* (Milan: Giuffrè, 1972), 92; A. Melucco Vaccaro, *I Longobardi*, 139; D. Claude, "Die Handwerker der Merowingerzeit nach den erzählenden und urkundichen Quellen," in H. Jankuhn ed., *Das Handwerk in vor-und frühgeschichtlicher Zeit* (Göttingen: Vandenhoeck & Ruprecht, 1980), 207, the latter emphasizes the key role of commissioners for the Merovingian craftsmanship.

[196]See A. Tagliaferri, *Strutture sociali*, 94; V. La Salvia, "L'artigianato metallurgico," 17.

[197]M.Cl. Bournand, "Les artisans," 15.

[198] See M. Brozzi, "Strumenti di orafo," 167-74; V. La Salvia, "L'artigianato metallurgico," 16.

[199]A. Melucco Vaccaro, *I Longobardi*, 139.

[200]See A. Melucco Vaccaro, *I Longobardi*, 48 and ff.; moreover, on the characteristic of I zoomorphic style see G. Haseloff, *Die germanische Tierornamentik der Völkerwanderungszeit* (Berlin-New York: de Gruyter, 1981).

[201]H. Roth *Die Ornamentik der Langobarden in Italien* (Bonn: Theiss, 1973), *passim* and 9; H. Kilbride-Jones, *Celtic craftsmanship*, 21.

[202]H. Roth, *Die Ornamentik*, 9. It is important to note that the period during which this stylistic development took place, starting from the very end of the AD 6th and continuing through the entire AD 7th century, does correspond to the period of definitive adjustment of the Lombard newcomers with the Italian situation, a fact which preludes to the general pacification of the end of AD 7th century. N. Åberg, *Die Goten und Langobarden*, 43 and A. Tagliaferri, *Strutture sociali*, 84.

[203]See, O. von Hessen, "Tecniche di Lavorazione," in G.C. Menis ed., *I Longobardi*, 208-09; N. Åberg, *Die Goten und Langobarden*, 47 where the author says about the comprehention of the Bizantine stylistic culture: "ist es wahrscheinlich, dass sie zunachst von nicht langobardischen Künstler und Handwerken angeführt worden ist, und dass die Langobarden erst allmählich sie verstehen und schätzen gelernt haben."

[204]P.M. De Marchi, "Note su produzione e scambi nella Lombardia di età longobarda: l'esempio degli scudi da parata," in G. P. Brogiolo ed., *II Congresso Nazionale di Archeologia Medievale. Musei Civici, Chiesa di Santa Giulia (Brescia, 28 settembre-1 ottobre 2000)* (Florence: All'Insegna del Giglio, 2000), 289: "La diffusione di questa arma da difesa [the shield] ha il massimo sviluppo a partire dai primi decenni del VII secolo, quando sono poco documentate tombe di fabbro/orefice, in questi anni è probabile si precisasse una nuova organizzazione del lavoro, di cui sfuggono le caratteristiche, ma che doveva vedere

cultural osmosis between these two ethnic components was deepened through the work and "emigration" of merchants, as well as the same craftsmen of Roman origin who were attracted to new centers of power, as Grave # 37 from *Castel Trosino* suggests. Grave # 37 possibly belonged to a goldsmith, dated to circa AD 600. It contained a crucible, two iron tools – of which one was not identified – and funerary goods of a certain high level including gold and silver objects. Thus, even if the grave lacks weapons the quality of the grave goods shows that this was the inhumation of a person of high status. Therefore, the finds from this grave and from *Crypta Balbi* in *Rome* are important for the reconstruction of the material basis of this cultural interaction since they attest the transfer of craftsmen and artifacts of Roman origin into Lombard controlled territories.[205]

The find of the *Crypta Balbi* in *Rome* yielded evidence of a large centralized workshop which was possibly connected to a monastic center reproducing various luxury goods such as textiles, bone and ivory objects, precious stones and metal goods and wooden objects. The atelier also supplied an extensive trade network as shown by finds in the Lombard necropolises of north and central Italy, especially in *Castel Trosino*. The similarity between precious metal objects in this necropolis with finds from the *Crypta Balbi* is, actually, remarkable. It seems that the products of this *Rome* workshop were fashioned for Barbarian clients. Thus, the frontier between the Lombard and the Byzantine parts of Italy were not at all impermeable. Moreover, through these artifacts it is possible to follow the process of mutual acculturation: the Roman-Byzantine society had acquired many Barbarian features, becoming somehow Longobardized, and both societies could make use of the same objects. In fact, at the same site of *Crypta Balbi* many polished objects decorated with *agemina*, and thus not *ipso facto* produced by the same workshop, were also found. Stylistically, this *agemina* displayed both the typical Byzantine decoration with points and commas as well as the characteristic zoomorphic patterns of Germanic origin. This evidence once more demonstrates the way artifacts typical of Lombard Italy circulated within the Byzantine territory.[206] A similar situation may also have existed in northern Italy, especially in *Firuli*, where several luxury goods deposited as grave goods may have been produced in Italian-Byzantine workshops and then distributed throughout Lombard areas.[207]
The Lombard kingdom consequently created a new market. The Byzantine workshops, on the one hand, undoubtedly received new impulses for production and, on the other hand, they strongly contributed to the diffusion of patterns and models of decoration which were then elaborated according to the new commissioners' tastes. These stylistic features were subsequently "digested" by craftsmen within the Lombard kingdom. Here, from the beginning of the 7th century, several workshops began their operations. They were soon able to provide the market with silver, copper alloys, and *agemina* iron objects which fell not too far from Byzantine production standards.[208]

While there is no doubt that the Italian Peninsula suffered from recession at the end of the AD 6th and the first half of the AD 7th centuries, society during the Lombard period appears to have been dynamic and flexible. Its Lombard component was open to influences, contacts and exchanges with the Byzantine civilization that played the role of "guardian" of the Mediterranean heritage. Yet, it is precisely from the AD 7th century that among Lombards craftsmanship was characterized by a strong standardization of production, typical examples of which are the *agemina* sword-belts with animal motifs, bronze belts with triangular plaquettes and belts with multiple *agemina* ornaments in a harmonious animal style. This type of production slowly replaced products of Byzantine origin in Lombard funerary goods. All these workshops diffused their products widely through Italian territory, including its Byzantine areas, without any break in continuity throughout the entire 7th century. After this period the rite of depositing funerary goods in graves was clearly abandoned, rendering the study of the development of these goods much more difficult for archaeologists.[209]

compresenti nella stessa bottega artigiani di origini etniche e culturali diverse, fattore che spiegherebbe le varianti nei simboli e nella lavorazione degli scudi;" and 290: "La trasmissione delle tradizioni artigianali tardoromane avvenne anche con la mediazione gota." In addition to this, see as well C. Citter, "I corredi funebri nella Toscana," 199; G. Fasoli, "Aspetti di vita economica," 151; U. Monneret De Villard, "L'organizzazione industriale," 30; on workshops as the heart of craftsmen production see as well, H. Kilbride-Jones, *Celtic craftsmanship*, 10.

[205]See, L. Paroli, "Aspetti e problemi," 14-15; and M. Ricci "Relazioni culturali e scambi commerciali," 253-56, 270; L. Paroli, "La Necropoli di Castel Trosino: un laboratorio archeologico per lo studio dell'età longobarda," in L. Paroli ed., *L'Italia centro-settentrionale*, 91 particularly note 5 for the quoted bibliography.

[206] V. La Salvia, "Aspetti dell'economia dell'Italia alto medievale. Artigianato e commercio fra Longobardi e Bizantini," in O. Merisalo and P. Pahta eds., *Frontiers in the Middle Ages* (Louvain la neuve: Brepols, 2006), 348-72.

[207]About *Castel Trosino, Nocera Umbra* and the general framework of this portion of central Italy see R. Mengarelli, "La necropoli barbarica di Castel Trosino presso Ascoli Piceno", *Monumenti Antichi, pubblicati a cura dell'Accademia dei Lincei*, 12 (1902): coll. 145-380; A. Pasqui, R. Paribeni, "Necropoli barbarica di Nocera Umbra," *Monumenti Antichi, pubblicati a cura dell'Accademia dei Lincei*, 25 (1918): coll. 137-352; L. Paroli, "Aspetti e problemi," 14-15; L. Paroli, "La Necropoli di Castel Trosino: un laboratorio archeologico," 91, 92, 102, 105, 109; M. Ricci, "Relazioni culturali e scambi commerciali," 253-56, 270; P. Delogu, "Considerazioni conclusive," in L. Paroli ed., *L'Italia centro-settentrionale*, 428-30; L. Paroli, "Introduzione," in L. Paroli ed., *La necropoli altomedievale di Castel Trosino. Bizantini e Longobardi nelle Marche* (Milan: Silvana Eitoriale, 1999), 22, 23, 25; L. Paroli, "La Necropoli di Castel Trosino: un riesame critico", in L. Paroli ed., *La necropoli altomedievale di Castel Trosino*, 199, 206, 208-10; M.C. Profumo, "Le marche in età longobarda: aspetti storico-archeologici," in L. Paroli ed., *La necropoli altomedievale di Castel Trosino*, 128, 130; A. R. Staffa, "Un quadro di riferimento per Castel Trosino: presenze longobarde fra Marche e Abruzzo," in L. Paroli ed., *La necropoli altomedievale di Castel Trosino*, 111, 116-17; L. Jørgensen, "Castel Trosino and Nocera Umbra. A chronological and social analysis of family burial practices in Lombard Italy (6th-8th cent AD)", *Acta Archaeologica* 62 (1991): 2-10, 10-18, 42-44, 44-48; C. Rupp, "La necropoli longobarda di Nocera Umbra,"23-40.

[208]L. Paroli "La cultura materiale," 270, 285

[209]L. Paroli "La cultura materiale," 283, 285, 296-97

4.1.1. New objects from new people? Technology transfer within early medieval Italy.

Among the numerous finds dated to the Lombard period, are several iron objects that apparently, reflect non Mediterranean characteristics. On the contrary, they present features typical for an alloctonous derivation from a northern and/or central-eastern European cultural milieu. Since some of these objects, such as some particular types of plough, iron and copper alloys stirrup, were unknown in the Italian Peninsula before the Lombard invasion, it has to be considered that this event, beside having political and military implications, had a deep influence on the development of this aspect of the material culture too. The Lombard contribution to the development of the tool assemblage of the Italian Early Middle Ages is possible to assess through evaluation of some of these finds and the production methods used to manufacture agricultural tools, swords and stirrups. Moreover, it seems clear that the Lombard kingdom was a vital part of the cultural and commercial network that connected and permeated the entire Eastern Merovingian area. In this way, a fairly uniform zone was shaped that, far from being simply on the periphery of the Frankish kingdom, played an important role in the transmission of technological and cultural patterns from Northern and Eastern Europe into the Mediterranean regions.

4.1.2 Finds of Agricultural tools from the Lombard period in Italy and within eastern Merovingian area

The evidence of early medieval agricultural tools in Italy is fairly scant. This is particularly true for those from the Lombard period. However, there are some objects whose typology clearly diverges from artifacts of classic Mediterranean origin. On the contrary, they apparently indicate a Central European origin emphasizing once again the importance of the eastern Merovingian area for the transmission of technology and tools from continental Europe to southwestern Europe.

At *Belmonte* (Piedmont), the most important group of agricultural tools from the Lombard period came to light. Here, a large settlement has been identified with houses of the elite and a defensive wall which was repaired three times. Six ploughshares, five picks, one hoe, one spade and a shovel were found within the site. Some of the unearthed tools were found in two different cubbyholes, while the remainder of the tools was found scattered all around the site. One was a sort of wooden barrel, of which only the metal hoops have survived and which was hidden between two walls. The other important find spot was a small storeroom linked to room C. Here, the remnants of a smithy together with some iron slag, indicates that this area may have been the storeroom for an iron making workshop. The activity of the latter, moreover, can be connected to the patron's presence as suggested for room C. The placement of tools in cubbyholes is rather common for early medieval contexts. Some scholars suggested that the people who hid the tools thought their abandonment of the settlement would only be temporary although we know that it must have become permanent. In both places the variety of tools is noteworthy. This bears witnesses to the presence within the settlement, of numerous craft activities. A pick, a bit, a trivet, a blacksmith's tongs, a *francisca*, an axe typical for the period between AD 6th and 7th centuries were placed in a wooden barrel. Several mining tools (a pick, a crowbar and a bit) as well as tools used in agricultural such as another pick and three ploughshares were brought to light in another barrel near room C.[210]

The typology of the six *Belmonte* ploughshares, all dated between the late AD 5th and the first half of the AD 7th century is extremely interesting (**see figure 23**). This type of object have triangular points and a long bar with a rectangular cross-section. They are called *Speerförmige*, that is to say, they are spear-shaped ploughshares. Their average length is between 73 and 90 cms, while their average width (measured at the triangular point) is between 13.3 and 22.5 cms.[211] Similar tools have also been found in Italy at *Carignano* (Province of Turin, Piedmont),[212] *Parma* (Emilia-Romagna)[213] and *Masegra* (Province of Sondrio, Lombardy), [214] unfortunately without any reference to their original archaeological contexts. However, the ploughshare from *Parma* is relatively larger compared to those from *Belmonte*. Its triangular point is 23.5 cm long and 26 cm wide. Its bar is 80 cm long. The origin of this form of ploughshare has been ascribed to the Central European Celtic tradition.[215] Let us now consider the specific character of these tools and in what ways they differ from those of classical

---

[210]On the site of *Belmonte* see, E Micheletto and L. Pejrani Baricco, "Archeologia funeraria ed insediativa in Piemonte tra V e VII secolo, in L. Paroli, *L'Italia centro-settentrionale*, 318-25; M. Cima, "Le Origini della Metallurgia del Ferro nel Canavese," *Rivista di Archeologia*, 11 (1987): 113 and 119. Moreover, the settlement of Belmonte, where agriculture and a specialized craft such as iron metallurgy coexisted, finds its parallel in many early medieval written sources that often present a framework within which iron production is managed in the countryside by a sort of "peasants-smiths," beside urban contexts where specialized craftsmanship takes back its usual role. These smiths possessed, already for the AD 9th and 10th, huge patrimony in real estates. For iron making also the production of the smiths working within monastic estates must have been important especially to fulfil the needs of the *curtes*. On this topic see, V. La Salvia and F. Zagari, "Cultura materiale e tradizione tecnica," 971; M. Baruzzi, "I reperti in ferro dello scavo di Villa Clelia (Imola). Note sull'attrezzatura agricola nell'Altomedioevo," in R. Francovich and G. Noyè eds., *Archeologia e storia del Medioevo italiano* (Rome: NIS, 1987), 160; V. La Salvia, "Gap or Continuity?" 266.

[211]In details the lenght and width of four of these six ploughshares is 90.2 and 21 cm; 89 and 22.5 cm; 73 and 13.3 cm, and 89 and 14 cm.

[212]F. Scafile, "Di alcuni oggetti in ferro rinvenuti a Belmonte," *Ad Quintum*, 3 (1972): 28; E. Micheletto and L. Pejrani-Baricco, "Archeologia funeraria," 322.

[213]M. Catarsi Dall'Aglio, "Materiali longobardi di provenienza ignota presso il Museo Archeologico di Parma," in M. Catarsi Dall'Aglio ed., *I Longobardi in Emilia occidentale* (Parma: Editoria Tipolitotecnica, 1993), 74.

[214]G. Forni, "Gli aratri dell'Europa nord-occidentale dalla preistoria al Mille," in R. Comba and F. Panero eds., *Il seme, l'aratro, la messe. Le coltivazioni frumentarie in Piemonte dalla preistoria alla meccanizzazione agricola* (Cuneo: Società per gli Studi Storici, Archeologici ed Artistici della Provincia di Cuneo, 1996), 38 ff.; E. Micheletto and L. Pejrani-Baricco, "Archeologia funeraria," 322.

[215] G. Forni, "Strumenti aratori in Aquileia Romana. Loro origine, evoluzione, tipologia. Il contesto socio-economico," *Antichità Altoadriatiche* 35 (1989): 324; V. La Salvia and F. Zagari, "Cultura materiale e tradizione tecnica," 970.

Mediterranean origin. The main difference lies in the system of connecting the iron part to the wooden support: the *Speerförmige* ploughshares of northern Italy have neither holes nor metal wings to mortise them to the wooden moldboard. Therefore, it must be imagined that the connection between their long bar and the wooden moldboard was made through ropes or iron strips, drawn together firmly by iron nails or check pins. Thus, considering both the position and the inclination of the ploughshare, this clasping system would have been completely different from what was typical for the Roman world – evidence of a diverging techno-cultural tradition. As a matter of fact, for the ploughshares of the Mediterranean area, the junction with the moldboard was produced by metal wings, later to be replaced by an iron bolt. However, there are also numerous ancient ploughshare found in northern Italy, more precisely in the vicinity of the town of *Aquileia*, which also do not fully belong to the Roman tradition. For these objects as well, either direct import from *Noricum* or local production under the strong influence of the blacksmithing methods of that region has been proposed. In both cases, the connection with Celtic metal-working traditions is evident.[216]

The possible Central European origin of these instruments may be confirmed by examining other finds from this same region. A *Speerförmige* ploughshare for example, may also be found among the Imperial period agricultural tools from the provinces of *Raetia, Noricum and Pannonia* – where the Celtic substratum seems to have resisted Romanization. This object was found in *Bregenz* in 1906 and was dated to the AD 1st century. It had a length of 76 cm and a width at the point of 13.6 cm (**see figure 24**).[217] The dimensions and the shape of this object are very similar to similar ploughshares discovered in Italy. Moreover, the place where it was found seems to coincide with ancient *Brigantium* on Lake Constance, in *Raetia* province. According to some scholars this province can be considered the cradle of the wheeled plough mentioned by Pliny the Elder.[218] J. HENNING also mentions that this type of plow was well rooted in *Raetia* from the AD 1st to 4th centuries. Single examples of *Speerförmige* ploughshares have also been found in the area immediately south of the western flow of the Danube, together with very elongated coulters (**see figure 25**). According to this scholar, this type plow resulted from the "meeting" of Late Antique and Celtic-Germanic traditions in the Migration Period. Similar ploughshares, called *Stielschäftung* (type B, according to J. HENNING's typology), have been also documented for south-eastern Europe, along the Danube. Together with them, types such as *Tüllenschäftung* (type A) and *Pflugschar mit tüllenartigem Stiel* (type C), shovel-like or leaf-shaped ploughshares (**see figure 26**), whose common feature is the method of clasping using metal-wings may also be

216 G. Forni, "Strumenti aratori," 324-27; V. La Salvia and F. Zagari, "Cultura materiale e tradizione tecnica," 972-73. On the importance of the mutual cultural relationship, in its wider sense, between Aquileia and the Celtic world see R. Bianchi Bandinelli, *Roma. La fine dell'arte antica* (Milano: Rizzoli, 1999), 110: "Importanza del tutto particolare assume la fondazione, in territorio già celtico e in posizione di dominio sulle vie di comunicazione verso l'Illiria ed il Norico, della colonia di latini ed alleati italici di Aquileia (181 a.C., cfr. Liv. XL, 34, 2) ... Aquileia appare, infatti, determinante per il repertorio artistico del Norico e dell'Illiria, della Pannonia e della Dacia."

217 R. Pohanka, *Die eisernen Agrargeräte der römischen Kaiserzeit in Österreich – Studien zur römischen Agrartechnologie in Rätien, Noricum und Pannonien* (Oxford: BAR, I.S. 298, 1986), 36-38, tav. 7, 24; V. La Salvia and F. Zagari, "Cultura materiale e tradizione tecnica," 973-74. For the survival of a celtic cultural substratum in these regions, especially among the peasantry, see S. Mazzarino, *La Fine del Mondo Antico* (Milano: BUR, 1999), 30, 140; R. Bianchi Bandinelli, *Roma. La fine dell'arte antica*, 123, 127: "L'unica di queste regioni ad avere alcuni caratteri iconografici ed ornamentali peculiari è proprio il Norico, dove a una antica civiltà del ferro era succeduta una intensa produzione metallurgica nella Carinzia (Carnuntum, Magdalensberg) abitata da popolazione celtica. Vi sono tipiche vesti e larghi berretti a due punte nelle stele delle donne indigene. Il repertorio funerario ellenistico-romano viene (per esempio, Eros e Psiche in un sarcofago di Sirmium) racchiuso entro una tipica incorniciatura di elementi ricurvi ('cornice norica') che risente del gusto celtico, e che continua dal I al IV secolo trovandosi ancora si sarcofagi con raffigurazioni cristiane" and 134 "...in altri documenti del Norico e dell'alta Pannonia, dove si era estesa l'area celtica, possiamo probabilmente cogliere un riemergere del sostrato locale più antico che porta a prediligere forme schematiche, di gusto primitivo, quali era diffuse nella civiltà detta di La Tène (stele di Aurelius Januarius a Budapest; stele di Aptomarus, dalla Carinzia, a Deutsch Altemburg). Dalla Carinzia ci viene proprio il documento più evidente di questo innesto di concezione romana su forme di tradizione celtica nella singolare stele di un Popaeius Senator, ancora in sito a Matrei."; T.L. Markey, "Germanic in the Mediterranean," 55: "... Of course, to underscore the matter, Roman Noricum was founded on an earlier Celtic substrate, cultural parts of which persisted, albeit in new guises: for example, Celtic Noreia=Roman Isis, an effective translation of the inscription on the Roman building stone inserted, however surreptitiously, above the main entrance to the church at Michelsberg outside present-day Klagenfurt."; A. Mócsy, *Pannonia and Upper Moesia. History of the middle Danube provinces of the Roman Empire* (London: Routledge & Keagan, 1974), 338: "Roman culture and presumably the Latin language as well were confined to those who by office or by wealth were closely connected with the government and with the administration;" and 358: "One might expect that the local populations were able to resume the use of their Celtic Illyrian and Thracian languages in the 5th century, for the tenacious survival of the local tongues of the original population can be proved in Pannonia as well as in Moesia." Still on the situation in Pannonia see, L. Borhy, "Romani e Pannoni. Aspetti dell'acculturazione in una provincia di frontiera," in G. Hajnóczi ed., *La Pannonia e l'Impero Romano. Atti del Convegno Internazionale. Accademia d'Ungheria e l'Istituto Austriaco di Cultura (Roma 13-16 Gennaio 1994)* (Milano: Electa, 1994), 72: "... la produzione artigiana locale si sforzò di soddisfare le pretese dell'esercito: la ceramica celtica del tardo periodo La Tene presenta iscrizioni latine (*da bibere*) o stampi a gemma italici, forme ceramiche vengono imitate, ma decorate, secondo il gusto celtico, con una pittura a fasce rosse e bianche. Ciò comportò un intreccio fra le locali tradizioni tardo celtiche e quelle romano provinciali in sviluppo."

218 Pliny the Elder, *Natural history*, vol. V, bks. 17-19, H. Rackham ed. (London/ Cambridge, Mass.: Heinemann/Harvard University Press, 1950), bk. 18, 171-72: "*culter vocatur inflexus praedensam priusquam proscindatur terram secans futurisque sulcis vestigia praescribens incisuris quas resupinus in arando mordeat vomer. alterum genus est volgare rostrati vectis. tertium in solo facili non toto porrectum dentali sed exigua cuspide in rostro. latior haec quarto generi et acutior in mucronem fastigia eodemque gladio scindens solum et acie laterum radices herbarum secans. non pridem inventum in Raetia Galliae ut duas adderent tali rotulas, quod genus vocant plaumorati; cuspis effigiem palae habet.*" This is the only written sources directly mentioning such a ploughing tool. Its interpretation is though controversial. It is possible that this wheeled plough, already diffused in AD 1st century *Retia*, was similar to that atteste in the area of Mantova (Lombady) already in the AD 4th cent. at the time of Servius. On this topic see, G. Forni, "Strumenti aratori," 327, 328; G. Cherubini, "Le campagne italiane dall'XI al XV secolo. L'intensificato sfruttamento dei terreni coltivati," in O. Capitani, R. Manselli and G. Cherubini eds., *Storia d'Italia*, IV, *Comuni e Signorie: istituzioni, società e lotte per l'egemonia* (Turin: UTET, 1981), 296-303.

found. The A and B types, according to J. HENNING, assured deeper ploughing.[219]

Besides ploughshares, there are other instruments found on the same site that present peculiarities in their form of clasping that, again, are different from that found on objects of Roman-Mediterranean origin. Among the various types of clasping used for agricultural tools in medieval Italy, that of "eye-shaped" or "canon-like" clasping is prevalent. This kind of clasping technique may be found in any period and is independent from the ways these implements were used or their geographical origin. In this clasping method, the wooden support was inserted into the "eye-shaped" or "canon-like" metal portion and could be better fixed or adjusted by immersing the wood in water or by inserting small wooden pins. The reasons for such homogeneity in the clasping system of Roman-Mediterranean ploughshares should probably be attributed to the strong linking of the metal points and the wooden handles that could be achieved by it.[220] On the other hand, the presence of a pointed tang in a hoe and a spade found on the same Piedmont site and dated to somewhere between the AD 6th to 7th centuries provide evidence for a strong connection between this type of variation in the form of clasping and the agricultural traditions of early medieval *Belmonte* (**see figure 27**).[221]

The hoe is 15.3 cm long and 11.7 cm wide and is also dated between the AD 6th and 7th centuries. Its triangular blade is thinner towards the edge and has a 9.5 cm long tang for joining the blade to the wooden handle. The tang has a hook-like bend in it, thus, forming a fairly acute angle. Once again, the clasping system denotes a diverging techno-cultural tradition. As far as the spade is concerned, its form is rather similar to those *palae* of Roman origin. However, as it appears to be typical for *Belmonte*'s tools, it has a pointed tang with a hook-like bend. Moreover, it has a *zig-zag* decoration all along its medium axis, a feature which makes it something of an *unicum*. This clasping system, based on a pointed tang inserted into the wooden handle, appears to be quite common at *Belmonte* and seems to have been preferred over the clasping system of Mediterranean origin. However, in order to fasten the two parts together where they join, metal ropes or strips may just as well be used. As a matter of fact, within this settlement this allochthonous method for joining metal points and wooden handles appears to be used for different tools used in different ways.

[219]On the question of ploughshares typology and development between Late Antiquity and Migration Period see, J. Henning, "Zum Problem der Entwicklung materieller Produktivkräfte bei den germanischen Staatbildungen," *Klio*, 68 (1986): fig. 1; Id., *Südosteuropa zwischen Antike und Mittelalter* (Berlin: Akademie Verlag, 1987), 49-51, 58-59, 60, 61, 63, 64, 65; V. La Salvia and F. Zagari, "Cultura materiale e tradizione tecnica," 973-79.

[220]F. Zagari and V. La Salvia, "Aspetti della produzione," 875-76; V. La Salvia and F. Zagari, "Cultura materiale e tradizione tecnica," 972-73.

[221]L. Pejrani Baricco, "Schede," in G.C. Menis ed., *I Longobardi*, 347 and ff; V. La Salvia and F. Zagari, "Cultura materiale e tradizione tecnica," 973.

Beside ploughshares, hoes and spades, other tools found in early medieval northern Italy apparently had a Central European origin having been developed between Late Antiquity and the Migration Period and may have reached Italy through Lombard mediation. Among the picks discovered at *Belmonte* and *Villa Clelia* there are two objects of a type with analogies again in Central Europe, more precisely in Imperial Roman *Pannonia*.[222] Both these picks have two sharp opposite and perpendicular points (**see figure 28**). Such tools could easily have been used for both agricultural work and mining activities. The points of one pick from *Belmonte*, dating from between the late AD 6th and 7th centuries, measure 36.5 x 2.7 x 2.8 cm.[223] The points of the other pick, dated to the AD 6th century measures 30 x 2.8 x 1.5 cms. Both tools are extremely close in type and chronology to each other.[224] The same type of picks is found in the area between *Noricum* and *Pannonia* provinces as shown by an object discovered in present-day Austria and dated to the Roman Imperial period.[225] This pick is very similar to the one from *Belmonte* but its back point is much thicker and larger than the Italian one. This is, in any case, the only major difference between the three almost completely analogous tools. The origin of their typology is apparently rooted, not in the "Roman-Mediterranean" world, but in Central Europe since the only comparable evidence comes from the province *Pannonia*. Therefore, within the body of the techno-cultural tradition that the Western Germanic people inherited from the Celts during the time they were settled in Central Europe they picked up a body of agricultural knowledge as well.[226]

In addition to the North Italian finds, other archaeological evidence from the eastern Merovingian area attests a profound transformation in agricultural tools and technology between Late Antiquity and the Migration Period and to the importance of Central Europe as a crossroads in technological transfer. In fact, data from the old Roman *castrum* of *Osterburken*, located on the Roman *limes* between *Würzburg* and *Heilbronn* do confirm this hypothesis. It lies within the Reatic-Alamannic region. This area, similar to North Italy, is critical as a zone of cultural integration thanks to its geographical position in the middle of Europe. Actually, being located on the edge of the Alps, it was both protected by them and an important crossroad for trade and exchange. There is evidence for both a strong continuity in the provincial population, as shown by the well known *Lex Romana Curiensis*, a Roman law issued for the Bishopric of *Chur* (in present day Switzerland), and for the presence of newcomers, the Alamans along with the techno-cultural background they brought with them. The Alamans were a tribe with a deeply mixed

[222]R. Pohanka, *Die eisernen Agrargeräte*, tab. 20, 87.

[223]L. Pejrani Baricco, "Schede," 345-46.

[224]M. Baruzzi, "I reperti," 153, 154.

[225]R. Pohanka, *Die eisernen Agrargeräte*, 115-17, tav. 20, 87.

[226]V. La Salvia and F. Zagari, "Cultura materiale e tradizione tecnica," 979-80; V. La Salvia, "L'artigianato metallurgico," 15.

character, not altered by the Frankish conquest of the late AD 5th century.[227]

A large hoard of iron tools was unearthed in a container within the rampart ditch of the Roman fort in 1897 at *Osterburken*. At first, many scholars suggested that these instruments had been hidden in the AD 3rd century by a Roman blacksmith in reaction to Alamannic raids along the *limes*. However, recently these finds have been assigned to the AD 5th century as a part of the set of tools that would be found on an Alamannic estate.[228] Among the various tools, some of the harvesting blades are of a noteworthy type. These tools are clearly transitional forms, providing a link between Roman and medieval types. The shape of these tools lies somewhere between a scythe and a sickle. Scythes, known and used since Celtic times, were employed only for harvesting hay while crops were collected by sickles. The latter, being shorter, had less flexible blades and, thus, could assure more accurate cuts. The Roman scythes were similar to enlarged sickles but, between the blade and the wooden handle, they formed an angle allowing peasants to hold the blade parallel to the soil while standing upright. The *Osterburken* specimens have the very same angle and, at the same time, have thinner blades with the flanged back parts, assuring that there was no loss in rigidity. In this way, the blades were noticeably lighter but the quality of performance was then similar. Actually, this form maintained both the precision of the sickle in cutting and permitted the worker to maintain an upright position, as was the case while working with the scythes. Therefore, although these tools are rooted within Roman culture in terms of both their type and the way they were made while at the same time they are also an example of technological progress. Their form already points to tools typical of the full Middle Ages.[229] Moreover, ploughshare dated to the AD 5th century came to light from the very same hoard. The form of this ploughshare again lies somewhere between Roman and Medieval ones, anticipating models found from the 8th century onwards. The Roman plough was basically a simple improvement on the hook-shaped ploughard, a scratching metal point, which only broke up the ground. Therefore, the soil had to be turned several times a year. However, during the Roman Imperial period, a new type of ploughshare was developed where the socket and working blades were placed apart. This new type was considerably bigger than the previous type. These ploughshares have been found particularly in the context of the great villa farms in *Pannonia* province and Gaul. The *Osterburken* one is of this latter type but its dimensions are smaller than usual. Comparable examples from Late Antiquity have been also discovered outside the Roman Empire. It is important to note that coulters were discovered among the *Osterburken* tools as well. Thus, this seems to have been a plough that not only broke up the earth but also turned the clods. Such ploughs may as well have been equipped with wheeled front stools. In this case, they would have been drawn by teams of oxen. However, such a possibility existed only for the large landed estates.[230] It was precisely within these landed villa estates in Central Europe, especially in the area of the upper Danube, that the new type of ploughs developed during Late Antiquity. The Celtic technical tradition was, altered to attain new technical heights.[231] As already mentioned, Roman agriculture first only used hook-shaped ploughard. This tool was suitable for working on the small and square fields typical of the Mediterranean region where soils were dry. The dimensions of the fields and the quality of the earth required frequent turning over and criss-cross ploughing. In this way the problems linked to superficially turning the earth were partially avoided. In contrast to this, the Central European ploughs were heavier and more suitable for ploughing long and narrow fields with heavier, wetter soils.[232]

Within these regions, and particularly for the provinces of *Pannonia*, *Dacia*, *Moesia*, and *Thracia*, the economic organization of Roman agricultural production forced intensive improvement in the way crops were grown. Therefore, the old Roman farming traditions underwent profound changes throughout Late Antique times up to the AD 6th century.[233] The change in plough type and function was not obviously an abrupt but rather involved a gradual shift towards new forms. During this phase there must have been transitional types which occasionally possessed coulters as well (**see figure 29** for main trends of development of agricultural tools).[234]

The archaeological evidence also supports the Central European invention of this heavy plough in the time between Late Antiquity and the Migration Period. Such finds are mostly associated with the great *villae rusticae*. During the AD 3rd and 4th centuries, together with the formation of the *latifundia*, there is evidence that the landed estates grew significantly in size with concomitant increases in the size of barns and other service buildings. This development led to a massive growth in agricultural production. In fact, for the same period there is also evidence of a general improvement in the quality of all agricultural tools from sickles to spades, from shovels to billhooks. It might well be chance that this growth in countryside production coincided with a crisis in urban

[227] K. Brunner, "Continuity and Discontinuity of Roman agricultural Knowledge in the early Middle Ages," in D. Sweeney ed., *Agriculture in the Middle Ages* (Philadelphia: University of Pennsylvania Press, 1995), 23-24.

[228] K. Brunner, "Continuity," 21; J. Henning, "Zur Datierung von Werkzeug- und Agrargerätenfunden im germanischen Landnahmegebiet zwischen Rhein und oberer Donau (Der Hortfund von Osterburken)," *Jahrbuch des römisch-germanischen Zentralmuseums Mainz* 32 (1985): 570-94.

[229] K. Brunner, "Continuity," 23-24; V. La Salvia and F. Zagari, "Cultura materiale e tradizione tecnica," 987-88.

[230] K. Brunner, "Continuity," 25-26.

[231] B.A. Slicher van Bath, *The agrarian history of Western Europe. AD 500-1850* (London: Edward Arnlod Ldt., 1963), 62-65.

[232] K. Brunner, "Continuity," 25; V. La Salvia and F. Zagari, "Cultura materiale e tradizione tecnica," 983.

[233] J. Henning, *Südosteuropa*, 48-61, 99, 101, tab. 21-26; V. La Salvia and F. Zagari, "Cultura materiale e tradizione tecnica," 984.

[234] V. La Salvia and F. Zagari, "Cultura materiale e tradizione tecnica," 983-84.

crafts (**see figure 30** for the development of real estates and *villae* during Late Antiquity).[235]

As for the import of this wheeled plough during the Early Middle Ages into the Mediterranean regions, and particularly into Italy, neither written sources nor the archaeological evidence provide us with a definitive answer. Both types of sources are, in fact, too scanty to permit a full reconstruction or understanding of the plough's morphology and function. However, the hypothesis that this type of plough was directly imported from Central Europe to Italy through Lombard mediation is well grounded. A tool called a *plovum* and its parts are mentioned in the *Edictus Rothari Regis*. Moreover, in many Italian dialects the word for plough handle in Italian is *sterzo*, a Lombard loan word.[236]

Thus, as far as the development of early medieval agricultural tools is concerned, the Central European area around the Roman *limes*, far from representing an impenetrable barrier, acted as a major zone where different cultures such as the Celtic, the Germanic and that of provincial Roman origin could integrate with each other. The clear link of the *Speerförmige* ploughshares found in northern Italy with a distinctively Central European type and the possible origin of the wheeled plough from the same regions, suggest a remarkable transfer of tools and technology from these areas to the Mediterranean region. Certainly, the North/East-South/West Germanic migrations must have played an essential role in the circulation of this technical knowledge all around the old continent. In particular, the evidence from northern Italy, on the one hand, emphasizes the role of the Lombards in shaping the set of agricultural tools and technology used on the Italian peninsula during the Early Middle Ages, and, on the other hand, underlines once again the special attitude of the Lombards towards iron metallurgy.[237]

To sum up, the investigation of some aspects of the early medieval history of agriculture has shown the composite character of the rural equipment used in this period. This situation was the result of assimilation and integration of various techno-cultural traditions taking root within a new cultural and social context in which the Celtic-Germanic agricultural tradition and cereal agriculture of the Mediterranean were integrated into a new and more rewarding pattern.[238]

4.1.3 On sword production

As far as the production of weapons and consequently the use of iron is concerned, the situation appears to have been much more complex than that for non-ferrous metals, a field of metallurgy where the Mediterranean influence appears to be more evident.[239] With respect to weapons in early medieval Europe, two important things come immediately to mind and must be emphasized. Both are aspects of production of craftsmen directly linked to the Germanic metallurgical and iron making traditions, thus related to the manufacture of objects outside classical tradition, and emphasizing the importance of Lombard contribution to the development of iron metallurgy on the Italian Peninsula. First, the Germanization of early medieval weapons[240] must be considered not so much the result of importing of finished and functional objects and/or new fighting strategies, as of new methods of production brought by craftsmen working in the Germanic technical tradition. In fact, metallographic analysis of Merovingian and Lombard swords have demonstrated a high level of uniformity and continuity in the use of specific production processes not only generally among Germanic peoples, but have shown that among the Lombards in particular there existed strong traditions of blacksmithing

---

[235] J. Henning, *Südosteuropa*, 48-61, tabs. 21-26, 99, 101; Id, "Frühgeschichtliche Landwirtschaft Südosteuropas: Vom Großgrundbesitz zur Großgrundwirtschaft," in F. Horst and B. Krüger eds., *Produktivkräfte und Produktionsverhältnisse in ur- und frühgeschichtlicher Zeit* (Berlin: Akademie-Verlag, 1985), 301-08; again on the general improvment of latifundia in this period see, A. Mócsy, *Pannonia*, 297-99; J. Fitz, "Economic life," in A Lengyel and G.T.B. Randan eds., *The Archaeology of Roman Pannonia* (Budapest/Lexington: Academia/University Press of Kentucky, 1980), 324; M. Biró, "Roman villas in Pannonia" *Acta Archaeologica Academiae Scientiarum Hungaricae* 26 (1974): 52-53; C. Edward Stevens, "Agricoltura e vita rurale nel Tardo Impero Romano," in *Storia Economica Cambridge. L'agricoltura e la società rurale nel Medioevo*, 141, 150-52; R. Koebner, "Popolamento stanziale e colonizzazione dell'Europa," in *Storia Economica Cambridge. L'agricoltura e la società rurale nel Medioevo*, 33, 37; V. La Salvia and F. Zagari, "Cultura materiale e tradizione tecnica," 979, 987.

[236] *Edictus Rothari regis*, *M.G.H. Legum*, tom. IV, chap. 288, 69; C.A. Mastrelli, "La terminologia longobarda dei manufatti," in *Atti del Convegno Internazionale: La Civiltà dei Longobardi in Europa*, 262; F. Zagari and V. La Salvia, "Aspetti della produzione," 883-84, White L. jr., *Medieval technology*, 53; J. Henning, *Südeosteuropa*, 53 where is recorded the find of an asymetric plough in the Hungarian site of Tác within a phase dated between the last Lombard and the ealry Avar period.

[237] F. Zagari and V. La Salvia, "Aspetti della produzione" 871; V. La Salvia and F. Zagari, "Cultura materiale e tradizione tecnica," 988-89.

[238] Lynn White, Jr., *Medieval religion and technology*, 139-40. On the importance of the period between Late Antique and Migration Period for the development of agricultural techniques see B.A. Slicher van Bath, *The agrarian history*, 59ff. on the early medieval origin, documented from the AD 8th cent., of three-course rotation: winter corn (wheat or oats) was sown the first year, spring corn (barley or oats) the second, and the third year the ground lay fallow; 64: "The chronological development, far from being certain, probably proceeded in the following way. In the Roman era there were hamlets with square fields. On these fields a two-course rotation was followed; they were ploughed with a 'sliding' plough with no moldboard, drawn by oxen. In about the 6th cent. AD, when the climate had already become much rainer, the people of western Europe went over from the prehistoric walled fields (a type of square field) to strip cultivation, by which better drainage was secured. Still better results were achieved with the help of wheel-plough with fixed moldboard, with which it was possible to plough fields in a markedly ridged form. Now, for the first time, the boggy districts ... could be tilled. In the 8th cent. the peasantry ...changed over to a three-course rotation, which resulted in raising the production of food for human consumption." See also, Ch. Parain, "L'evoluzione delle tecniche agricole," in *Storia Economica Cambridge, L'agricoltura e la società rurale nel Medioevo*, 156, 173, 179, 185.

[239] M. Ricci, "Relazioni culturali e scambi commerciali," 253-56, 270.

[240] On this question see, R. Forrer, *Die Schwerter*, 9-10, 12, 16; E. Salin, *La Civilisation merovingienne*, 6, 81-82; W. Menghin, *Das Schwert im frühen Mittelalter* (Stuttgart: Theiss, 1988), 15; F. Cardini, *Alle origini della cavalleria medievale* (Florence: La Nuova Italia, 1991), 60.

which reached Italy when the Lombards entered the Italian peninsula.[241]

This trend in continuity becomes particularly evident when comparing Italian and pre-Italian products through archaeometallurgical analyses. The metallographic analyses, carried out under the supervision of Prof. L. MIHOK of the Politechnical University of *Kosice* (Slovak Republic), were performed on two swords (swords # 65.59.1 and # 65.43.1) from the Hungarian necropolis of *Hegykő* dating to the Lombard period. The results were compared with the data published by M. ROTILI in 1977 on two swords (no. 5 and 6) and one scramasax (no. 1) from the Lombard necropolis of *Benevento*. This comparison conclusively showed that, on the one hand, the technological level of Lombard blacksmithing was just as high as the average Merovingian standards, and, on the other hand, marks the establishment of the technical tradition for sword manufacturing among the Lombards.

In fact, the evidence from the metallographic analyses indicates clear continuity in the methods used for sword making. The composition of the core of the blade, created by welding together various metal sheets with different carbon contents, and the addition of the edges through hammering are the basic common features. Moreover, the edges were apparently separately worked as half-products with completely different structural and, therefore, functional qualities. In fact, these were the very parts of the sword where enhancement in the carbon content was crucial. Nonetheless, such an increase would have been useless without uniform distribution of the carbon inside the structure.

Concerning the two samples from the *Hegykő* necropolis, the fact that it was not possible to analyze samples cut near the tips of the swords could be misleading with respect to the evaluation of the quality of the artifacts. Actually, the sample from sword # 65.59.1 yielded evidence of a mildly and unhomogeneously carburized structure at the edge. Nevertheless, one has to keep in mind that the blacksmith had to control many variables (the quality of the iron ore, the reducing conditions, the quality of fuel) in order to produce a sword. The entire working process was carried out without the aid of any kind of precision instrument. Indeed the work was based only on the skillful experience of the craftsmen. Thus, it is evident that the quality of an artifact could range widely between extremes. The second sample from *Hegykő* (# 65.43.1) indicates that this was a masterfully worked sword. Even if it did not undergo the quenching and tempering processes, a profound homogeneity in the diffusion of carbon was achieved in the structure of the edge through annealing which entirely released all internal stresses. Thus, the level of iron technology used in the production of this sword was equivalent to what was produced in Italy.

Pattern welding was also used to produce the swords and scramasax from *Benevento*, that is to say, different iron materials were hammered and assembled together. The metal sheets which constituted the core used to produce these swords were not worked in a spiral shape and then flattened by hammering in order to obtain the damascening structure; however, it is evident that the technique of production was of a very high level. In fact, for both objects the microscope analysis presents clear evidence of the succession of different elements into the core of the swords. This difference is tied to the different carbon content of the various metal sheets. Generally, through the analysis of the cross-sections, it is possible to observe that the structure of the metal passes from the internal areas, that is the core with a mostly *ferritic* structure and little carburization, to areas with higher carbon content, notwithstanding the presence of sections containing *Ferrite*. Towards the blades, there is only *Pearlite*, and the carbon content rises to a high level.

The evaluation of the results of the analyses, even if not univocal, yielded conclusions which were not contradictory. The not always uniform succession of *Pearlite* from the inside to the outside ending in an almost exclusive *Pearlite* content in the blades, could be regarded either as a consequence of a differentiated work worked-out on a single layer of metal or the result of the pattern welding of different metal sheets. However, in this case, the vertical alternation inside the *Ferrite* core areas and the carburized ones indicates that the pattern welding of various metal sheets of different carbon content through hammering must have been the normal practice. Thus, the vertical alternation inside the core with its *Ferrite* and carburized areas possibly indicates that the various metal sheets had been pattern welded.
The analyses of the cross-sections and the macro photographs stress the fact that the blades were added on by hammering. This is suggested by vertical lines appearing on the swords' surfaces of which indicate the line along which the *Pearlite* structure was dissolved by the reheating of the metal because of welding by hammering in a core with a structure in which the sheets, outside the thick, carburized, and central part, had an entirely *Ferrite* structure or little carburization. Aside from the inclusions of oxides, which may have generated some problems in terms of the quality of the artifacts, the hardness tests remove all doubts about the level of the production processes. Actually, the blades of these swords in the cross-sections sampled near the edge exceed the level of 400 Vickers, often reaching indexes of up to 800. The cross-section sampled near the tang of sword 5 revealed indices of around 290 and around 120-140, but only for the structure of *Ferrite* in the core, while the carburized metal sheets displayed Vickers indices of 200. This analysis indicated that a quenching and tempering process was used in the production.
The carbon content is analogous to what is found in Merovingian production, if not in certain cases even superior. In Merovingian swords the average carbon

[241]For the metallographic analyses on Merovingian and Frankish swords see, E. Salin, *La Civilisation merovingienne*, 59-60, 62, 64-68, 72; for those on Lombards ones see, M. Rotili, *La Necropoli longobarda di Benevento* (Napoli: Università di Napoli. Istituto di Storia Medioevale e Moderna, 1977), 40-50, 129-31, 132-33; V. La Salvia, *Archaeolmetalurgy*, 33-66; Id., "La fabbricazione delle spade,"41-50.

content in the core carburized areas is on the order of 0.1-0.2% and in the blades between 0.4% and 0.6%. In the samples from *Benevento* there are indexes exceeding 1% with respect to the core and around 2% for the blades. Since these are average evaluations, they bear witness to the top quality level of the working technology employed in their manufacture.

The continuity in production methods is, apparently, emphasized by the peculiar structure of scramasax 1 from *Benevento*. Here, the two metallographic cross- sections (one taken near the tang and the other close to the tip) have shown that towards the spine there was a highly carburized area which continues along the whole sax as shown by its presence in both the samples. Its structure is almost *pearlitic* and in some areas *pearlitic* and *ferritic*. This steel element may perhaps constitute the very core of the sax. Thus, the core may comprise a steel bar sandwiched inside a single and larger sheet, the borders of which apparently touch each other along the longitudinal line recognizable in both the samples. This structure appears to have been produced by pattern welding of metal sheets of different quality. The edge was welded to this core with a cat's paw joint. This exact structure is extremely similar to that found in sword # 65.34.1 from *Hegykő*. Through the evaluation of these blacksmithing methods, it is possible to show the importance of the Germanic traditional metallurgy. These production methods and standards became characteristic and separable from the iron making technologies found in both Mediterranean and Celtic areas.[242]

In addition to the archaeometallurgical data already presented, X-rays investigation of the entire set of swords from the Lombard necropolis of *Nocera Umbra*, recently conducted by C. RUPP, have shed new light on both the production methods used in their manufacture as well as their circulation and diffusion.[243]

The X-rays investigation of these swords yielded evidence of a general use of western welded damascening, a method through which the blacksmith assembled together thin layers of soft iron sandwiched between layers of carburized iron. The resulting bar was entirely or partly reduced to a spiral and finally flattened out by hammering. This bar constituted the exact body of the sword to which the edges would be added and welded on by hammering.[244]

The entire set of the analyzed swords from *Nocera Umbra* necropolis can be further subdivided into subgroups according to the stylistic characteristics of the damascening of the blades.

1) The first group is the larger one and consists of 10 pieces. These swords came from graves number 1, 16, 20, 27, 49, 76, 11b, 115, 145 and 163. The decorative motif on the western welded damascening is, in this case, something known in the literature as *dreibahnige Winkeldamast* (**see figure 31**, group 1).

The same decorative type was also found on swords found in the Alamannic necropolis of *Schretzheim*. Here, this type of sword is associated with the first phase of the cemetery utilization, the late Merovingian period. As for the chronology of the Italian swords, they could just as well have come from an early phase of the necropolis according to C. RUPP. In fact, the grave goods from grave 145 can be dated to *Zeitsufe I* (AD 570-590). However, grave 76 is a later one and can be placed in *Zeitsufe III* (AD 610-620/30), the last phase at the cemetery. I. BÓNA, however, stated that similar swords were also found among the Lombard finds from their Hungarian period. The production of this group of objects can, at least

[242]On the difference between Roman and Celtic-Germanic traditions in sword-making see, R. Maddin, A. Hauptmann and D. Baatz, "A metallographic examination of some iron tools from the Saalburgmuseum," *Saalburg Jahrbuch* 46 (1991): 21: "Although the Roman metallurgists did not invent or discover fundamentally new methods for treating iron, they adopted the techniques developed by cultures to the East perfecting them so as to yield more consistent results. It has to be stressed that the techniques achieved have been used even in the most remote areas of the Roman Empire." Moreover, the following statement, substantiated by an overall evaluation of the archaeometallurgical evidence, precisely defines the differences in the working procedures in the two different given periods: P.T. Craddock, *Early Metal Mining and Production* (Edinburgh: University Press, 1995), 260: "Clearly the carburization and heat treatment of iron to produce steel was known from the beginning of the European Iron Age, even if treatment was rare and did not always produce the best possible products." For a closer investigation with respect to archaeometallurgical analyses, see the bibliography in the volume. Concerning the metallographic analyses of iron objects from the Lombard necropolis of Benevento see M. Rotili, *La Necropoli longobarda di Benevento*, 40-50, 129-31, 132-33; for those made on the swords from the necropolis of Hegykő, performed under the supervision of Prof. L. Mihok of the Faculty of Metallurgy of the University of Kosice (Republic of Slovakia), see. V. La Salvia, *Archaeometallurgy*, 49-54, 63-64, 67, and I. Bóna, J. Gömöri, L. Mihok and V. La Salvia, "Metallographic Analyses," 180-86; on the general methods for sword production during the Migration Period see, Salin, *La Civilisation merovingienne*, 51-53, 59-60, 62, 64-68, 72, 98; V. La Salvia, "La fabbricazione delle spade," 28-54; W. Menghin, *Das Schwert, passim*; N.T. Belaiev, "Sur le damas oriental et les lames damassées," *Métaux et Civilisation* 1 (1945): 10-15; C. Böhne and H. Dannheimer, "Studien an Wurmbuntkliegen des frühen Mittelalters," *Bayerische Vorgeschichtblätter* 26 (1961): 107-22; A. Bruhn De Hoffmeyer, "Introduction to the History of European Swords," *Gladius* 1 (1961): 30-75; J. Emmerling, "Die Metallografische Untersuchungen der Schwertklingen aus Münsingen," *Jahrbuch des Bernischen Historischen Museums in Bern* 47-48 (1967-1968): 147-90; Id., "Technologische Untersuchungen an eisernen Bodenfunden," *Alt-Thüringen* 12 (1972): 267-320; F. Wewer, "Das Schwert," 7-71; J. Ypey, "Drei römische Dolche mit tauschierten Scheiden aus niederlandischen Sammlungen," *Berichten van de Rijksdienst voor het Oudheidkundig Bodemonderzoek* 10-11 (1960-61): 347-62; Id., "Au sujet des armes avec damas soudé en Europe," *Archéologie Medievale* 11 (1981): 147-65.

[243]All information concerning the swords of *Nocera Umbra* contained in this paragraph are entirely based on C. Rupp, *Das langobardische Gräberfeld von Nocera Umbra* (unpublished Dissertation, University of Bonn, 1994), 74-79. As reported in this dissertation, only the swords from graves 61 and 71 together with the fragments of the swords from graves 9 and 119 were not analyzed. For the dating of *Nocera Umbra* graves and the consequent structuration of different *Zeitstufe* and for the grave goods quality groups definitions see, C. Rupp, "La necropoli longobarda di Nocera Umbra," 23-40.

[244]J.F. Finó, "Notes sur la production du fer," 51. Moreover, with regard to western welded damascening and the differences between these working procedures with the oriental damascening method, see A. Leroi-Gourhan, "Notes," 46; E. Salin, "Les techniques metallurgiques après les grandes invasions," in *Les fer atravers les ages, Actes du Colloque International. Nancy 3-6.10.1955* (Nancy: Université, 1956), 51; F. Wewer, "Das Schwert," 37; F. Cardini, *Alle radici*, 61; C. Böhne, H. Dannheimer, "Studien an Wurmbuntkliegen," 111-22.

partially, be attributed to a pre-Italian period or to an early phase of the utilization of the necropolis.[245]

2) The second group consists of 7 swords from graves 5, 6, 47, 74, 122, 139, 143. Here, the decoration comprises the *zweibahnige Winkeldamast* (**see figure 31**, group 2). In *Nocera Umbra*, this kind of western welded damascening decoration is found over the entire period of the use of the cemetery. Its closest Italian equivalent was found at the necropolis of *Trezzo d'Adda*, in grave 3 which is dated to around the second quarter of the AD 7th century. Outside Italy, weapons made with similar techniques can again be found in the Alamannic region, always within the necropolis of *Schretzheim*.[246]

3) The third group consists of 5 swords decorated with *vierbahnige Winkeldamast* (**see figure 31**, group 3). The objects come from graves 7, 52, 67, 127, 132. They could be dated to between *Zeitstufe I* and *II*, that is to say, around the end of the 6th and the first thirty years of the 7th century. Comparable items are known from Italy from the excavations at *Sirmione* where two similar swords have been found.[247] Unfortunately these finds do not come from a well-preserved funerary context and, therefore, are difficult to date precisely. In contrast to the previous groups described above, these latter swords have no parallels in the necropolis of *Schretzheim* where this type of decoration is quite rare.

4) A fourth large group consists of eight swords. They come from graves 36, 38, 79, 98, 111\A, 131, 134, 137. The damascening on the peculiar form of their blades is a *drei-und vierbahninge Winkeldamast* which changes\dissolves into *Streifendamast* (**see figure 31**, group 4). This type may be found throughout all three phases (*Zeitstufe I, II, III*) at the burial site. Their chronology can be reconstructed through the evaluation of the grave goods: graves 36 and 38 come from *Zeitstufe I* (AD 572-590); graves 79, 98, 111\A, 131 from *Zeitstufe II* (AD 590-610) and; 134 and 137 from *Zeitstufe III* (AD 610-620\30).

This group of swords is extremely important because the morphology of this type of damascening was apparently wide spread over all of the Italian peninsula during the Lombard period. Comparable swords, come from the site of *Boffalora d'Adda*; *Trezzo sull'Adda*; *Sirmione*; *Borgo d'Ale*; *Fornovo S. Giovanni* e *Nasate*.

The homogeneous geographical and chronological diffusion of this type of sword into Italy during the Lombard period testifies to a production technique mastered by craftsmen in Italian workshops.

Outside Italy, similar objects have also been found in the Alamannic cemetery of *Schretzheim* where they date to the first phase of utilization of the burial area.[248]

5) Two other swords from grave 15 and 32 respectively, make up a fifth group. These artifacts are characterized by a *gleichgerichtete dreibahnige Winkeldamast* (**see figure 31**, group 5). Within the blade of sword 32 there is a small section with *Rosendamast*. The grave goods from this tomb can be dated to the last third of the AD 6th century. Unfortunately, the grave goods from grave 15 do not provide enough clear evidence in to enable a precise chronological order to be established.

This damascening technology is extremely rare at the necropolis of *Schretzheim* and is associated to phases 3-5. However, a similar object also came to light in grave 36 at the Bavarian necropolis of *Irlmauth* which is dated to the first half of the AD 6th century.[249]

Here, it is important to emphasize that these two swords are, as far as we know at present, the only finds of this sort in Italy.

6) The sword from grave 86 dates to *Zeitstufe I* (AD 572-590) and is unique in Italy. It has a grill-like damascening pattern on it (**see figure 31**, group 6). The object which is closest to it is a sword discovered in the Netherlands by *Maas* near *Stevensweert* (*Limburg* province) and dated around the year AD 400. This chronological horizon underlines the long lasting tradition of sword production technology.[250]

7) The swords from graves 48 and 125 have a peculiar wavy\corrugated *Streifendamast* (**see figure 31**, group 7). Grave 48 is dated to *Zeitstufe I* (AD 572-590), while the other one from grave 125, may be assigned to *Zeitstufe II* (610-620\30). These objects should be regarded as isolated specimens in Lombard Italy. To the south of the Alps, in fact, these are the only artifacts of this kind. It seems likely that they were not actually produced within a Lombard milieu but rather imported from the western Merovingian area where objects comparable to them have been found. Typologically the closest analogy to the Italian swords was found in grave 66 at the Frankish necropolis of *Iversheim* from the second half of the AD 7th century. Two other similar objects are chronologically quite far from those from *Nocera Umbra*. However, they are mentioned here because it is important to emphasize the long duration of the sword-making technological

[245] U. Koch, *Das Reihegräberfeld bei Schretzheim*, 98 and note 46; I. Bóna, "Die Langobarden in Pannonien," in R. Busch ed., *Die Langobarden*, 66.

[246] E. Roffia, "La necropoli longobarda di Trezzo sull'Adda," *Ricerche di Archeologia Altomedievale e Medievale* 12-13 (1986): tab. 16, 1a; 17, 2; U. Koch, *Das Reihegräberfeld bei Schretzheim*, 98.

[247] G.P. Brogiolo, S. Lusuardi Siena, P. Sesino, "Ricerche su Sirmione longobarda," *Ricerche di Archeologia Altomedievale e Medievale* 16 (1989): tab. V, 2c; VII, 2f.

[248] P.M. De Marchi, "I ritrovamenti di Boffalora d'Adda," in *Nuovi Contributi agli studi longobardi in Lombardia. Atti del convegno (Arsago Seprio 29 Settembre 1984)*, (Busto Arsizio: SAP, 1986), 21 ff. tab. II, 1; E. Roffia, "La necropoli longobarda, " tab. 2, 2a; G.P. Brogiolo, S. Lusuardi Siena, P. Sesino, "Ricerche su Sirmione," tab. 5, 2e; L. Breccairoli Taborelli, "Tomba longobarda da Borgo d'Ale," *Quaderni della Soprintendenza Archeologica del Piemonte* 1 (1982): tab. LIX, 3; P.M. De Marchi, "Catalogo dei materiali altomedievali delle civiche raccolte archeologiche di Milano," *Notizie dal Chiostro del Monastero Maggiore*, 4 (1988): tabs. XXIII, XLVII, 5, 6; U. Koch, *Das Reihegräberfeld bei Schretzheim*, 98.

[249] U. Koch, *Das Reihegräberfeld bei Schretzheim*, 98, notes 50 and 51.

[250] J. Yepey, "Damaszierung,"in *Reallexikon der Germanischen Altertumskunde*, Band 5: Chronos – dona, H. Beck, H. Jankuhn, K. Ranke and R.Wenskus, eds., (Göttingen: 1984, De Gruyter) 298, pic. 14.

tradition: a sword from *Rhenen*, in grave 821, dated to the first half of the AD 5th century and a sword discovered by *Waal* near *Nimegen* and dated to some time between the 9th and 10th centuries.[251]

Beside the general features of their damascening decoration, the swords from graves 6, 15, 122 and 143 have other noteworthy characteristics just below the tang. In addition, there are other small designs and forms on these swords within the general framework of their decoration (**see figure 32**). The sword from grave 15 displays a little circular\spiral shaped motif on one side (a) which is unique in Lombard Italy. The object from grave 122 also has a small section *Schleifendamast* visible on it, a wire-like ornament, which is very different from other *Winkeldamast*.

However, the objects with the most important and unusual decorative patterns are the swords from grave 6 and 143. Both of them have an S-shaped design below the tang, inserted within the body of their damascening. There are two S-s, one over the other, on the sword from grave 6 making it a unique piece. Although the decorations are similar they are not identical and may reflect the hands of different masters. Nonetheless, slight differences occur frequently in products from the same craftsman. Within this system of production it was difficult to achieve a repeated and standard uniformity in the final products. Thus, it is difficult to judge on the basis of only two samples whether or not they are products of the same workshop. Moreover, it has to be stressed that this S-form motif is rare south of the Alps while north of the Alps, especially in southwest Germany and northern Switzerland, that is, in Alamannic areas, this motif is much more common.[252] D. ANKNER has suggested that such designs served as workshops trademarks. According to the same scholar, it has a chronological horizon ranging between the last third of the AD 6th and beginning of the 7th centuries.[253] The swords from *Nocera Umbra* could also be dated to the last third of the AD 6th century and, thus, they fit into the chronology proposed by D. ANKNER. However, the discovery close to *Waaxens* (Netherlands) nearby *Petersen* of a blade from the Carolingian period apparently with the same type of S-shaped ornament complicates the identification of the S-shaped motif with trademarks.[254] The appearance of the same motif after such a long period probably has a different meaning leaving the question open.

To conclude, the evaluation of the swords of *Nocera Umbra* has shed new light on:

a) The presence and the activity of a lively workshop production in Italy during the Lombard period, as emphasized by certain features on the objects described in groups 2, 3, and especially 4. These organizations of craftsmen operated in specialized ateliers that were particularly active between the last third of the AD 6th and the first half of the AD 7th centuries.

b) A commercial network closely linked the Lombard kingdom with the rest of the eastern Merovingian area, particularly with the Alamannic region. This can be seen by the many formal similarities between the materials from *Nocera Umbra* and *Schretzheim*. As shown by the objects in group 1, contact with the Alamannic region could be dated to the seventies of the AD 6th century. Sometime, the contact seems to have stretched as far as the west Merovingian zone, basically the Frankish area, as apparently shown by the swords in groups 5, 6 and 7. These objects, as single pieces, appear to be quite unusual in the Italy during the Lombard period and, therefore, may easily be considered finished products which had been directly imported. The same may be said about the swords from graves 6, 15, 122 and 143.

c) The long lasting tradition of sword making in the western Germanic milieu and, more specifically, the possible transfer/continuity of blacksmithing technology from the Pannonian phase to the Italian phase of Lombard settlement. Moreover, both I. BÓNA suggestions concerning the production of welded damascene weapons already in *Pannonia* and the comparison between the swords from *Hegykő* with those from *Benevento* indicate that sword manufacturing among the Lombards was a specialized craft.

The time between the last third of the AD 6th and the beginning of the 7th centuries was crucial for the development in weapon making. In the words of Paul the Deacon: *Arma quoque precipua sub eo* – King Alboin – *fabricata fuisse ...*[255] As already mentioned, this is also the period when the blacksmithing tradition of the Lombards met those of the Mediterranean. During the AD 6th century, local production traditions remain undisrupted in the Danubian provinces. Thus, the beginning of "new production" in this chronological horizon may well not be simple coincidence but the result of integration between, the Germanic and the Late Antique manufacturing traditions.

4.1.4.0 The Introduction and Spread of the Stirrup in Lombard territories

4.1.4.1 Preliminary questions: Historiography, Hypothesis and Suggestions

The presence of the stirrup in both Italian territory and in the Alaman-Bavarian area, that is to say, the very

[251] Chr. Neuffer-Müller, *Das frankische Gräberfeld von Iversheim*, in *Germanische Denkmäler der Völkerwanderungszeit*, Band 6 (Berlin: Gebr. Mann Verlag, 1977), tab. 9 and J. Yepey, "Damaszierung," 198, 199 pic. 16.

[252] For a detailed description of the location of such sword types see, C. Rupp, *Das Langobardische Gräberfeld von Nocera Umbra* (unpublished Dissertation, University of Bonn, 1994), 78 particularly note 41.

[253] D. Ankner, "Zur Damaszierung der Spathen aus Alterding," in H. Helmuth, D. Ankner and H.-J. Hundt eds., *Das Reihengräberfeld von Alterding in Oberbayern*, vol II (Mainz: P. Von Zabern Verlag, 1996), 150 and ff.

[254] J. Yepey, "Vroegmiddeleeuwse zwaarden uit Nederlandse verzamelingen,"*Berichten van de Rijksdienst voor het Oudheidkundig Bodemonderzoek* 14 (1964): 89, pic. 1c.

[255] *Pauli Historia Langobardorum*, bk. 1, ch. 27.

southeastern Merovingian area,[256] undoubtedly opens an interesting and important chapter in the relationship between these territories and the Carpathian Basin. The latter, thanks to the Avar conquest, had "acquired" a new territorial unity and a certain socio-economic coherence, which lasted at least until sometime after the failed siege of Constantinople in AD 626.[257] The Avars also introduced the stirrup in Europe (**see figure 33**), along with other military innovations, such as the composite bow and *Lamellenpanzer*. Thus, the area of the Danube River, especially in its upper part, played a central role as a gateway through which objects and techniques of Eastern origin penetrated into Western Europe.[258]

The main object of this work, however, is not to investigate and come to a definite conclusion about the *vexata quaestio* of the origin and diffusion of the stirrup in Europe and in Italy in particular. Rather, the purpose is simply to emphasize the role of the Lombards in its dissemination, within the eastern Merovingian area, thanks to their favorable relationship with both the Avars and the Byzantine territories. In addition to the archaeological evidence, linguistics also supports this hypothesis. Of the total of 280 words of Lombard origin that entered various Italian dialects only 70 (thus, 25%) denote objects. Nonetheless, the etymology of the Italian word 'staffa' clearly and undoubtedly indicates that it is a Lombard loan word, a term which, therefore, entered the lexicon of early medieval Italy solely through Lombard mediation. It is important to stress that this word was not a calque from either Latin or Greek (in the Byzantine *Strategikon* stirrups are known as *sideroi skala*). Moreover, all the component parts of this object were given technical terms of Lombard linguistic origin as well.[259] This is quite remarkable since the Lombards are known for their "disponibilità alla ricezione del modello culturale e religioso bizantino," as stated recently by F. BURGARELLA.[260] It is, thus, difficult to believe that the Lombards adopted this particular object directly from a Byzantine milieu.

On the origin and diffusion of the stirrup, the traditional historical-archaeological paradigm has always asserted that these iron artifacts, of earlier far eastern origin (possibly Chino-Korean), were imported into Europe through Avar mediation during the last thirty years of the AD 6th and the beginning of the 7th centuries. This idea has been held on the basis of scant but significant set of archaeological evidence by J. HAMPEL indicating that, in the Avar necropolis of *Szentendre*, stirrups and coins of the Emperors Justinian I (AD 518-527) and *Phocas* (AD 602-610) were found buried together.[261] However, the dating of stirrups deposited in Avar graves between the second half of the AD 6th and the very beginning of the 7th century has already been challenged by D. CSALLÁNY.[262] Recently, the idea introduced by L. WHITE in *Medieval technology and social change* crediting a major role to the Byzantine milieu in introducing the stirrup into Europe has been taken up again by CS. BÁLINT.[263] Through a sound revision of archaeological evidence he attempted to demonstrate that there are not enough well-dated finds to identify an Inner-Asian route by which the stirrup would have traveled from the Far East to Europe via the Avars. In my opinion BÁLINT's main contribution has been to deconstruct what had become an unchallenged paradigm, making it clear that

---

[256] The stirrups, apparently, had already begun to diffuse into this area from the late AD 6th cent. although the process was not uniform, see U. Von Freeden, "Das Grab eines awarischen Reiters von Moos-Burgstall, Niederbayern," *Bericht des römisch-germanischen Kommission* 66 (1985): 16 and Ead., "Das frühmittelalterliche Gräberfeld von Moos-Burgstall," *Bericht des römisch-germanischen Kommission* 68 (1987): 523.

[257] I. Bóna, "Gli Avari. Un popolo dell'Oriente nell'Europa altomedievale," in *Gli Avari. Un popolo d'Europa* (Tavaganacco (UD): Arti Grafiche Friulane, 1995), 18, 29.

[258] I. Bóna, "Neue Nachbarn im Osten. Die Awaren," in H. Dannaheimer and H. Dopsch eds., *Die Bajuvaren. Von Severin zu Tassilo 488-788* (Munich/Salzsburg: Freistaat Bayern/Amt der Salzburger Landesregierung, 1988), 108.

[259] Concerning these linguistic aspects see, C.A. Mastrelli, "La terminologia longobarda dei manufatti," 258 and 266: "Particolarmente significativo é, poi, il caso del longobardo *staffa*, 'staffa,' che probabilmente si riferisce a una novità tecnica nell'arte del cavalcare che era ignota nei secoli anteriori; infatti, come é noto, la prima menzione della staffa é soltanto del 602 AD. Il vocabolo é solo longobardo e, quindi, la presenza di questo vocabolo comporta delle conseguenze di estremo interesse nella storia dell'equitazione e nel confronto con le altre popolazioni germaniche. Non é da escludere che i Longobardi avessero appreso questa tecnica da popolazioni più orientali, come ad esempio gli Avari. La longobardicità del termine mi pare poi ulteriormente rafforzata dal fatto che ancora oggi l'elemento trasversale della staffa si chiama 'panca,' con termine quindi anch'esso longobardo, il quale può riflettere un impiego semantico già antico e non dunque secondario rispetto al significato più comune di 'panca,' sedile. Sempre nell'ambito della terminologia relativa alla bardatura del cavallo si può citare il longobardo **bridel* 'redine,' che sopravvive nel toscano predella 'briglia;' ma questo termine non é altro che la longobardizzazione di una situazione anteriore: infatti, già i Goti avevano introdotto il vocabolo **brigdil*. Del resto, anche i Franchi avevano introdotto il medesimo termine nel gallo-romanzo. Il che sta a significare che, a differenza della staffa, la briglia doveva essere un finimento proprio dei Germani in genere e con caratteristiche abbastanza simili rispetto alle redini dei Latini, i quali possedendo già il termine *habenae* sostituito da *retinae* (= all'italiano redini), e potevano evitare la mutazione del vocabolo germanico." The same point of view is confirmed by F. Sabatini, *Riflessi linguistici della dominazione longobarda nell'Italia mediana e meridionale* (Florence: Olschki, 1965), 190-92: " ... l'italiano *staffa*, 'montatoio, predellino:' anch'esso introdotto dai Longobardi e perciò, come *staffilo*, 'palo,' in campo romanzo limitato sostanzialmente al solo italiano;" see also 227; and A.Castellani, "Capitoli d'una introduzione alla grammatica storica italiana, II: L'elemento germanico," in *Studi Linguistici Italiani* 11 (1985): 173: "È curioso che fra i vocaboli sicuramente di origine longobarda del lessico italiano non ce ne sia quasi nessuno di carattere militare ... Si può forse aggiungere staffa, data l'importanza della cavalleria negli eserciti germanici, (da compararsi con l'anglosassone *stapa*, 'staffa') e si veda anche staffile, 'correggia' a cui sta appesa la staffa."

[260] F. Burgarella, "Bizantini e Longobardi nell'Italia meridionale," in *I longobardi dei Ducati di Spoleto e Benevento*, 186.

[261] J. Hampel, *Alterthümer des frühen Mittelalters*, vol. 1, 217, 223; I Kovrig, "Contribution au probleme de l'occupation de la Hongrie par les Avars," *Acta Archeologica Accademiae Scientiarum Hungaricae*, 6 (1955): 175 mentioning graves 121 and 173 from the Avar cemetery of Jutas; and Ead., "Deux tombes avares de Törökbálint," *Acta Archeologica Accademiae Scientiarum Hungaricae* 9 (1957): 131-33 for an early 7th century Avar grave with stirrups; more recently, among others see, E. Garam, "Sepolture di cavalli," in *Gli Avari. Un popolo d'Europa*, 146.

[262] D. Csallány, *Archäologische Denkmäler der Awarenzeit in Mitteleuropa: Schriftum und Fundorte* (Budapest: Hungarian Academy of Sciences, 1956), 77-220, 240 and Id., "Grabfunde des Frühawarenzeit," *Folia Archaeologica* 1 (1939): 171.

[263] See L. White, *Medieval technology and social change*, 14-28.

the origin of the stirrup is still up in the air.[264] On the other hand, it has to be said that if the Byzantine cultural milieu had such an impact on neighboring societies (the Germanic and nomadic ones) one would expect to find stirrups in graves of horsemen whose grave goods were heavily influenced by Byzantine fashion and imports. However, this is not the case for two Lombard Italian graves. The burials of the knight of *Castel Trosino* and grave No. 5 from *Nocera Umbra*, both containing grave goods (belts or saddle decorations) of Byzantine, if not precisely of Constantinopolitan, provenance, lack stirrups.[265] In contrast, all the earliest finds of stirrups are characterized by extremely complex archaeological contexts within which it is not always easy to identify and clearly differentiate all the various cultural influences. The analysis of written sources, basically all of Byzantine origin, is problematic as well. In fact, comparing texts from the first half of the AD 6th century with those from the first half of the AD 7th does not help to narrow down the time range suggested heretofore on the basis of archaeological evidence (i.e. AD 570/600). As an example, *Procopius*, beside several times mentioning the influences that oriental fashions had on both Germanic and Roman warfare, devoted several lines in his *History of Wars* to training cavalierly (for example bk.1, ch.1, 12-15 where the role of the ἱπποτοξότης is enthusiastically described). There, he also speaks about the use of the composite bow and straight daggers, both of "Hunnic origin," but does not mention the use of stirrups. However, one may infer the existence of some sort of footrest from the description of the types of movements made by mounted soldiers.[266] As with many other Late Antique sources (i.e. *P. Flavi Vegeti Renati Epitoma Rei Militaris*),[267] *Procopius* is, thus, referring to both to the way nomadic bands were able to penetrate the ranks of the Roman army and to consequent changes made in Roman military strategies. In order to find a direct and clear mention of the stirrup one must refer to the *Strategikon* of Emperor *Maurikios* bk.1, ch. 2, 7 and bk. 2, ch. 8, 3.[268] The military handbook of *Maurikios* contains the golden rules for the Byzantine knight; bk.1, ch. 2, 7 prescribes that a pair of stirrups must be attached to the saddle. The word *scala*, used in the handbook, meaning step or stair is a loan from Latin. As has been proposed by S. SZÁDECZKY-KARDOSS the textual context in which the discussion of the stirrups is included favors the interpretation that these artifacts were an Avar import. In fact, the paragraphs immediately before (bk. 1, ch. 2, 6) and after (bk. 1, ch. 2, 8) that discussing the stirrups, present the organization and equipment characterizing Avar warfare as one to which the Roman Army should refer.[269] These bands of horsemen, equipped according to Avar fashion (bk. 2, ch. 2, 6), were believed to constitute the invincible core of the Imperial army. The stirrups were also used by mounted attendants in the Imperial army to help collect wounded soldiers from the battlefield.[270] Based on the data presented above, it is still difficult to draw a conclusion about the origin of the stirrups. Both theories about the origin of the stirrup, the Avar and the Byzantine one, rely heavily on indirect proof and inferences. As suggested by B. GENITO, the "invention" of the stirrup should be considered the final act of a great change in cavalry and warfare, beginning in the AD 4th/5th centuries on, to which, in addition to the Avars, various people and cultures such as Byzantines and Arabs may have contributed.[271]

4.1.4.2 The stirrups in Italy according to the Archaeological evidence

To date, stirrups were discovered in the following places in Italian territory:

1) In the large necropolis of *Borgomasino* (Piedmont), a few kilometers away from the early medieval graves of *S. Germano* cemetery, reserved for the Lombard aristocracy, and which probably comprised more than

[264]On the argument see, Cs. Bálint, "Byzantinisches zur Herkunfsfrage des vielteigen Gürtels," in Cs. Bálint ed., *Kontakte zwischen Iran, Byzanz und der Steppe im 6.-7. Jahrhundert*, *Varia Archaeologica Hungarica* 10 (2000): 99, 101, 102, 103, 104, 120-21, 136-38 and notes 117-127; Id., "Zur Geschichte und Archäologie der osteuropäischen Reiterhirten im Frühmittelalter," in *Reitervölker aus dem Osten. Hunnen + Awaren* (Bad Vöslau: Grasl, 1996), 202-04; Id., "Über die Datierung der osteuropäischen Steppenfunde des frühen Mittelalters (Schwierigkeiten und Möglichkeiten)," *Mittelungen des archäologische Insitut der Ungarische Akadamie des Wissenschaften* 14 (1985): 137-47; Id., *Die Archäologie der Steppe. Steppenvölker zwischen Wolga und Donau von 6. bis 9. Jahrhunderts* (Vienna: Böhlau, 1989); A.D. Bivar, "The strirrup and its origin," *Oriental Art* n.s. 2 (1955): 61-65; A.K. Ambroz, "Stremena i sedla rannego srednevekovja kak xronologičeskij pokazatel' (IV-VIIIvv.)" [Early medieval stirrups and Saddles as chronological indicators, AD 4th-8th cents], *Sovjetskaja Arkheologija* 4 (1973): 81-98 and I.L. Kyzlassov, "O proisxoženii stremjan" [On the origin of Stirrup], *Sovjetskaja Arkheologija* 4 (1973): 25-36; M.A: Littauer, "Early stirrups," *Antiquity* 55 (1981): 99-105; J. Werner, "Nomadischen Gürtel bei Persern, Byzantynern und Langobarden," in *Atti del Comvegno Internazionale: La Civiltà dei Longobardi in Europa*, 117; W. Pohl, *Die Awaren. Ein Steppenvolk in Mitteleuropa* (Munich: Beck, 1988), 311-12; I. Bóna, "Gli Avari," 21.

[265] On the subject see, L. Paroli, "The langobardic finds and the archaeology of central Italy," in K. Reynold Brown, D. Kidd and C.T. Little eds., *From Attila to Charlemagne* (New York: Yale University Press, 200), 140-52; F. Vallet, "Une tombe de rich cavalier lombard découvert a Castel Trosino," in F. Vallet and M. Kazanski eds., *La noblesse romaine et les chefs barbares du 3e au 7e siécle* (Saint-Germain-en-Laye: AFAM, 1995), 335-49; C. Rupp, "La necropoli longobarda di *Nocera Umbra*," 36 and Ead. "Catalogo," in L. Paroli ed., *Umbria Longobarda*, 119-26.

[266]As stated by C.M. Mazzucchi, " ΚΑΤΑΓΡΑΦΑΙ dello Strategikon di Maurizio," *Aevum* 55 (1981): 131-32 and note 80.

[267] *P. Flavi Vegeti Renati Epitoma Rei Militaris*, ed. A. Önnerfors, (Stuttgart-Leipzig: Teubner, 1995), bk. 1, ch. 20: 34 and bk. 2, ch. 15: 76-77.

[268]On the dating of this work to the late AD 6th or beginning of the 7th centuries see, among the others, M. Springer, "Germanisches und anderes in der 'Kriegskunst' des Maurikios und anderswo," in Jh. Irmischer ed., *Die Literatur der Spätantike—polyethnisch und polyglottisch betrachtet* (Amsterdam: A.M. Hakkert Verlag, 1997), 183-84, note 2 and S. Szádeczky-Kardoss, "Der Awarisch-Türkische Einfluss auf die byzantinische Kriegskunst um 600 (Anmerkungen zum Strategikon des Maurikios)," *Acta Universitatis de Attila József Nominatae. Acta Antiqua et Archaeologica* 24 (1986): 205, note 2; 209, note 15, and 210.

[269]S. Szádeczky-Kardoss, "Der Awarisch-Türkische Einfluss" 208-11.

[270]C.M. Mazzucchi, "ΚΑΤΑΓΡΑΦΑΙ," 126.

[271]See, B. Genito, "Sepolture con cavallo da Vicenne (CB): un rituale nomadico di origine centroasiatica," in S. Gelichi ed., *I Congresso Nazionale di Archeologia Medievale. Auditorium del Centro Studi della Cassa di Risparmio di Pisa (Pisa, 29-31 maggio 1997)* (Florence: All'Insegna del Giglio, 1997), 287.

100 tombs with rich grave goods now, unfortunately, almost completely lost. Here, the presence of burial places for horses together with accompanied horse-equipment such as bridles and stirrups has been attested. It is possible that one of these tombs may have contained iron stirrups and human and horse bones together. Disappointingly, the general preservation of these finds and the almost complete loss of undamaged grave goods do not permit accurate reconstruction of the archaeological features of the necropolis, even from a chronological point of view. Nevertheless, the presence of certain objects from this cemetery suggests that these graves have an early date at least in a section of the necropolis. Vessel remains with stamped and *stralucido* decoration, in fact, do have remarkable formal and typological similarities with those from the Pannonian phase, particularly as concerns the jugs with cylindrical necks as well as the sack-shaped vases and bottles, all worked on a slow wheel. Moreover, the discovery of two golden coins, one an imitation of a coin by Byzantine Emperor Maurice Tiberius (AD 582-602) and another minted by *Phocas* (dated to the middle of the AD 7th century) may mean that that part of the necropolis can be dated between the end of the AD 6th and the first half of the 7th centuries.[272]

2) Important evidence comes from the area of *Friuli*. Here, 'sporadic' finds, that is to say, stray objects not found in an exact funerary or settlement context will be first considered. These finds comprise two iron stirrups found in the excavation conducted during restoration of the southern side of the *Palazzo del Pretorio* in *Cividale*. These may be considered direct evidence for Avar raids into Italy at the beginning of the AD 7th century. The typology of these stirrups falls within that of the circular type with oblong eyes, dated exactly to the early Avar period.[273] Moreover, the graves with horse burials found in *Illegio* and *Visnale dello Iudrio* as well as in *Cividale* are extremely significant. These burial places in *Cividale* are particularly important since there stirrups were found in a grave south of the Church of *S. Pantaleone* and in grave n. 43 in the necropolis of *S. Mauro*.[274] The latter grave is particularly interesting. Its finds correspond to Avar style graves even if the form of grave goods (shield's umbo, long sword, sax, spear heads and bows) is more reminiscent of an east Merovingian cultural milieu, which in its turn was heavily influenced by Roman-Byzantine borrowing. This is particularly evident in the type of the points of the bows, typical for the Late Roman *castra* situated along the Friulian *limes*, and in the tip of the laurel leaf-shaped spear. At the same time, the fittings of the horse bit are comparable to those found in the Alamannic territory, for example, those from tomb 9 of *Niederstotzingen*, thus, clearly demonstrating direct contacts with southern Alamannic territories. There are two other important aspects concerning grave n° 43 of *S. Mauro*. First, iron tools as part of the grave goods (a graver and a chisel with an incomplete tip) should be mentioned. The occurrence of tools, together with a non identifiable Late antique bronze coin and a quadrangular Byzantine weight with the mark of its worth incised on it (N, *nomisma*) may mean that this was the grave of someone involved in trade and/or an artisan. The second important aspect is that this grave, dated to the last third of the AD 6th century belonged to the first Lombard migrating generation.[275] Thus, this grave also shows that the Lombard migration into Italy implied both the penetration of Lombard origin tradesman/artisans and the introduction of the stirrup.

3) Two different iron stirrups have been found with *agemina* decorations in grave No. 86/2 of the early medieval necropolis of *La Selvicciola*, in the vicinity of *Ischia di Castro* (Latium, Viterbo County) dated to the middle of the AD 7th century (**see figure 34**). This type of stirrup is comparable to ones found in Hungary. At this site, traces of at least 100, more or less, clearly visible graves have been discovered. Many of those were destroyed due to agricultural work, ancient robbery and to the relocation and/or re-use of burial places consequent to the building of a cult edifice within the necropolis while it was still in use. This building, stretching from north to south, was a one-nave church with the apses in the southern wall and a quadrangular room along the longitudinal wall. The construction of this building influenced the orientation of the graves so that those, most likely the older ones oriented east - west, were cut or covered by the foundation of the church, while those graves dating after the construction of the church are oriented north - south. The oldest objects from this necropolis can be dated to the end of the AD 6th to the first half of the 7th century while the newer ones date to a period around the end of the AD 7th century and possibly to the beginning of the next one. Besides the *agemina* decoration on the iron stirrups of grave 86/2, the *agemina* decorative motif on a belt buckle, a projectile point and five other elements from the same grave are also important for dating and understanding of the cultural environment of these objects. This decoration is similar to that found on a belt from grave 86/11 comprising a buckle, a loop, five small tongue-like elements and four bigger ones, plus another group of five elements, possibly part of the same set, composed of another four tongue-like elongated pieces and a long point. These decorated objects find their exact counterpoints in the Alamannic-Bavarian territories of the second half of the AD 7th century. Such correspondence has also been confirmed by X ray analysis on stirrup no. 8/3. The analysis revealed

[272] E. Ferrero, "Borgomasino. Sepolcreto barbarico scoperto nell'abitato," *Notizie degli Scavi di Antichità* (1893): 189; G. Berattino, "I reperti della necropoli a Reihengräber di Borgomasino," *Bollettino di storia e arte canavesana* 7 (1981): 87, 88, 91; E. Micheletto and L. Pejrani Baricco, "Archeologia funeraria," 307; on the pottery types belonging to the Pannonic phase, see O. Von Hessen, "Tecniche di lavorazione," in G.C. Menis ed., *I Longobardi*, 209.

[273] I. Ahumada Silva, "Testimonianze archeologiche avare a Cividale," *Forum Julii* 14 (1990): 63-67.

[274] I Ahumada Silva, "Necropoli longobarde a Cividale ed in Friuli," in *Paolo Diacono ed il Friuli alto medievale*, 339, 355-56.

[275] P. Lopreato, "La necropoli di S. Mauro-Cividale," in E. Arslan and M. Buora eds., *L'oro degli Avari* (Milan: In-Form, 2000), 196; I Ahumada Silva, "Cividale del Friuli. Necropoli di S. Mauro. Tomba 43 di cavallo e cavaliere," in E. Arslan and M. Buora eds., *L'oro degli Avari*, 198-205.

the presence of *agemina* decorations which could be placed in the same chronological and productive space as those on the belts from graves 86/11 and 86/2, that is to say, the middle of the 7th century in the Alamannic-Bavarian territories.[276]

4) A couple of cast bronze stirrups came to light in grave 41 of the necropolis of *Castel Trosino* (Marche, County of *Ascoli Piceno*). This burial is composed of a simple pit with its wall lined with stone slabs, a grave type widely diffused throughout the cemetery from the first half of the AD 7th century. The grave goods, beside the stirrups, consisted of a pectoral cross, two necklace beads and a glass bottle. The dimensions of these cast bronze stirrups, certainly not suitable for an adult male as well as their typology and material are uncommon. Their closest analogy may be found in grave 33 of the necropolis of *Vicenne* where another bronze stirrup was unearthed which was remarkably similar with those from *Castel Trosino*. This correlation is also a confirmation that this find can be dated to around the years 50/60 of the 7th century.[277] However, the peculiar typology of these stirrups, together with their particular material, shape and funerary context, call for a more detailed investigation for attaining a more precise attribution of grave goods and their relative chronology. Once again, findings from Central-eastern Europe provide some archaeological evidence suitable for confrontation with the Italian items. These are stirrup-like artifacts from the Byzantine settlement of *Caričin Grad* (Serbia) and from grave 1 of the early Avar necropolis of *Szegvár-Oromdülő* (Hungary) (**see figure 35**). The couple of stirrup-like objects from *Caričin Grad* was made of iron, possibly through the smithing of a quadrangular iron bar. They have an average height of 8 cm. and the footrest bar is almost 9, 5 cm long. Thus, these artifacts do not properly resemble, for their shape and structure, the stirrups as described in the *Maurikos' Strategikon*. Actually, the footrest seems to be too short and thin to able to support a male mounted soldier, and, as it was already hypothesized by J. WERNER, the very part of this object could also be covered with fabric or fur. In grave 1 of *Szegvár-Oromdülő* cemetery, a horse was buried together with a human skeleton, attributed to a female. Here, a small stirrup-like artifact made of bone came to light together with a little golden cross which possibly pertains to the horse fittings. This bone artifact is only 7 cm long and 4 cm high. In this case as well, this item does not appear to be able to sustain a mounting adult male, thus confirming the attribution of the grave to a female. The rest of grave goods is relatively rich: beside the horse fittings, 3 glass paste beads, 4 amber beads, 4 copper alloy medallions decorated with granulation, a larger copper alloy disc shaped medallion, a golden and a silver earrings, various elements of a belt, and a hand made jar are worth mentioning. According to G. LŐRINCZY, the entire set of objects points to a very early phase of Avar occupation of *Pannonia*, that is to say, to the last third of the AD 6th century or to the beginning of the 7th. The objects from *Szegvár-Oromdülő* and *Caričin Grad*, apparently, prove that between the end of the AD 6th and the beginning of the AD 7th centuries there existed different sort of footrests for mounting horses of possible Sassanid origin, as also suggested by the frescos of the palace of *Kahn Varkhman* dated to the mid AD 7th century. These tools were probably of far eastern origin as well but enjoyed different functional and structural characteristics in respect to those depicted in *Maurikios' Strategikon* and found in the Avar and Germanic necropolises of Europe. These were mainly used by men for military purposes whereas the others would better fit, given their shape and dimensions, for supporting feet of women while riding. The couple of copper alloy stirrups from grave 41 of *Castel Trosino* would then fall into the latter category, being an important Italian example of these artifacts. Such a conclusion, however, could also date grave 41 to an earlier phase of the Italian necropolis.[278]

5) Various iron and bronze stirrups have been found and at least 10 graves of riders and horses (graves no. 16, 29, 33, 66, 73, 79, 81, 85, 110 and 102 in the nearby locality of *Morrione*) at the necropolis of *Vicenne* (Molise, County of *Campobasso*) again in a context from the second half of the 7th century. Instead of one big necropolis, the cemetery organization consisted of various groups of tombs around several nucleuses. This typology seems to indicate a model more typical of nomadic settlement. The graves are lined up in relatively parallel rows oriented north to south. Each grave keeps strictly to an east-west orientation, with the face of the buried person turned towards the east. There are also inhumations with coffins, since nails were found in some burial pits (tombs 117, 115, 76, 73, 85). Generally, the rider occupied the southern side of the grave while the horse lay on the opposite side with the animal lying on its side with his mouth turned towards his lord. However, it is difficult to reconstruct any uniformity in the mode of deposing the horse and the horsemen based on archaeological data alone.[279]

---

[276]M. Incitti, "La necropoli altomedievale della Selvicciola ad Ischia di Castro (VT) ed il territorio castrense in età longobarda," in L. Paroli ed., *L'Italia centro-settentrionale*, 225-33; Id., "La necropoli longobarda della Selvicciola," in E. Herring, R. Whitehouse, J. Wilkins eds., *Papers of the Fourth Conference of Italian Archaeology, 4* (London: Accordia Research Centre, 1992), 213. The Hungarian items are in N. Fettich, *Das awarenzeitliche Gräberfeld von Pilismarót-Basaharc* (Budapest: Akadémiai kiadó, 1965), 20, pic. 25, nn. 9-10; A. Salamon and I. Erdely, *Das völkerwanderungszeitliche Gräberfeld von Környe*, (Budapest: Akadémiai Kiadó, 1971), 82 pl. 4, nn. 1-2 and 4-5; 83, pl. 5, nn. 32-33; 84, pl. 6, n. 5; 86, pl. 8, nn. 9-10 and 25-26; 88, pl. 10, n. 22; 93, pl. 15, nn. 20-22; 96, pl. 18, nn. 35-36; 98, pl. 20, nn. 34-35; 100, pl. 22, nn. 12-13; 101, pl. 23, nn. 17, 37, 41; 106, pl. 28, nn. 1-4, 6-7; 107, pl. 29, nn. 3-5; see also, I. Bóna,"Neue Nachbarn im Osten,", 113.

[277]L. Paroli, "La necropoli di Castel Trosino: un riesame critico," 206, 301.

[278]On the paragraph's subject see, L. Paroli, "La necropoli di Castel Trosino: un riesame critico," 206, 301; G. Lőrinczy, "Vorläufiger Bericht über die Freilegung des Gräberfeldes aus dem 6.-7. Jahrhundert in Szegvár-Oromdülő," *Communicationes Archaeologicae Hungariae* (1992): 81-124, in particular on tomb 1 see 81-90, 103, 105-09, 110; on the iron stirrups like artifacts from *Caričin Grad* in Serbia see, J. Werner, "Ein byzantinischer Steigbugel aus Caričin Grad," in N. Duval and V. Popovic eds., *Caričin Grad. I, Les basiliques B et J de Caričin Grad* (Rome: Ecole Française de Rome, 1984), 147-48, 150, 153.

[279]V. Ceglia and B. Genito, "La necropoli altomedievale di Vicenne a Campochiaro," in S. Capini and A. Di Niro eds., *Samnium. Archeologia*

del Molise (Rome: Quasar, 1991), 329-34 and, in the catalogue of unearthed materials, 354, 358; V. Ceglia, "Campochiaro (CB)località Vicenne. La Necropoli altomedeievale," *Bollettino di Archeologia* 5-6 (1990): 213-17; B. Genito, "Archaeology of the early medieval nomads in Italy: the horse-burials in Molise (7th century) south-central Italy," in Cs. Bálint ed., *Kontakte zwischen Iran, Byzanz und der Steppe*, 229-47; Id., "Sepolture con cavallo da Vicenne (CB)," 286: "Alcuni tratti culturali, in qualche modo riconducibili all'Asia nomadica, sono stati già in passato identificati in Italia centro-settentrionale e meridionale; ma nonostante ciò, molto poco delle relazioni tra la penisola e quel grande movimento migratorio di cui gli antichi nomadi d'Asia sono parte integrante, è stato, mai, archeologicamente messo in evidenza prima della scoperta della necropoli di Vicenne (Campochiaro, Molise). Questa necropoli, infatti rappresenta un *unicum* culturale, ad un tempo "asiatico" e "nomadico" caratterizzato dalla presenza di numerose tombe con cavallo che rimandano ad analoghe forme di sepolture rinvenute proprio tra i popoli nomadi delle steppe Eurasiatiche ... Il seppellimento contemporaneo, nella stessa tomba di un uomo e di un cavallo costituisce un esempio di rituale funerario con una lunga e complessa tradizione storica alle spalle che risale alla cultura dei cavalieri nomadi delle steppe Eurasiatiche. Questo tipo di seppellimento, per la prima volta documentato in Europa meridionale, ma generalmente rinvenuto in Asia centrale e in Europa centro-orientale, costituisce l'aspetto più significativo della necropoli di Vicenne e, all'interno della documentazione archeologica europea esso rappresenta davvero una significativa eccezione ... D'altra parte sulla base delle analisi archeologiche e archeo-zoologiche [S. Bökönyi, "Analisi archeozoologica dello scheletro del cavallo nella necropoli di Vicenne," *Conoscenze* 4 (1988): 69-76 and Id., "Two more horse graves from Vicenne," in *Samnium*, 342-43], queste tombe costituiscono senza alcun dubbio seppellimenti contestuali di un uomo e del suo cavallo, unitari nel tempo e nello spazio; ciò va particolarmente enfatizzato considerando l'estrema varietà tipologica delle numerose fosse con cavalli o parti di essi (con o senza corredo), rinvenute in Europa occidentale dalla Scandinavia alla Germania e databili dall'epoca Merovingia a quell'Ottoniana (V-XI secolo A.D.) [M. Müller Wille, "Pferdegrab und Pferdeopfer im frühen Mittelalter," *Berichten van de Rijksdients voor het Oudheikundig Bodenmonderzoek* (1970-71): 119-248. J. Oexle, "Merowingerzeitlichen Pferdebestattungen oder Beigaben?," *Frühmittelalterliche Studien. Jahrbuch des Instituts für Frühmittelalterforshung der Universität Münster* 18 (1981): 122-71] ... " and 287: "Il fatto che queste staffe di tipologia avarica siano state rinvenute in un contesto archeologico caratterizzato da molti oggetti di corredo di carattere "germanico" piuttosto che costituire un'anomalia rappresentano la conferma piena del fatto che tali oggetti, completamente sconosciuti nel mondo classico sono arrivati in Occidente proprio attraverso il tramite di popoli "diversi" che vivevano in un contesto multi-culturale e multi-etnico." and finally 288: "L'area più occidentale nelle quali le tombe con cavallo sono state rinvenute sono l'Ungheria [Balint C., "Les tombes a ensevelissement de cheval chez les Hongrois aux IXe-XIe siecles," *Archivum Eurasiae Medii Aevi* 2 (1982): 5-32.], l'ex Cecoslovacchia [A. Točik, *Slawisch-awarisches Gräberfeld in Stúrovo*, (Bratislava: Vydavatel'stvo Slovenskej akadémie vied, 1968)] e l'ex Jugoslavia dove contesti archeologici con un forte carattere "asiatico" sono stati attribuiti agli Slavi, gli Avari o gli Ungheresi. Dall'Ungheria si possono citare gli esempi di Szekszárd [Gy. Rosner, "A Szekszárd-Bogyiszlói úti avar temető ló és lovas temetkezései," [Burials of horses and horsemen in the Avar necropolis of Szekszárd at Bogyiszló street] *Szekszárdi Béri Balogh Ádám Múzeum Évkönyve* 6-7 (1975-76): 103], Ivancsa [I. Bóna, "Avar lovassir Iváncsáról," [The Avar horseman burial of Iváncsá] *Archaeologiai Értesítő* 97 (1970): 243-63, fig. 1]; dall'Ukraina, vicino alla regione di Nikolaev, quello di Portovoe [A.I. Ajbabin, "Pogrebenie hazarskogo voina," [The grave of Khazar warrior] *Sovjetskaja Arkheologija* 3 (1985): fig. 9] e altri dall'Asia Centrale, dall'Oasi di Tuva [B.B. Obcinnokova, "Pogrebenie drevnetjurskogo voina v central'noj Tuve," [The Grave of an ancient Turkic warrior in central Tuva] *Sovjetskaja Arkheologija*, 3 (1982): fig. 1; R. Kenk, "Früh- und hochmittelalterliche Gräber von Kudyrge im Altai, Frühmittelalterliche Gräber aus West-Tuva," in *Materialen zur allgemeinen und vergleichenden Archäologie*, Munich: 1982, figg. 9, 11, 14, 19, 20, 31, 41, 43] e dall'Altai superiore [Ivi, figg. 14, 15, 17, 18, 19, 20, 22, 23]. Con questi confronti non si intende, qui, però affermare l'esistenza di una diretta continuità tra le tombe con cavallo dell'Europa dell'Est e l'Asia Centrale da un lato e quelle dell'Italia meridionale dall'altro, né di stabilire precise correlazioni etniche tra gli inumati e i relativi resti archeologici così lontani tra loro. Queste ipotesi di confronto possono per il momento, solo fornire l'evidenza di connessioni tipologiche, strutturali di un rituale con al centro il sacrificio del cavallo che non appartiene al mondo germanico, sebbene un culto di quell'animale è comunque documentato anche tra quelli [C. Balint, "Les tombes a ensevelissement de cheval chez les Hongrois aux IXe-XIe siecles," *Archivum Eurasiae Medii Aevi* 2 (1982): 5-32]. Ritrovare un rituale funerario di tipo centro-asiatico in un contesto germanico, come certamente è quello di Vicenne, è cosa particolarmente significativa, ... Su questa base credo che i tre livelli interpretativi proposti qualche tempo fa per l'analisi dei resti della necropoli di Vicenne, l'asiatico, il germanico ed il locale trova la sua più piena conferma: un contesto multicultuale e multietnico non può che essere analizzato e interpretato se non con criteri interpretativi a più livelli." On the variety of horse burials typology see, E. Garam, "Sepolture di cavalli," 144: "la forma più diffusa di sepoltura equestre di epoca avara era quella che poneva il cavallo presso il suo proprietario. Il cavallo poteva giacere nella stessa fossa dell'uomo, alla sua destra oppure alla sinistra, a croce davanti ai suoi piedi, oppure coricato longitudinalmente, con lo stesso orientamento dell'uomo oppure con orientamento opposto. Talvolta veniva scavata una fossa distinta per il cavallo nei pressi della tomba dell'uomo. I cavalli con poche eccezioni venivano sepolti completamente bardati;" and for a wide presentation of examples see of I. Kovrig, "Contribution," 164-74.

6) Extremely interesting finds came to light from the locality of *Casino Vezzani-Vassarella* near *Crecchio* (province of *Chieti*, Abruzzo). Here, several weapons and iron horse equipment were brought to light. Among other things, four iron stirrups were unearthed. This entire assemblage possibly belonged to mounted soldiers and was attributed by A.R. STAFFA to a Byzantine troop in charge of defending the border with the Lombard kingdom, a frontier which lasted in this part of *Abruzzo* only until circa AD 580. Therefore, the timeframe for this evidence was established by the same author as the end of the AD 6th century. This attribution was hypothesized by STAFFA based on the presence, in the same stratigraphic level, of a copper-alloyed Coptic bowl, in addition to various *fibulae* recalling Late Antique fashions; of a specific pottery type, known as *tipo Crecchio*, considered by A.R. STAFFA to be an indicator of the Byzantine presence; and a wooden *cathedra* again of possible Coptic origin.[280] This assumption is, in my opinion, rather questionable from both a chronological and "ethnic/cultural" point of view. Grave goods of North African, that is to say, Coptic origin, and copper alloyed Coptic bowls in particular, represent a widely distributed type of luxury grave goods in non Byzantine territories as well and were, thus, present in Lombard necropolises between the end of the AD 6th and the first half of the 7th centuries. These types of objects attest to both the continuation of trade between Italy and North Africa in this period as well the existence of commercial relations between the two sides of Italy (i.e. the Lombard Kingdom and Imperial territory), something that has also been demonstrated by the recent discovery of the *Crypta Balbi* in *Rome*.[281] It is difficult to identify the presence of such objects with a particular ethnic, or rather, cultural identity. Even the *Crecchio* type ceramics cannot be

[280] See, A.R. Staffa, "La persistenza di logiche tardoantiche nella difesa dell'Abruzzo dai Longobardi. Reperti inedito da *Castrum Truentinum* e Crecchio," in M. Buora ed., *'Miles Romanus' dal Po al Danubio nel Tardoantico* (Pordenone: Lucaprint, 2002), 251-72.

[281] On this question see, A. Melucco Vaccaro, *I Longobardi*, 150-52; J. Werner, "Italienisches und koptisches Bronzegeschirr des 6. und 7. Jahrhundert nordwärts der Alpen," in *Memnosymon Theodor Wiegand* (Munich: F. Bruckmann Verlag, 1938), 78. V. La Salvia, "Aspetti dell'economia dell'Italia alto medievale," *passim*.

easily considered as "exotic" Byzantine production since it has also been found in funerary contexts within *Abruzzo* such as in the catacombs of *Castelvecchio Subequo* and *S. Vittorino*,[282] in areas that were soon to be outside of direct Byzantine control. The typology of horse equipment, stirrups and of some of the weapons, as for a sickle-like spear point (known as *harpago* or *lupus* and dated to the mid of AD 7th), does not easily fit into the chronology of the end of the AD 6th century. A.R. STAFFA himself admits that the spear point and one of the iron bids match similar items from the necropolis of *Vicenne* (a funerary context from the middle of AD 7th century). Moreover, the stirrups resemble a type found throughout the Carpathian Basin during the very late AD 7th century, such as at *Tiszaluc* and *Gyenesdiás* in Hungary.[283] The other weapons, three knives, a scramasax are somewhat closer to the chronology proposed by A.R. STAFFA since similar objects have been discovered in Lombard necropolises for the period between the end of the AD 6th and the beginning of the 7th centuries. Nevertheless, such artifacts are known from later contexts as well, such as graves 53 (dated between AD 610-630), 60 (dated around AD 660/690), and 69 (dated after the middle of the AD 7th century) from the necropolis of *Collegno* (province of Turin, Piedmont).[284]

To summarize, the highly composite character of the finds from *Crecchio* does not allow any precise ethnic/cultural and/or chronological attributions to be drawn. Many objects match those found within earlier Lombard territories and funerary contexts. Moreover, the chronology and typology of some other artifacts (the spear point, the horse bit and the stirrups) point to a situation rather far from a "pure" Byzantine context in the late AD 6th century. At the same time, these objects greatly resemble finds from *Vicenne* with its composite cultural structure mentioned above. It is therefore difficult to agree with A.R. STAFFA's idea that these objects came from a group of Byzantine soldiers defending the frontier against the Lombard kingdom.

[282] See A.M. Giuntella et alii, "Recenti indagini nella catacomba di Castelvecchio Subequo (Aq)," *Rivista di Archeologia Cristiana* 2.67 (1991): 249-321, especially 301-06; A.M. Giuntella, "Il ducato di Spoleto: persistenze e trasformazioni nell'asseto territoriale (Abruzzo e Sabina)," in *I longobardi dei Ducati di Spoleto e Benevento*, 796-97.

[283] On the changes occurring in stirrup typology in the late Early Avar period see, E. Garam, "Sepolture di cavalli," 146: "La forma delle staffe nel bacino dei Carpazi cambiò e dalla fine del VII secolo quando l'occhio delle staffe fu annodato e l'appoggio dritto (che alla fine dell'epoca avarica divenne più largo) rafforzato con coste." See as well P. Tomka, "Il Costume," in *Gli Avari. Un popolo d'Europa*, 84-85 where it is emphasized that the Euroasian nomads always used felt riding boots with smooth soles and that its invention in Europe can be ascribed to the early Avar period since this type of boots would only fit into the roundish rod stirrups. Moreover, he stressed that only after the coming of other nomadic people of Turkic\Mongol origin, around the year AD 670, and the consequent diffusion of stirrups with a rigid rod was it then possible to mount with hard leather sole boots as well. The *harpago* or *lupus* is known from grave 119 at *Castel Trosino* while another one came to light in *Pisa*, see G.C. Menis ed., *I Longobardi*, 188; G. Ciampoltrini, "La falce del guerriero e altri appunti per la Tuscia fra VI e VII secolo," *Archeologia Medievale*, 20 (1993): 596-97.

[284] See C. Giostra, "Catalogo," 97-109, 111-15, 127-32. For the objects attributed to an earlier phase see, G.C. Menis ed., *I Longobardi*, 46, 188, 426, 444; C. Rupp, "Catalogo," 118; L. Paroli, "La Necropoli di Castel Trosino: un riesame critico", 262.

Based on the evidence presented above, it seems more plausible to date these objects to the middle of the AD 7th century.

4.1.4.3 Conclusions

The presence of horse burials, unquestionably a tradition of oriental origin,[285] and of funerary contexts influenced

[285] On the oriental origin of the fashion of horse burials and of the importance of the horse as a status symbol see, B. Genito, "Materiali e problemi," *Conoscenze* 4 (1988): 56; E. Garam, "Sepolture di cavalli," 143: "Nel primo millennio d.C. cominciò a diffondersi nelle steppe asiatiche l'uso di seppellire con i guerrieri i loro cavalli bardati. Prima dell'arrivo degli Avari tale usanza era un'eccezione nel bacino dei Carpazi;" see also R. Brulet, "La sépulture du roi Childéric à Turnoi et le site funéraire," in F. Vallet and M. Kazanski eds., *La noblesse romaine et les chefes barbares du III au VII siecle*, 314: the distribution map of horse burials in the Merovingian period shows that this mode of interment remained exceptional between the Rhine and Seine, and only began to evolve from the AD 5th century onwards in Thuringia and even later along the Elbe, Rhine and Danube rivers; see as well, M. Müller-Wille, "Pferdgrab und Pferdopfer," 122-72; S. Piggott, *Wagon, Charriot and Carriage* (London: Thames and Hudson, 1992), 108, 109 and especially 112: "When we turn to the horse as a prestige steed, the Scythian world of the fifth century BC on the Central Asian steppe shows a rather different pattern, illustrated by princely *Pazyryk* burilas in the Altai. Here we are in a horse-centred nomadic or semi-nomadic world ... where horses indicated status in herds rather than as individuals." B. Genito, "Sepolture con cavallo da Vicenne (CB)," 286-87: "Storicamente l'uso di seppellire cavalli è attestato da Erodoto (IV, 72) per l'epoca scita (V-IV secolo a.C.) e da evidenze archeologiche per l'età del Bronzo e del Ferro, nel Vicino Oriente, in Cina (nel periodo tardo Shang, XIII-XI secolo a.C.) ed in Asia Centrale con la significativa variante della presenza di carri. Per l'epoca medievale il rituale è testimoniato da numerosi autori e da una discreta documentazione archeologica con la deposizione contemporanea di uomo ed animale, variante assente nell'età del ferro. Una delle principali fonti relative a tale rituale si trova nelle note di viaggio di Ibn Fadlan, ambasciatore del califfo Abbaside, Muqtadir, inviato nel IX-X secolo presso il re dei Bulgari del Volga. L'autore nelle sue note ci ha fornito una grande quantità di informazioni storiche, geografiche ed etnografiche, inoltre, sulla popolazione dei Khazari, tra le quali anche la descrizione della morte, del seppellimento di un uomo e della pratica rituale di sacrificare contestualmente un cavallo. In tale rituale è evidente che l'uccisione di un cavallo e la sua collocazione in una tomba, consente, all'uomo deceduto di salire simbolicamente al Paradiso. Il rituale descritto da Ibn Fadlan, documentato anche presso i Mongoli ci fornisce l'evidenza anche di un pasto rituale prima che il cavallo sia imbalsamato e ci conferma quanto tali sacrifici fossero diffusi nel mondo dell'est-europeo dell'epoca. Il seppellimento di cavalli insieme agli uomini nelle tombe di periodo altomedioevale è archeologicamente attestato nel V secolo nella Russia Meridionale e nel Caucaso; gli Avari nel VI-VII secolo lo diffondono in Occidente nelle aree Danubiane e della Tisza, e gli Ungheresi, nel IX/X secolo lo adottano di nuovo nel bacino dei Carpazi. Indipendentemente da ogni attribuzione etnica di quel rituale per cui sono stati di volta in volta proposti popoli come i Saragouri, gli Ouguri, gli Onoguri, i Bulgari, i Finno-ugri, gli Unni, e gli Avari, esso fu, molto probabilmente, introdotto contemporaneamente da diversi popoli in aree diverse. Sebbene la questione appare molto complicata la più gran parte degli studiosi attribuisce l'uso di seppellire i cavalli al ramo occidentale dei su citati popoli di origine turca, e la probabile area originale del rituale alla Siberia di Sud-ovest. I seppellimenti più tardi con cavallo del X-XI secolo nella Russia meridionale e a Sarkel la famosa capitale dello Stato Kazaro sono, invece, attribuibili a nuove ondate di popoli turchi, come i Pecheneghi. Generalmente considerati relativi a credenze religiose, i seppellimenti con cavallo del medio-evo appartengono culturalmente in toto al mondo nomadico eurasiatico, e, sono chiare espressioni della particolare importanza del cavallo nel contesto socio-economico del nomadismo pastorale equestre che col tempo ha determinato lo sviluppo di particolari rituali connessi con il suo sacrificio. Questo rituale richiedeva che il cavaliere portasse con se dopo la morte il cavallo e la sua cavalcatura ed è, così, strettamente riferibile al valore sacro del suo spirito o della forza vitale, capace di sopravvivere attraverso le ossa

by Avar tradition within Lombard and Alamannic-Bavarian territories makes it clear that significant relationships existed between these areas and the Carpathian Basin. Actually, as reported by B. GENITO, this connection must have played a central role in shaping early medieval tools in Italy.[286] The Avar migration into Central Europe around the end of the sixties of the AD 6th century, on the one hand, implied a radical reorganization of the political structure, particularly in the Byzantine provinces in the Balkans. On the other hand, this reorganization did not result in any major or disruptive event as far as the local populace of the Carpathian Basin was concerned. There is strong evidence in the archaeological record of the persistence within the early Avar Kaghanate, of peoples both of Germanic, either western or eastern, and provincial, either Roman or local, origin, as the example of *Keszthely* culture demonstrates.[287] Apparently, this very situation lasted until the 80ies of the AD 7th century when the arrival of other nomadic tribes from the east lead to a crisis in Avar power in the Carpathian Basin.[288] The relationship between the Avar Kaghanate and the east Merovingian area remained regular and stable both from a military point of view, often the main one, and commercial exchanges.[289] Therefore, the existence of burials of western Germanic people and of objects of eastern Merovingian provenance in Avar territories may be primarily regarded as an indication of connections with Lombard Italy, granted that in the Italian peninsula as well there is plenty of material of Avar and Avar-Byzantine origin.[290] As an example, grave 168 at *Romans d'Isonzo* contained a straight sword, typical of the Avars, even if they are decorated in the II animal style.[291] However, this relationship with the Carpathian Basin was not one way as concerns either the direction of trade or the various *gentes* involved in it. Artifacts and burials of western Germanic origin in Avar lands demonstrate that there existed all-inclusive contacts with the populace in the Alamannic-Bavarian area and contacts were not exclusively with Lombard Italy alone. This is apparently, the case of grave 200 in the Avar necropolis of *Mezőfalva* (for the *agemina* type of belt decorations) and other finds in Transylvania [more precisely in the Avar necropolis of *Bánd*, *Veresmart*, *Nagylak* (*Nadlac*, Romania) and *Baráthegy* (*Bratei*, Romania)].[292] As far as Lombard burials within Avar territory are concerned, excluding the controversial cemetery of *Várpalota* in which, according to some scholars, Lombard and Avar burials followed each other without any discontinuity,[293] graves 16, 216, 350, 356, 390, 760 of *Szekszárd-Bogyiszlói* street necropolis and of that of *Zamárdi* may be considered. The first burial ground is important because it contained iron stirrups of the very same type as those from *Vicenne\Borgomasino*. This parallel find suggests

(cranio e quattro zampe) e la pelle dell'animale, generalmente lasciate intatte nelle tombe. Considerando l'alto costo di un cavallo, sempre maggiore di un bovino, l'uso di seppellire un cavallo completo potrebbe essere stato un privilegio dei livelli più alti della società, là dove la deposizione della sola bardatura potrebbe avere caratterizzato i livelli bassi. On the very subject see as well, A. Azzaroli, *Il cavallo nella Storia antica* (Milan: L.L. Edizioni equestri, distributore Dielle 1975); M. Canard, "La relation du voyage d'Ibn Fadlan chez les Bulgares de la Volga," *Annales de l'Institut d'etudes Orientales* 16 (1958): 200b, 75, 76, and note n. 136; L. Chi, *Anyang* (Seattle: University of Washington Press, 1977); K.C. Chang, *Shang Civilisation* (New-Haven: Yale University Press, 1980); Cs. Bálint, "Les tombes a ensevelissement de cheval chez les Hongrois aux IXe-XIe siecles," *Archivum Eurasiae Medii Aevi* 2 (1982): 19 ss., 25; K.F. Smirnov, "Arheologiceskie stat'i v Trudah Saratovskogo oblastonogo Muzeja kraevedenija," [Archaeological papers in the works of Saratov regional museum] *Sovjetskaja Arkheologija* 4 (1962): 270-73; I.P. Zaseckaja, "Osobennosti pogrebal'nogo obrjada gunnskoj epohi na territorii stepej niznego povolz'ja i severnogo pricernomr'ja," [The specific grave-goods of Hunnic period on the Steppes of the lower Volga basin and northern Black Sea area] *Arheologiceskih Sbornik* 13 (1971): 64-65, 69.

[286] B. Genito, "Materiali e problemi," 49. Moreover, *Lamellenpanzer*, *Spangenfederhelm* and stirrups appear in Lombard graves only after AD 568, therefore precisely from their Italian phase. This moment in their history, represented an occasion to establish new contacts and re-enforce older ones; relative to that see, J. Werner, "Nomadischen Gürtel," 119.

[287] The long lasting Keszthely culture certainly must be connected to the persistence of a *castrum* on the southern shore of Fenékpuszta peninsula on Lake Balaton. The fort flourished between the AD 4th and the middle of the AD 7th centuries. Since it was located close to the frontier of the early Avar Kaghanate its population could maintain connections with both the Balkans and northern Italy. At that time, its populace consisted, beside people of provincial origin, of persons of Byzantine origin, possibly prisoners of war captured during Avar raids, Germanic people, beside various craftsmen who would have been attracted there since the strategic and political role of the *castrum* was preserved. From the early 30s of the AD 7th c. the situation started to change, especially after the Avar defeat at Constantinople in AD 626 and the subsequent succession war within the Kaghanate. Around AD 630, the fort was destroyed and never rebuilt. During this period the Avars finally gained full control of the region and the Christian populace of Keszthely culture remained isolated; concerning this theme see, R. Müller, "La cultura di Keszthely," in *Gli Avari. Un popolo d'Europa*, 165-66 and 167-68 Id., "Die Keszthely-Kultur," in *Reitervölker aus dem Osten*, 265-74; L. Barkóczi, "A 6th century Cemetery from Keszthely-Fenékpuszta," *Acta Archeologica Accademiae Scientiarum Hungaricae* 20 (1968): 275-311.

[288] Gy. László, *Steppenvölker und Germanen. Kunst der Völkerwanderungszeit* (Vienna/Munich: Schroll, 1970), 52-55; F. Daim, "The Avars. Steppe people of Central Europe," *Archaeology* 37.2 (1984): 34-35; W. Menghin, T. Springer and E. Wamers eds., *Germanen, Hunnen und Awaren. Schätze der Völkerwanderungszeit* (Nurnberg: Verlag Germanisches Nationalmuseum, 1987), 261; W. Pohl, *Die Awaren*, 55, 89, 92, 308-12; A Kiss, "Germanen im awarenzeitlichen Karpatenbecken," in F. Daim ed., *Awarenforschungen. Studien zur Archäologie der Awaren* 4 (Vienna: 1992), 36-38; T. Vida, "Merovingische Spathagurte der Awarenzeit," *Communicationes Archaeologicae Hungaricae* (2000): 171-72; P. Stadler, "Das germanisches Substrat: Langobarden, Gepiden und anderen germanische Völkerschaften im frühawarischen Material," in *Reitervölker aus dem Osten*, 281-82; F. Daim, "Archäologie der Awaren," in *Reitervölker aus dem Osten*, 200: "Vor allem in Westungarn spielt die germanische Komponent im awarischen Material eine bedeutende Rolle und manche germanische Verzierungsweisen werden dann im awarischen Kunsthandwerk übernommen und weiterentwieckelt. Sicher spielen hier die besiegten Gepiden und zurückgebliebene Langobarden eine wesentliche Rolle."

[289] S. Szádeczky-Kardoss, "Histoire des Avars et leur heritage en Europe," in S. Csernus and K. Korompay eds., *Les Hongrois et l'Europe: conquete et integration* (Paris/Szeged: Publications de l'Institut Hongrois de Paris, 1999), 150, 151, 153, 155, 157; M. Brozzi, "Avari e Longobardi friulani," in *Gli Avari. Un popolo d'Europa*, 57, 58, 58-59, 60 ff.; I. Bóna, "Gli Avari," 28-29; A. Kollautz, "Awaren, Langobarden und Slaven in Noricum und Istrien," *Karinthia* 1 (1965): 619-45; A. Tagliaferri, "Il Friuli e l'Istria nell'Altomedioevo," *Antichità Altoadriatiche* 2/2 (1972): 273-94.

[290] I. Bóna, "Neue Nachbarn," 111-13.

[291] A. Giovannini, "La necropoli di Romans d'Isonzo," in *Paolo Diacono ed il Friuli alto medievale*, 641-42.

[292] I. Bóna, "Neue Nachbarn," 111.

[293] M. Martin, "Zur Interpretation des langobardischen Gräberfeldes von Várpalota, Komitat Veszprém," *Basler Beiträge zur Ur- und Frühgeschichte* (1976): 194-99.

acquisition *in situ* of this new and fundamental riding tool for those Lombards living in direct contact with the Avars. Moreover, 70% of the early Avar stirrups come from former Lombard *Pannonia* (**see figure 36**).[294] The necropolis of *Zamárdi* is located on the southern shore of Lake Balaton. It was probably the residence of Avar Khan Bajan's successors although the settlement itself has not yet been found. The excavations were conducted over a 20,000 square meters area and 2226 tombs were finally unearthed, dated to a period from the AD 7th to 8th century. However, the site flourished during the AD 7th century and ended in a crisis during the following century. Many of these graves had been robbed and it has been calculated that the entire number of burials would have been around 5000. An iron making site with reduction furnaces was discovered as well in the same area. The archaeological evidence from the necropolis testifies to the existence of long distance trade connections with the Alaman-Bavarian zone, Italy and Byzantine territory. Beside imported Byzantine jewelry, there are many other artifacts, such as belt ornaments, horse bits and fittings from Lombard Italy, Bavaria, Alamannia, Frankish kingdom or from the Romanized populace of eastern Alps. The overall situation of the finds structure and quality in *Zamárdi* is, thus, a demonstration of the mutual influence between the Avar lands and nearby territories.[295]

As far as the relationship with Bavaria is concerned, graves 40, 74, and 97 from the *Linz-Zizlau* cemetery, grave 35 from the necropolis of *Moss-Burgstall* and the burial grounds of *Budenheim* can be mentioned. In these two last necropolises the fashion of horse interment must certainly be considered an oriental import through Avar mediation. [296] The contacts between these two areas greatly intensified during the last thirty years of the AD 7th century because of the military instability on the frontier. This situation is, apparently, demonstrated by finds from some Avar necropolis such as *Záhorska-Bystrica*, *Sommerein* and *Zalakomár* where belt fittings of Bavarian origin are found together with Avar pottery; in the military cemetery of *Wien-Liesing* (grave 3 with a late 7th century "Bavarian" sword). In some north Hungarian necropolis such as *Hédervár*, *Vasasszonyfa*, *Lukácsháza*; where various long sax came to light; in some Lower Austrian cemeteries (at *Zwölfaxing*, graves 3 and 233) and in *Münchendorf*.[297] In relation to the Avar-Alaman connection, the stirrup imports have been fully investigated by J. OEXLE. Moreover, even if the Lombard mediation is emphasized in explaining the diffusion of stirrups in Alaman regions, the existence of contacts with the region in the upper courses of the Danube are no less significant.[298]

The archaeological features from the necropolis of *Moss-Burgstall* and grave 35 in particular, are important. From an anthropological point of view, it has a quite mixed composition: northern European types are widely diffused among women although there are Roman and east-European\Mongoloid types as well. The inhumations of grave 35 and 3 represent the latter physical type. The grave goods of grave 35 yields evidence, beside the stirrups, of typical Germanic weaponry (shield's umbo, long sword and sax) and, moreover, the belt decoration points to an eastern Merovingian origin for these artifacts, more precisely a north Italian one. However, grave 35 can be certainly ascribed to a rider of Avar origin who would have been in charge, on behalf of the Bavarian Duke, of controlling the relevant strategic zone around the confluence of the rivers Isar and Danube during the early 30s of the AD 7th century.[299] This Bavarian burial place displays a broad range of similarities with the Italian necropolis of *Vicenne*. In both cemeteries, as noted by B. GENITO, there is evidence of structural features, such as horse interments and objects (i.e. the stirrups) of clearly Asiatic derivation within a generalized Germanic context. [300] In addition, the presence in Bavaria and Lombard Italy of necropolis that are quite similar to each other in terms of their "oriental character" dating from the mid AD 7th century (as shown by horse burials and stirrups), sheds light on contemporary written sources mentioning the migration and settling of 'Proto-Bulgarian' peoples in these same areas. Thus, the Bulgars led by Duke *Alzec\Alcio* and mentioned by Paul the Deacon in *Historia Langobardorum* V, 29; by Fredegar in his *Chronica* IV, 72; by Theophanes in the *Chronographia* 6171; and by Nikephoros Patriarch of

---

[294] I. Kovrig, "Contribution," 177.

[295] On the necropolis of Zamárdi see E. Bárdos, "La necropoli di Zamárdi," in *Gli Avari. Un popolo d'Europa*, 151, 153, 163; I. Bóna, "Gli Avari," 34; about Szeksárd see G. Rosner, "Das awarenzeitliche Gräberfeld in Szekszárd-Bogyiszlói Straße," *Monumenta Avarorum Archaeologica* 3 (1999): 12, 34, 48, 51, 54-55, 96, 154-56.

[296] I. Bóna, "Neue Nachbarn," 113; see also F. Daim, "Die Bayern, die Nachbarn der Awaren westlich der Enns," in *Reitervölker aus dem Osten*, 308-15.

[297] I. Bóna, "Neue Nachbarn," 111-13, 113-14.

[298] J. Oexle, *Studien zu merowingerzeitlichem Pferdegeschirr am Beispil der Trensen*, in *Germanische Denkmäler der Völkerwanderungszeit, serie A band 16* (Mainz am Rhein: von Zabern, 1992), 82, 99, Kirchheim/Ries, Ostalbkreis, n. 80-81, tab. 38, 142-43; Mühlhausen im Täle, Kreis Göppingen, n. 95, tab.41, p. 147; Mannheim, unpublished, tb 313, n. 93, tav. 41, 146; Pfahlheim, City of Ellwangen, Ostalbkreis, n. 114, tab. 56, 157-166; Rißtissen, Staddt Ehingen, Alb-Donau Kreis, n. 139-140, tab. 73, 167-68; Wilflingen, Kreis Biberach, n. 172, tab. 85, 175-76; Aschheim, Kreis München, n. 176, tab. 86, 177; Au, Gem. Rehling, Kreis Aichach-Friedberg, n. 177, tab. 86, 177-78; Friedberg, Kreis Aichach-Friedberg, n. 190-191, tabs. 92-93, 182; Geltendorf, Kreis Landsberg/Lech, n. 196, tab. 92, 183; Giesing, Stadt München, n. 198, tab. 93, 183; Künzig, Kr. Deggendorf, n. 203, tab. 95, 184; Ottmaning, near Ainring, Kr. Berchtesgadener Land, n. 218, tab. 99, 188; Regensburg, n. 222, tab. 105, 190; Budenheim, Kr. Mainz-Bingen, n. 267, tab. 123, 203; Flomborn, Kr. Alzey-Worms, n. 272, tab. 125, 204; Staubing, Kr. Kelheim, n. 254, 198; Windecken, Stadt Nidderau, Main-Kinzig Kreis, n. 334, tab. 151, 224; Bremen, Gem. Ense, Kr. Soest, n. 380, tab. 176-77, 240. See as well R. Christlein, *Die Alamannen. Archäologie eines lebendigen Volkes*, 66-67 on the cultural stream that from Lombard Italy and from Avar Pannonia reached initially Bavaria and only after Alamannia: as a matter of fact, the Alamans settled in the northern part of their territories or in the mountainous regions of Switzerland kept their oldest ethnic style for a longer period; 75 on the change of riders graves style after 650 AD when stirrup is imported; 76: "Auch bei den Steigbügeln wird deutlich, daß sie eher auf italisch-langobardische denn auf awarische Vermittlung hin zu den Alamannen gelangten."

[299] U. Von Freeden, "Das Grab," 7, 8-16, 19-20; Ead., "Das Frühmittelalterliche Gräberfeld," 559, 567, 596.

[300] B. Genito, "Materiali e problemi," 55, 57.

Constantinople in his *Breviarium*, even if they cannot be directly identified with the people buried in the two necropolises do, in fact, represent the deep cultural and commercial exchanges between these regions. These relationships, therefore, also imply the movement and subsequent settlement of large parts of populations for military and strategic reasons.[301]

The necropolis of *Környe* (Hungary) contains material evidence of the complex relationships between peoples during the Migration Period in the Carpathian Basin, the eastern Merovingian area, especially Lombard Italy, and the Byzantine East. Moreover, in the case of this cemetery, it was possible to carry out archaeometallurgical analyses on ten different iron items. Such information helps illuminate connections between peoples, particularly as regards iron production, technology transfer and the mobility of craftsmen. The burials began at the necropolis at the end of the AD 6th century. The 26 swords found at the cemetery gives it a strong military character. The general archaeological structure of the necropolis attests, that besides the dominant Avar population, the populace consisted of tribes from the South Russian steppes, the so-called proto-Bulgarians within the region. Other groups that have been identified in this region include a small population belonging to the so-called Ant tribe alliance, like the Slavs from the area of the Dnieper river; and former inhabitants of the region including surviving Samatians and Gepids, the Romanized population of *Pannonia* and some of the western Germanic tribes that had lived previously in this area. There were also groups who have been nominally identified as Lombards, as well as smaller groups of Lombards who stayed behind after most of their people had left. Ten iron artifacts from this necropolis underwent archaeometallurgical analyses: spectrographic, metallographic and hardness tests were performed on these objects. These objects include 5 swords (no.1 from grave 78; no. 2 from grave 97; no. 3 from grave 149; no. 4 and 5 from field survey work); part of an *umbo* from a shield (from grave 66); two spear points (no. 6 from graves 129 and no.7 from field survey walks); an axe (*francisca*, from grave 125) and; a stirrup (from grave 129). These iron objects can be divided into five different groups based on their production technology and the quality/character of their raw material:

1) The first group consists of iron artifacts with low carbon content, for example spear point no. 7 and the partial *umbo* from a shield (consisting of almost all *ferritic* iron).

2) Within the second group, the artifacts were manufactured using iron that was not homogeneously and, possibly thus unintentionally carburized. This was the situation for the stirrup and spear point no. 6 from grave 129.

3) A third group comprises sword no. 1 from grave 78 made from a very well carburized and hard steel which had undergone intentional heat treatments, as revealed by its *sorbitic* structure.

4) The fourth group consists of the axe (the *francisca*, from grave 125) and sword no. 3 from grave 149. These artefacts were carburized through *Cementation* and later tempered in order to obtain an almost homogeneous structure. Actually, the metallographic analysis has shown the presence of *Martensite*, *Troostite* and later *Sorbite* within the iron structure.

5) The last group comprises artifacts with the most complicated production technology and consists of swords no. 2 from grave 97 and swords nos. 4 and 5 found during field survey. Here, beside heat treatments, there is also evidence of pattern welding of iron bands with different carbon contents.

Thus, the results of these analyses show that different production processes were employed to manufacture different objects and that all of the objects were produced from a raw material basically lacking any Phosphorus content with the highest proportion among the analyzed objects was, in fact, only up to 0.12%. The technology used in production was not, however, unconscious. Thus, the smith carefully used all his arts to produce objects that would properly fit their functions. Actually, there is no need for a stirrup or a shield handle to be subjected to heat treatments in order to be used efficiently whereas it would be essential for a sword. It is, however, difficult to determine whether all of these objects were produced in different workshops or in the same place but with different production strategies. The general technological features of these iron objects, as revealed during archaeometallurgical investigations, has numerous analogies with artifacts produced within the so-called Lombard milieu in the middle of the AD 6th century in Central Europe (i.e. items from the settlement of *Březno* and the necropolises of *Hegykő* and *Kajdacs-Homokbánya* and *Tamási-Csikólegelo*). The presence of *Martensite*, *Troostite* and later *Sorbite* within the iron structures of the axe and of swords no. 1, 3 and 4 (the most similar to the *Hegykő* type since pattern welding is found in those objects together with *troostitic* and/or *sorbitic* structures), indicates that these objects were manufactured with high standard production technology through heat treatments. In fact, the production processes used for the artifacts from the necropolis of *Környe* share a certain degree of technological correspondence particularly with samples no. 619, no. 620 (knives) from *Březno*, the sword # 65.34.1 from *Hegykő*, the handle from a shield (# 75.22.2) from *Kajdacs-Homokbánya* and *Tamási-Csikólegelo*. This artifact was not subjected to such intense smithing as with the object from *Környe*. Moreover, no sulphur content could be detected in this last tool. These technological analogies should be regarded as proof of a wide-ranging technological transfer within a vast area including the Avar kaghanate and the eastern Merovingian region and even, blacksmiths' mobility within the same territory. This hypothesis may also find support on comparisons based on the form of some tools found at *Környe* such as the

[301] In relation to this see H. Ditten, "Protobulgaren und Germanen im 5.-7. Jahrhundert," *Bulgarian Historical Review* 3 (1980): 69-73, 75-76; L. Capo ed., *Paolo Diacono. Storia dei Longobardi* (Vicenza: Lorenzo Valla, 1992), 548-49: N. Christie, "Longobard weaponry," 16-17.

"curved" knives (*Krummesser*). In terms of the ways these objects were made there was a fairly strong correlation with objects of both eastern and western Germanic origin. The *Krummesser* from graves 66, 75 and 130 are extremely important since aspects of their manufacture are very similar to productions modes from objects found in the grave of the smith from *Poysdorf*.[302]

To summarize, the eastern Merovingian area, that is to say Lombardy, Alamannia and Bavaria, apparently played a relevant role in the importation of artifacts and technologies of oriental origin into Western Europe, totally unrelated to the Graeco-Roman heritage. Far from being a simple periphery of the Merovingian world, this zone played an essential mediation role in exchanges with the East, including territories under Byzantine rule, between the end of the AD 6th and the mid of the 7th centuries. The frequency and the profundity of these commercial, cultural and military connections, also implying migration of whole peoples and family kinship, between and among Avars, Lombards, Alamans and Bavarians, lent this area a strong degree of homogeneity. The dynastic bonds between Lombards and Bavarians are too well known to deserve further explanation here. However, it is important to stress that during the AD 7th century, iron artifacts and jewelry consistently passed through Lombard Italy to Bavaria from whence they continued on via Danube and its southern tributary rivers. It is not by chance that all finds of weights and scales are strictly located all along the courses of the Danube, the Rhine and their tributaries. As far as Lombard-Alaman connections are concerned, the recent conclusions of G. GRAENERT are relevant. According to this scholar, contacts between these two peoples were not solely restricted to military or commercial relations and, on the contrary, involved the creation of family links as possibly mirrored by the presence in Alaman necropolis of women of 'Lombard' origin, on both side of the Alps. This situation may have been supported by an early Lombard migration, already starting at the end of the AD 6th century, directly from their Pannonian settlements into the most eastern parts of Alamannia such as the Lech valley.[303]

Finally, iron stirrups as well as weapons were produced by Avar artisans employing sophisticated smithing technologies: a rhomboidal iron bar was flattened by hammering thus acquiring its typically roundish form at the stanchion level. Evidence of the existence of this specialized craftsmanship among the Avars comes from grave 166 in the *Jutas* necropolis and at iron making sites in *Zamárdi* and within all of *Somogy* County. The necropolis of *Jutas* also has a mixed burial context containing skeletons of both Caucasian and mongoloid human types from a period between the end of the AD 6th and the beginning of the 7th century.[304] The direct import of this technology of production, beside the objects themselves, into eastern Merovingian regions is still difficult to prove because of the scarcity of Italian finds and the lack of archaeometric analyses on these objects which would be the only way to separate regional variations within a producing technology.

4.2.0 Iron making in Italy during the Lombard Period

It is possible to evaluate the Lombard influence on the system of exploitation and processing of metals based on archaeological evidence and written sources from the period of the Lombard Kingdom in Italy. There are direct mentions of this metallurgical activity in sources. These mentions are often concentrated in the same geographical areas that were traditionally considered mining zones. These include Piedmont, Lombardy and Tuscany, where traces of human influence were always a part of the cultural landscape. Both written and archaeological evidence show the complexity of technological

---

[302]On this topic see, J. Piaskowsky, "Metallkundliche Untersuchungen an Eisengegenständen aus dem Gräberfeld von Környe" *Acta Archaeologica Academiae Scientiarum Hungaricae* 26 (1974): 117-30, see particualrly 120 ff, 123; A.Salamon and A. Cs. Sós, "Pannonia – Fifth to Ninth Centuries," in A Lengyel and G.T.B. Randan eds., *The Archaeology of Roman Pannonia*, 407-08; A. Salamon and I. Erdély, *Das völkerwanderungszeitliche Gräberfeld von Környe*, 55, 57, 66-67.

[303] On this paragraph's topic see, F. Moosleiter, "Handwerk und Handel," in H. Dannaheimer and H. Dopsch eds., *Die Bajuvaren*, 219; I. Bóna, "Neue Nachbarn," 108; H. Steuer, "Handel und Ferbeziehungen," in *Die Alamannen*, 391, 400, 408-10, 410-14; V. Babucke, "Nach Osten bis an den Lech," in *Die Alamannen*, 256; I. Stork, "Friedhof und Dorf, Herrendorf und Adelsgrab," in *Die Alamannen*, 295-99, 301; G. Barni, "Alamanni nel territorio lombardo," *Archivio storico lombardo* n.s. 16 (1938): 137, 140, 144-50; G. Graenert, "Langobardinnen in Alamannien. Zur Interpretation mediterranen Sachgutes in südwestdeutschen Frauengräbern," *Germania* 78 (2000): 418, 422-23, 426ss., 431, 432-34; J. Werner, "Fernhandel und Naturalwirschaft im östlichen Merowingerreich nach archäologischen und numismatischen Zeugnisse," *Bericht des Römisch-germanischen Kommission* 42 (1961): 310 ff.; Id. "Zur Verbereitung frühgeschichtlicher Metallarbeiten (Werkstatt-Wanderhandwerk-Handel-Familien Verbindung)," *Early Medieval Studies* 1 (1970): 65 ff.; H. Schutz, *Tools*, 198: "Thus the presence of northern square-based Lombardic fibulas in the Bavarian cemeteries indicates that Lombards may have participated in the tribal genesis of the Bavarians." The homogeneity of many aspects of the eastern Merovingian material culture and the importance of the oriental influences, running along the Danube, is confirmed also by pottery production as noted in O. Von Hessen, "A proposito della produzione di ceramica nel periodo delle migrazioni nell'Europa centrale e meridionale," in *Artigianato e tecnica*, 757-58: "Vorrei parlare ora di un terzo prototipo di ceramica dell'epoca merovingia, proveniente dalla Germania meridionale e più precisamente dalle zone bajuvariche ed alemanne. Questo prototipo è particolarmente importante per la ceramica longobarda trovata in Italia … Si tratta … di bicchieri e di e vasi a forma di otre … Le forma di questo gruppo di recipienti … non trovano riscontro nella Germania settentrionale. Questo tipo di vasi non è di provenienza occidentale, bensì di origine orientale. La sua diffusionenella Germania meridionale segue più o meno il corso del Danubio. Questi vasi a forma otre … sono quasi sempre decorati da stampigliature e sono modellati a mano. I loro parenti più prossimi … formano il gruppo principale della ceramica longobarda in Italia, con la differenza però che quest'ultima, contrariamente a quella proveniente dagli scavi della Germania meridionale, è modellata al tornio. Ma ambedue i gruppi risultano probabilmente dalle stesse influenze orientali."

[304]I. Kovrig, "Contribution," 164; I. Bóna, "Neue Nachbarn," 109; I. Bóna, "Gli Avari," 25-26, 37; G. Rhé and N. Fettich, "Jutas und Öskü, zwei Gräberfelder aus der Völkerwanderungszeit," *ΣΚΏΘΙΚΑ* 4 (1931): tab. IV, pics. 12-20; J. Gömöri, "Nemeskér-type iron smelting workshops from the time of Onogur colonization of Pannonia. Excavations in Zamárdi," in J. Gömöri ed., *Traditions and Innovations in the early Medieval Iron Production*, 149-59; Id., "Az avar kori és X-XI századi vaskohászat régészeti emlékei Somogy megyében" [*Archaeometallurgical site of Somogy county during Avar and early Arpadian period*], *Somogy Megyei Múzeumok Közleményei* 14 (2000):163-218.

developments within which diverse technical-cultural traditions interacted.

In the territory of Piedmont, particularly in *Canavese*, archaeological research has revealed the presence of a significant quantity of iron slag from prehistoric times to the Roman period, then from the Early Middle Ages and beyond. [305] Some among these numerous settlements (such as *Peveragno*, *Centallo*, *Misobolo*, *Belmonte*) are important from the point of view of chronology as well as their general archaeological structure. Many archaeological traces of iron making (much slag and many half-finished products) and other evidence of metallurgical processes (lead ingots) as well as metallurgical tools in general (a copper alloy anvil, small hammers and a steel chisel), not to mention tools for carpenters and for agricultural work have been found at the settlement of *Peveragno*, dated from the end of the AD 4th to 6th century, especially in the later phases of the village (**see figure 37**). The tools found in this settlement are comparable to the set of instruments typical for the contemporary culture of Merovingian Europe, as it has been reconstructed on the basis of grave goods found in a Frankish-Merovingian grave in *Hèrouvillette* and 'Lombard' graves from *Poysdorf* and *Brno*. Nevertheless, the hammers belong to a Mediterranean morphological type.

A grave of a 45-50 year old man found inside the cemetery basilica of *Centallo* (second half of the AD 7th century) marks an interesting case. The deceased had iron tools at his feet including an anvil, a hammer and an unidentified iron tool, all implements used for processing metals. Here, we can see again the presence of Germanic and Christian, that is to say, local elements existing side by side. Thus, while the disposition of tools at the feet of the buried man points to a typically Germanic ritual, confirmed by examples from *Hèrouvillette* and *Poysdorf*, the inhumation within a church and the clearly Mediterranean typology of the hammers attest the influence of the local culture. Moreover, the features of the inhumation seem to indicate the importance of the buried craftsman in his community. [306] In fact, metal workers and blacksmiths in particular, enjoyed a privileged position in the Germanic world as can be deduced not only from ancient Nordic sagas but also from Roman-Barbarian legislature. Thus, the Lombard situation, where metal-workers often appear as free men involved in real estate transactions, is not exceptional.[307]

In 1985, a bowl furnace for iron production and several pieces of slag were found near *Misobolo*. The structure of the furnace (**see figure 38 and 39**) and the settlement area where it was located can be dated between AD 7th and 8th centuries, thus fully within the period of the Lombard domination. The settlement of *Misobolo* is located in the *Canavese* valley, 20 km far from iron ore sources (**see figure 40**). Given the geological configurations of the place (a quaternary substratum) and its surroundings, it would have been impossible to get iron from a closer location. Therefore, there must have been such a strong market for metal that the transfer of minerals from the ore source to where it was used was encouraged. The bowl furnace of *Misobolo* must have been the central element in a craftsmen's production system that was closely linked to local needs.

An even more significant archaeological example reflecting the relationship of Lombard settlements with iron making may be found at the *castrum* of *Belmonte*. Two successive excavations (1968-75 and 1986-1994) have shown the site to be extremely important in terms of its archaeometallurgical features, particularly for several implements which seem to be absolutely new to the Mediterranean tradition. Numerous finds from the site, settled from the middle of the AD 5th to the middle of the 7th century, attest a Lombard phase. The quality and the characteristics of all the ceramic remains indicate, on the one hand, the reduced scale of contacts through commerce, and the existence of local ceramic production, on the other. This specific production process is also attested by specific discard material found at the same site. Altogether, archaeological finds and the presence of slag from production activities confirm that this site enjoyed a wide-spread and variegated culture of craftsmanship, including metallurgy, wood, textile and ceramic production, contributing to an economic well-being based on agro-pastoral activity. Iron tools, which are exceptionally numerous and varied, have been partly found in two cubby holes and partly spread over the entire site, inside as well as outside houses. Among the objects discovered here, there were mining instruments, (picks, a mining bit, a crowbar), a trivet, a horse bit, a pair of smithing tongs, six ploughshares, and an iron axe with a hammer-like back, the typology of which is well attested in Italy around the end of the AD 6th and the beginning of the 7th century, a sax, a spear tip, an arrow, four iron rivets from a belt, a fifth rivet with *agemina* decoration, and two iron buckles. A glass paste bead and a pierced coin were found elsewhere on the site. A bronze *fibula* (brooch) in the form of a cross placed on a dove, typical for female graves of the AD 6th-7th centuries was found in the southern part of the settlement, together with the already mentioned ploughshares. The abundance of iron implements, weaponry and slag at this site indicate the presence of a workshop, probably connected to a patron, as can be deduced from a particular residential area fit for aristocratic habitation.[308] There was a strong

[305] M. Cima, "Metallurgia," 180; E. Micheletto and L. Pejrani Baricco, "Archeologia funeraria," 300-01; on the Piedmontese situation see as well V. La Salvia, *Archaeometallurgy*, 27-29; Id., "L'artigianato metallurgico," 18-20.

[306] E. Micheletto and L. Pejrani Baricco, "Archeologia funeraria," 315-16, 334-35.

[307] On the importance of the smith in Germanic and in general in Medieval societies see, V. La Salvia, "Archaeometallurgy as a Source," 110-11; Id., "L'artigianato metallurgico," 19 especially note 51; Id., *Archaeometallurgy*, 30-31; F. Zagari and V. La Salvia, "Aspetti della produzione," 870-71; C. Giostra, *L'arte del metallo*, 13-22; P. Galloni, *Il sacro artefice* (Roma/Bari: Laterza, 1998), 26-27, 243-44; Cima, *Archeologia del Ferro*, 37, 51.

[308] M. Cima, "Metallurgia," 173, 177, 179, 181, 187, 188, 189; Id., "Le Origini della Metallurgia," 113, 119; E. Micheletto and L. Pejrani Baricco, "Archeologia funeraria," 318-24.

Lombard military presence in the *Canavese* area (*Misobolo* and *Belmonte*), which was incorporated in the Lombard Duchy of *Ivrea*, since it was contained within Frankish territory in Valle d'Aosta.[309]

The Piedmont finds indicate the presence of iron making activity with Germanic characteristics which was far from diminishing in the Lombard period. The *Belmonte* case, where the finds include tools for extracting minerals, slag and finished products, permits two important observations. First, activities related to a complete iron production processes from extraction of the mineral to the finished products. Secondly, this production was directly related to Lombard technical elements, a fact that is confirmed by the presence of iron tools, that seem "exotic" with respect to Mediterranean traditions.[310]

Lombardy, particularly the valleys of *Bergamo* and *Brescia*, is home to the most important iron ore sources (carbonate of iron).[311] The exploitation of these resources was already of great local importance in Late Antiquity, as can be surmised from a AD 5th century hagiographical source. According to the *gesta* of St. Virgil of Funto, this saint destroyed the statue of God *Tillino* in the *Trompia* valley. The statue was made of the local iron.[312] The valleys of eastern Lombardy were intensively used and occupied by the Lombards. According to tradition, Lombards had reinstituted *res publica Cammunorum* around the city of *Civate*, where there are, in fact, important indications of the Lombard presence. These include a rustic portal made of sandstone extracted from local quarries and the remains of a church rebuilt in the AD 7th century.[313] Moreover, the system of Lombard settlements covered the entire area between the rivers of *Oglio*, *Mella* and *Chiesa*, and ended at the natural border of the Alps. Their presence here is attested by the abundance and widespread distribution of archaeological evidence in the form of settlements, graves and necropolis. Thus, the Lombard element deeply penetrated this area after the invasion of AD 568 and throughout the AD 7th century. This area was connected to the Brescian valleys and the lakes of *Iseo* and *Garda*. In addition, two of the most important Brescian monasteries – those of *St. Giulia* in *Brescia* and *Leno* – held estates located within this territory. Even if they are documented only from the AD 9th to 10th centuries, the quality and distribution of their lands must have played an important role in the resurgence of an area that already had defined economic and commercial characteristics in the Lombard period.[314] The analysis of the formation of *St. Giulia* monastery's patrimony, allows us to date the relationship between it and the mines of the Brescian valley to at least the second half of the AD 8th century, based on the presence of *ferramenta*, already mentioned in the document of foundation issued by King *Desiderius*. The monastic possessions in the mining area of the *Mella* River must date from the same period as well.[315] Moreover, two

[309] F. Gabotto, "Un millennio di storia eporediese," *Eporediensia, Biblioteca della società storica subalpina* 4 (1900): 9-10.

[310] See F. Zagari and V. La Salvia, "Aspetti della produzione," 871-86; V. La Salvia and F. Zagari, "Cultura materiale e tradizione tecnica," 970-73.

[311] About Lombardy in the given period see, V. La Salvia, "L'artigianato metallurgico," 20-22.

[312] G. Rosa, "Sulle miniere di ferro della Lombardia," *Commentari dell'Ateneo di Brescia* (1843): 161-70.

[313] G. Bonafini, "Due reperti altomedievali in Valle Camonica," in *Miscellanea di Studi bresciani sull'Alto Medioevo* (Brescia: Comitato Bresciano per l'ottavo congresso internazionale dell'arte dell'alto medioevo, 1959), 117, 124; U. Vaglia, "Tracce di vita longobarda in Val Sabbia," in *Miscellanea di Studi bresciani sull'Alto Medioevo*, 127-28; F. Oodorici, *Storie bresciane dai primi tempi fino all'età nostra* (Brescia: 1854), vol. 2, 237-38, vol. 3, 116 ff.

[314] P.M. De Marchi, "Calvisano e la necropoli d'ambito longobardo in località Santi di sopra. La pianura tra Oglio, Mella e Chiese nell'altomedioevo," in L. Paroli ed., *L'Italia centro-settentrionale*, 388-89. Charters of a later period mention the possessions of several monasteries in this region tied with iron production: for Bobbio, a *breve* by Abbot Wala of 835 and a later one dating to 862 mentions the iron supplied to that monastery by its *curtes* at Luliatica and at Sarlasco. Moreover, an *auctoritas* by king Berengarius 1st dated to 903 confirms to the abbot of Bobbio Teodelassio the apostolic privileges and the *praecepti* obtained from Lombard kings, Charles the Great, Lotharius 1st, Lotharius 2nd, Carlomanno, Charles 3rd and Arnolfus; all that stresses how long-lived was the ownership the monastery had on that productive territory, listing among the confirmed possessions both of the *curtes* mentioned in the 9th century *breves* above as suppliers of iron to the cenobium established by S. Colombano. In the 10th century the monastery of Sant'Ambrogio in Milan imposes a supply of iron per year to its servants in the *curte* of Limonta, given to it since 835 by Lotharius 1st; G. Tiraboschi, *Storia dell'augusta Badia di Nonantola, aggiuntovi il codice diplomatico della medesima*, t. 2, 90 doc. LXVII (a. 907), mentions a document in which the blacksmith Godeperto, a Lombard name, submits himself to the abbey's authority. For the data presented above see, *Codice diplomatico del monastero di S. Colombano di Bobbio*, vol. 1 (Rome: 1918), respectively docc. n. XXXVI, 137 ff. and n. LXIII, 184 ff., doc. n. LXXXI, 272; V. Fumagalli, "Strutture materiali e funzioni dell'azienda curtense. Italia del Nord: sec. VIII e IX," *Archeologia medievale*, 7 (1980): 26; F. Carli, "Il mercato nell'età del Comune," in *Storia del commercio italiano* (Padova: CEDAM, 1934) vol. 2, 69-70 and G. Rosa, "Metallurgia storica bresciana," *Commentari dell'Ateneo di Brescia* (1877): 92; *Lotharii I et Lotharii II diplomata*, T. Schieffer ed., in *M.G.H , Diplomata karolinorum*, 3, n. 23, 93.

[315] P.M. De Marchi, "Calvisano e la necropoli," 388-89. The monastery of Santa Giulia from the 9th to the 10th centuries A.D. received about 170 kg. per year from its *curtes* in the valleys around Brescia. A large fragment of a polyptych dating among 879 and 906 tells us that the *curtes* at Grilliano, Bogonago, Andalvico, Casivico, Mariano, Vassanigo and Bradella in Valcamonica, supplied the monastery with both iron tools and raw-iron. Moreover, an analysis of the very structure of the monastery's patrimony at Santa Giulia as well as the results of recent excavations in its area make largely acceptable the hypothesis of an even earlier exploitation of the mineral resources in the region. Since 760 indeed, among both movable and immovable properties confirmed by the Lombard king Desiderius are mentioned - beyond iron tools *(ferramenta)* - among the monastery's possession ... *Brada, curte ducales, prope fluvio Mella ...*, in the mining district of Val Trompia. In relation with the *curtes* mentioned in the above polyptych as metal suppliers, very relevant seems to be the *diploma* by Lotharius 1st of 837, mentioning possessions at Grilliano and in Valcamonica. This document does not establish a new situation but simply records a reorganization of the monastery patrimony as carried over around 813-814 by two abbots and two bishops; incidentally, the assessment is the result of an on-the-spot *inquisitio* of the patrimony rather than of a mere documental survey. With regard to these questions see, V. Fumagalli, "Strutture materiali e funzioni dell'azienda curtense. Italia del Nord: sec. VIII e IX," *Archeologia medievale*, 7 (1980): 26; G. Pasquali, "Santa Giulia di Brescia," in M. Luzzatti, G. Pasquali and A. Vasina, *Inventari altomedievali di terre, coloni e redditi*, (Rome: Istituto Storico Italiano per il Medio Evo, 1979), 36-37, 49-50, 54, 63-65, 69, 71-73; *Codice diplomatico Longobardo*, 3, 1, n. 33, (Pavia 4 ottobre 760), 206-207; *Lothari I et Lothari II diplomata*, ed. T. Schieffer, in *M. G. H., Diplomata Karolinorum*, 3, n. 35, 112-15; G. Pasquali, "Gestione economica e controllo sociale di S.Salvatore-S.Giula dall'epoca longobarda all'età comunale," in C. Stella and G. Brentegani eds.,

pieces of slag from iron production have been found in *Ortaglia* area (sector Y2) during the excavation campaign of 1986/87, within the perimeter of the Monastery of *St. Giulia* at *Brescia*. Those can be dated to a period between the middle of the AD 6th to the beginning of the 7th century. These slag specimens attest the presence of a blacksmith, rather than iron reduction activity, even if the workshop was not identified.[316]

The discovery of another metallurgist's grave within the Lombard necropolis of *Porzano di Leno* (Province of *Brescia*, Lombardy), more exactly grave 224, is equally important. The cemetery is located along a five kilometers path in the countryside linking the villages of *Porzano* and *Leno*. These two rural communities are already attested in the AD 8th and 9th centuries as possessions of the monastery of *S. Salvatore*, *S. Maria e S. Michele* founded by King *Desiderius* in AD 757 and as a *curtis* of the royal monastery of *S. Salvatore* and *S. Giulia* of *Brescia*. The area within the cemetery is generally well organized. The presence of empty spaces between diverging zones with different concentration of graves indicates that the cemetery was originally established by separate groups of individuals or families that only gradually began to mix with each other. Here, 249 graves have been brought to light. Other 93 tombs were found close by in *Campi S. Giovanni*. The life of the necropolis was rather long. Actually, it stretches from a time around the end of the AD 6th to the second half of the 7th centuries. There are 15 graves (6% of the total) belonging to the earliest phase. These graves have a peculiar structure known as "house of death." These graves were constructed as large, deep rectangular holes in the ground with wooden posts vertically driven into each corner so that on the surface they may have served as a support for a fairly large wooden construction. Such structures are typical of the Pannonian phase of the 'Lombard' migration whereas such structures in Italy are only known at the Friulian cemetery of *Romans d'Isonzo*. However, among the different materials constituting the mass of the grave goods, iron is the most represented of the raw materials. Of the over one thousand objects, 625 comprise iron swords, scramasaxes, knives, arrow heads, spears and parts of shields, spurs, *armillae* and other small, often decorated, artifacts. This impressive amount of iron, thus, clearly indicates that iron making played an important role in the economy of this area. In this respect, grave 224, as the sepulture of a smith, is important as an indicator of metallurgical activity. The richness of the grave goods reflects the high status of the deceased. The grave goods included a sword and a scramasax, plus the fragments of their respective cases; a spear point; the iron handle and the umbo from a shield; multiple iron fittings of a belt, and other objects of the same sort from a bronze belt; four iron knives; a file; a pair of iron scissors; an iron anvil and an iron ingot. The anvil is quite similar to the one discovered in the grave found inside the cemetery basilica of *S. Gervasio* of *Centallo* (second half of the AD 7th century). The iron ingot matches those found at *Castelvecchio di Peveragno*, even though these latter are much thinner and elongated.[317]

It is true, on the one hand, that such finds indicate the "destructuration" of iron making activity as a result of upheavals in the economy, system of transport, as well as the social and political organization of the period. On the other hand, it should not be forgotten that metallurgy demanded an uninterrupted transmission of technical-operational notions, especially in areas with ancient metallurgical traditions. Such an uninterrupted transmission being documented, half-finished metal objects must also have been used as a means of exchange, at least in periods of economic restrictions and restructuring.[318]

In the Lombard area, the most interesting results were obtained from the excavation of two iron furnaces from the Lombard period near *Bienno* (Province of *Brescia*) and dated by the ^{14}C method. Here, tapped and compact slags were found as well, along with the remnants of a charcoal pit as part of the production site. It is significant that in the same area, Late Antique furnaces have also been discovered and analyzed. The analysis of slag from the area permitted comparison of different production techniques used in the same area for the same raw material in two different periods. The complete set of archaeological and archaeometric data have, thus, enabled scholars to reconstruct the operational chain of iron production in two distinct historical phases. As a result, it has been possible to state that the technology used in the Lombard period was significantly different from the Late Roman period. The goal of Lombard technology was to achieve higher temperatures rather than the improvement of reducing conditions inside the furnace.[319]

Among written Lombard sources a charter from Milan (6-6-725), related to a certain weapon maker *Theotperto – vir honestus* – is extremely interesting, especially since, as already noted by U. MONNERET DE VILLARD, it is from this city that comes the first early medieval document

S.Giulia di Brescia. Arte storia di un monastero dai Longobardi al Barbarossa (Brescia: Comune di Brescia/Grafo, 1992), 133-37.

[316]T. Mannoni, A. Cucchiara and A. Rabbi, "Scorie e forni di S.Giulia e la metallurgia nel medioevo," in C. Stella and G. Brentegani eds, *S.Giulia di Brescia*, 212-14.

[317] P.M. De Marchi and A. Breda, "Il territorio bresciano in età longobarda e la necropoli di Leno", in C. Bertelli and G.P. Brogiolo eds., *Il futuro dei Longobardi. L'Italia e la costruzione dell'Europa di Carlo Magno* (Ginevra\Milano: Skira, 2000), 473, 479, 488-90, pic. 333.

[318]T. Mannoni, A. Cucchiara and A. Rabbi, "Scorie e forni," 212-14; G.P. Brogiolo, "Trasformazioni urbanistiche nella Brescia longobarda: dalle capanne in legno al monastero regio di S. Salvatore," in C. Stella e G. Brentegani eds., *S.Giulia di* Brescia, 184.

[319]See, C. Cucini Tizzoni, "Un forno da ferro longobardo nelle Alpi italiane: Ponte di Val Gabbia - Val Camonica (Bienno, Brescia)," *Notiziario di Archeologia Medievale* 65 (1995): 8-9; C. Cucini Tizzoni and M. Tizzoni, "Early Ironworking in northern and central Italy. A critical review of the evidence," in *Proceedings of the International Conference Early Ironworking in Europe. Abstracts 19-25/09/1997* (Plas Tan y Bwlch (Wales, U.K.): 1997), 40-42.; C. Cucini Tizzoni and M. Tizzoni, *La miniera perduta* (Bienno: Comune di Bienno, 1999), 93-139 on Late Antiquity and 141-67 on the Lombard one, see particularly, 162-63.

(10th c.) attesting to a market where weapon makers played a significant role. It is possible that such a market organization goes back to the AD 8th century given that at that time Milan already had a lively market, able to satisfy the needs of the numerous pilgrims who visited the city, as mentioned in *Versum de Mediolano civitate* from AD 739. Other blacksmiths from Lombardy are mentioned as living in *Pavia* and *Monza.*[320]

Finally, we must consider the vast quantity of iron objects excavated in Lombard necropolises in the Brescian area. Given such a quantity of this metal, it is difficult to believe that we are always dealing in any way with the reuse of scrap iron material. On the contrary, it is more plausible that iron finds represent a direct indicator of mining activity. To remove great quantities of metal from the living and deposit it among funerary goods would not make sense as a logical choice for any population unless people were convinced that their "utensil metal property" would benefit from increased efficiency due to new conditions either in processing raw material or in the finished products[321]

Based on archaeological data, it appears that metal production in the northern parts of the Italian Peninsula was sufficiently developed during the Lombard period and capable of meeting market demand. This is confirmed by the quantitative analysis of iron products from this area carried out by LIBORIO ET.AL. (**see figure 41**). This work does not support the hypothesis that there was a decreased availability in iron in any given period from the period of Emperor Augustus until the 11th century.[322] The contribution of the Lombard metallurgical tradition seems to have been an integral part of this process of restructuring of iron production cycles in northern Italy in the Early Middle Ages. This is particularly true given that it was most likely in the Lombard period that this whole area – the valleys of *Bergamo* and *Brescia* – was restructured into a single economic unit centered around mineral exploitation.

The signs of past mineral extraction in Tuscany –largely attested archaeometallurgical from the 2nd century BC until the AD 5th century – must evidently have been an important center of metal working[323] linked to the renewed demand in metals stimulated, by the new Lombard "purchasing power."[324] From this region, or more precisely from the territory of *Lucca*, there exist four documents written between AD 742 and 773, that mention *calderarii*, i.e. copper workers. One of them deals with a property situated *in loco Cicina*, an area rich in copper ores near *Volterra* and not far from *Campigliese*, where there are possibly the only important tin veins in the Italian Peninsula (near *Monte Valerio*). There is important archaeometallurgical evidence from the territory of *Lucca* which confirms that in this particular location metal production was mainly based on copper alloys, something which the *Colline Metallifere* of the surrounding area possesses in abundance. Other relevant examples include the waste material from a workshop in *Lucca* that produced copper alloys between the AD 6th and 7th centuries in a location that already in the 4th century was involved in metallurgical activity. Traces of copper alloy production have also come to light in the stratigraphic features of the last phases of the *Domus dei mosaici* at the site of *Roselle* (Province of *Grosseto*). Again, on the basis of typological analysis of metal finds such as fibulae and earrings, it appears possible to recognize the work of one or more workshops that operated between the AD 6th and 7th centuries based on Mediterranean technologies and models aimed at the new Lombard or Lombardized commissioners. Thus, it may not be an accident that a 7th century manuscript concerned with the metallurgy of non-ferrous metals, is preserved in the archives of the Church of *Lucca* contained in a part of *Liber Pontificalis* from *Codex Lucensis 490* published by L.A. MURATORI in 1739.[325] In light of this, the inclusion of the territories that stretch from *Monteverdi* to *Massa Marittima* – an area rich in ores and woods of essential importance to metallurgical production – within the *iudiciaria lucense* demonstrates the interest of the Lombard aristocracy of *Lucca* in controlling this area for the use of its natural resources. The Lombard presence is clearly attested by the presence in this area of the monasteries of St. *Pietro di Monteverdi* and St. *Regolo*, constructed by the aristocracy of *Lucca*. Thus, the annexation of this area to the juridical district of the city of *Lucca* could not have been because of its possible strategic importance since these territories have no direct link to the territory of *Lucca*. In addition to this, the discovery of a coin of the Lombard period in the locality of *Montebamboli* close to *Massa Marittina*, in the core of the *Colline Metallifere* region is of importance

---

[320]For the documents on smiths of the Lombard period see, *Codice Diplomatico Longobardo*, vol.1, n. 36, 127; vol. 2 n. 218, 252; vol. 2 n. 231, 292; for *Versum de Mediolano civitate*, E. Duemmler ed., in *MGH, Poetae Latini aevi carolini* vol. I, 24-26. *5. Erga murum pretiosas novem habet ianuas,/**vinculis ferreis** et **claves** circumspectas naviter,/ante quas cataractarum sistunt propugnacula. ...16. Questu congrue ditantur venientes incole,/nudi quoque vestiuntur copioso tegmine,/pauperes peregrini saciantur ibidem./17 Rerum cernitur cunctarum speciebus inclita,/generumque diversorum referta seminibus,/vini copia et carne adfluenter nimiae./.* On the market of Milan during the early Middle Ages see U. Monneret De Villard, "L'organizzazione industriale," 32-33.

[321]On the iron findings of the area of Brescia see, G. Panazza, "Note sul materiale barbarico trovato nel bresciano," in A. Tagliaferri ed., *Problemi della civiltà e dell'economia longobarda,* 137-70. K.D. Withe, *Greek and Roman technology* (London: Clarendon Press, 1984), 113 considers a huge amount of iron artifacts as an indicator of iron making activities.

[322]C. Liborio, T. Marinig and A. Storti, "*Regiones augusteae X et XI*: ritrovamenti di manufatti in ferro databili dall'età augustea all'età altomedievale," in *Dal basso fuoco all'altoforno*, 23-37, especially 26.

---

[323]The idea tha the Lombards could have started again the exploitatation of Tuscan natural resources was alerady expressed by M.L. Simon, "De l'explotation des mines et de la metallurgie en Toscane pendant l'antiquité et le Moyen âge," *Annales des Mines ou recueil de memoires sur l'explotation des mines*, 5th ser., 14 (1858): 606-07: "les barbares d'ailleurs sortaient presque tous des forêts de la germanie, où, s'il faut en croire Tacite, les mines étaient en explotation dès la plus haute antiquité. De là l'idée naturelle aux barbares de reprendre le mines toscanes."

[324] On the importance of Lombard commisioneers, V. La Salvia, "L'artigianato metallurgico," 16-18; C. Giostra, *L'arte del metallo*, 10; L. Paroli, "La cultura materiale," 270.

[325]See *Antiquitates Italicae Mediaevi* t. II, *dissertatio* XXIV coll. 373-374.

because it sheds new light on both the use of coins in the countryside, in areas wrongly considered peripheral, and on the penetration into and possible control of the Lombard aristocracy of *Lucca* in this mining district.[326]

The overall results of the survey of both archaeological and written evidence strongly suggest that during the Lombard period non-ferrous metallurgy played a role of paramount economic importance within the territory of *Lucca*. Moreover, out of eight documents on mints from the Lombard period, no less than six are related to Tuscany. The examination of coins from the Lombard period confirms the presence of heavy coin-minting activity in Lombard Tuscany. It is logical to link coin-minting with the exploitation of mineral resources in the area.

As far as iron production is concerned, we may say that the Roman iron making system ended with the Lombard conquest of *Lucca* and *Populonia* and the consequent Byzantine withdrawal to the island of *Elbe*, which became the most important naval base in western Mediterranean. As a consequence, this principal and traditional source of iron became thus inaccessible for the Lombard Kingdom. Nevertheless, it is known that in AD 726 there was a blacksmith in *Pistoia*, a center linked to *Lucca* while the activities of a blacksmith on Mount *Amiata*, attested for the earliest Carolingian period (AD 803), may have had their origin in the Lombard period.[327] In the case of mining activity in Tuscany in the Lombard period a territorial restructuring can be observed that takes into consideration the importance of mines in an area where the Lombard element may have already penetrated, as with territories stretching from *Monteverdi* to *Massa Marittima*.[328]

In conclusion and on the basis of the above data, some areas of Piedmont (especially around the *Canavese* area), Lombardy (particularly around *Brescia* and *Bergamo*) and to a certain degree Tuscany, can be considered exemplary cases of the development of systems for the exploitation of iron and its production in the immediate aftermath of Germanic invasions and settlement in former Roman Europe. This phenomenon is generally characterized by the association of metallurgical activity with agricultural production on the one hand and by the diffusion of small or modest production centers in the countryside, linked to local economic dynamics on the other. At the same time, from archaeological and written sources the importance of the heritage of metallotechnical knowledge of the Germanic peoples in general, and in Italy of the Lombards in particular, in restructuring iron production cycles becomes evident. It is not coincidental that such a restructuring occurred simultaneously with Germanic settlement within the boundaries of the former Roman Empire. The iron making production system evolved into something very similar to that which developed in the first centuries of the Christian era at numerous other locations with strong iron making traditions beyond the Imperial frontier and also inhabited by Germanic tribes.

Certainly, the phenomenon of a restructured production system is not solely confined to Italy. On the contrary, similar archaeological material exists elsewhere in Roman-Barbarian Europe. There is the case of the Swiss *Jura*[329] and other documents, such as one from the monastery of *Lorsch* from AD 788, which indicates that it received 30 kg of iron from the village of *Weilnau* every year.[330] This indicates a certain type of iron production organization existed everywhere in the continent.[331] Thus, the economy of former Roman territories of Europe immediately after the settlement of various Germanic populations was characterized by a tendency to regionalization. This entailed a change to extensive exploitation of iron in many centers of production everywhere over the territory. This production, however, never became centralized though sending large quantities of half-finished material to one production point. The coincidence of such restructuring of production cycles with the influx of the material culture, craftsmanship techniques and traditions of the Germanic peoples into the Classical blacksmiths' heritage, was no accident. This tendency, which in Italy is already attested in the Late Antique period,[332] seems to be endemic to the Roman West. The consequent diffusion of iron production centers over the entire territory, all of probable Germanic origin, are part and parcel of the whole process and fit

[326] L. Tondo, "Moneta longobarda di Montebamboli," *Archeologia medievale* 17 (1990): 761.

[327]On the economy of early medieval Tuscany see, C. Citter, "I corredi funebri nella Toscana," 185-211; Id., "I corredi nella Tuscia," 179-95; R. Farinelli and R. Francovich, "Potere e attività mineraria," 449; V. La Salvia, "Gap or Continuity?" 264; Id., *Archaeometallurgy*, 70; Id., "L'artigianato metallurgico," 22-23.

[328]P.M. De Marchi, "Note su produzione e scambi nella Lombardia di età longobarda," 284: "In Toscana l'epicentro della produzione di preziosi è Lucca e circondario, i documenti scritti vi ricordano 5 orefici, 3 calderai e monetieri, mentre a Pisa è attestato un orafo e a Pistoia un fabbro. La distribuzione dei mercanti è molto interessante, se confrontata con le attività produttive, perché sono presenti soprattutto a Lucca e a Pavia, e sembrano quindi convergere verso i centri di produzione e/o di smercio."

[329]This Swiss region, occupied by the Franks during the early Middle Ages, yielded evidence of iron production for a period stretching from the Iron Age to the Middle Ages. Within the site of Boulies, two bowl furnaces have been unearthed. These furnaces, dated between 550 and 650 AD, had a diameter around 60-70 cm. and were 1.5 m. high. Apparently, they attest a single production phase and the activity is linked to local needs, however, within a system of workshops widely diffused in the territory. In fact, similar examples from the same area came from at Merishasen, Bellaires, Prins-Bois, and Monteherand. Moreover, the local necropolises of the Merovingian period have given back numerous iron tools and weapons; on this subject see, L. Eschenlor and V. Serneels, *Les bas fourneaux Mérovingiens de Boécourt, les Boulies* (Porrentruy: Office du Patrimonie Historique Société Jurassienne d'émulation, 1991), 12, 13-15, 117, 120-21.

[330]See M. Cima, *Archeologia del Ferro*, 40, on the same argument, U. Zimmermann, "Mittelalterlicher Bergbau auf Eisen, Blei und Silber - Begrenzte Mittel und zahlreiche Veränderungen," in G. Magnusson ed., *The Importance of Ironmaking*, 275.

[331] See, M. Cima, *Archeologia del Ferro*, 121; and M. Cima, "Metallurgia," 189.

[332]See, J.C. Edmondson, "Mining in the later Roman empire," 84-102; an italian case is described in G. Murialdo, E. Bonora, C. Falcetti, F. Ferretti, A. Fossati, T. Mannoni, G. Vicino and G. Imperiale, "Il Castrum Tardo-Antico di S. Antonino di Perti, Finale Ligure (Savona): fasi stratigrafiche e reperti dell'area D. Seconde notizie preliminari sulle campagne di scavo," *Archeologia Medievale* 15 (1988): 355-96.

well into the general economic-productive context, characterized by a tendency to a "schismatic" economic structure, mainly centered on the demands of regional markets.[333]

[333]On the end of Late Antique economic system see, S. Mazzarino, *L'Impero romano*, 812-15

Conclusions. From acculturation to integration of material cultures.

On the basis of archaeometallurgical data presented in previous chapters, metal working, particularly iron craftsmanship in Italy, which flourished during the Lombard domination, sheds light on two fundamental issues. First, artisans in the region were organized so efficiently that they were able to supply the market with standardized products, at least from the middle of the AD 6th century. As a result, the Lombard Kingdom was able to take a leading role in the whole eastern Merovingian area in such production. It is true that commercial relations with southern Germany's Alamannic-Bavarian regions were carried out through a network based on exchanges with the Byzantines, who occupied most of the costal areas of the Italian peninsula, and on the import and diffusion of North-African and oriental products. However, local (Italian)[334] craftsmen's' products were a constant with production volume of such a high level as to make this trade an important component of all commercial exchanges on the peninsula, particularly as far as the production of shields, weaponry and golden objects were concerned. [335] Archeological finds from Lombard necropolises demonstrate that Byzantine-oriental commercial traffic penetrated deeply into the Italian peninsula also immediately after AD 568. Finds from *Cividale*, where the oldest Lombard necropolises in Italy are located, of vessels made of Byzantine glass, amber and alabaster, should be interpreted within this context. A large number of other archaeological data also demonstrates the existence of international commercial exchanges for the period between the end of the AD 6th and the beginning of the AD 7th century. North-African bronze bowls found at the necropolis of *Castel Trosino*, testify that commercial relations with this region continued as well. Finds from the necropolises of *Nocera Umbra* and that of *Castel Trosino* – African red slip ware and ivory objects –from North Africa, need to be studied along the same lines. Finds in Lombard necropolises in Italy, i.e. amphora for transporting highly prized wine from Eastern Mediterranean (Gaza or Cyprus) or oil from Africa even in the 7th century together with the last examples of fine African ceramics for table-wear as well as the *spatia*, Coptic bowls, ivory objects, and globular earrings (possibly of Syro-Palestinian rather than Avar origin), are found among the richest grave goods carried on overseas boats not as the main product of exchange but for the less extensive luxury trade. These objects indicate that even if there was a general economic recession, oriental workshops continued to function and even engage in international trade.[336]

At the same time, the Lombard Kingdom created a new market. Byzantine objects undoubtedly benefited from the demands of Lombard commissioners contributing to the diffusion of decorative models. The models themselves were also modified depending on the tastes of new commissioners. These models, then, were absorbed and integrated into the work of craftsmen in Lombard regions, where since the first half of the AD 7th century there are workshops that are able to guarantee a vast array of new products in silver, bronze and *agemina* and which are not that different from what could be found in the politically Byzantine controlled parts of Italy. It is clear, however, that this process of transformation that Lombard technical/artistic culture was undergoing, as a result of contacts with the Mediterranean area, presupposes continuity of craftsmanship and a stable workshop organization which allowed continuous up-grading of technical procedures aimed at modifying the products as the tastes of commissioners evolved, commissioners who had become gradually accustomed to the rights, the language and customs of the Mediterranean regions. Thus, the absorption of Roman-Byzantine models and their use within a wider artisan context takes into consideration technical-productive skills of Lombard craftsmen as well and assumes that the workshops were able to keep functioning continuously.[337] At the end of

[334] The local, that is to say Italian culture, cannot simplisticly and immediately be assimilated to the Byzantine one; Italian and Byzantine were not at that time regarded as synonyms; on these arguments see, S. Gasparri, *Prima delle Nazioni* (Rome: Carocci, 2000), 113-14, 117, 120, 121, 128, 141, 157, 158; N. Zanini, *Le Italie Bizantine. Territorio, insediamenti ed economia nella provincia bizantina d'Italia (VI-VIII secolo)* (Bari: Edipuglia, 1998), 100; L. Paroli, "Introduzione", 22; D. Harrison, *The early state and towns* (Lund: University Press, 1993), 213; and V. La Salvia, "Aspetti dell'economia dell'Italia alto medievale," 371-72 .

[335] On the commercial network between Italy and southern Germany and the leading role of the Lombard kingdom, see L. Paroli, "La cultura materiale," 270-71, 283, 285, 296-97; G. Panazza, "Note sul materiale barbarico trovato nel bresciano," in A. Tagliaferri ed., *Problemi della civiltà e dell'economia longobarda*, 176; J. Werner, "Langobardischer Einfluss in Süddeutschland während des 7. Jahrhundert im Lichte der archäologische Funde," in *Atti del I Congresso internazionale di Studi longobardi. (Spoleto 27-30 settembre 1951)* (Spoleto: CISAM, 1952), 521; on the diffusion of golden crosses see, Id., *Das Alamannische Fürstergrab von Wittischlinge*, Band 2 of *Veröffentlichungen der Kommission zur vergleichenden Archäologie römischer Alpen- und Donauländer* (Munich: Verlag C.H. Beck, 1950), 33; A. Taglaiferri, *Strutture sociali e sistemi economici*, 105.

[336] For the analyses and descritpions of the archaeological objects from the necropolises of *Cividale*, *Nocera Umbra* and *Castel Trosino* quoted in the paragraph see A. Melucco Vaccaro, *I Longobardi*, 150-52; Ead., "Agere de arte, agere per arte: la trasmissione dei saperi tecnici fra tradizione colta e fonti materiche," in *Morfologie sociali e culturali in Europa fra Tarda Antichità e Alto Medioevo. Atti della Settimana di Studio del Centro Italiano di Studi sull'Alto Medioevo, 25 (Spoleto 3-9 Aprile1997)* (Spoleto: CISAM, 1998), 369; M. Brozzi, "La più antica necropoli longobarda in Italia — Cronologia relativa e osservazioni economiche", in A. Tagliaferri ed., *Problemi della civiltà e della economia longobarda*, 122; G. Panazza, "Note sul materiale," 176; J. Werner, "Italienisches und Koptisches," 78.

[337] H. Roth, *Kunst und Handwerk im frühen Mittelalter* (Stuttgart, Theiss, 1986), 9, emphasizes this process in respect to the stylistic deveolpment of *Bugelfibeln*. On the same subject see, N. Åberg, *Die Goten und Longobarden*, 47; Tagliaferri, *Strutture sociali e sistemi economici*, 94; L. Paroli, "La cultura materiale," 285; S. Lusuardi Siena and C. Giostra, "L'artigianato metallurgico longobardo attraverso la documentazione materiale: dall'analisi formale all'organizzazione produttiva," in *I Longobardi dei Ducati di Spoleto e Benevento*, 925-27 and 929-30: "Quanto allo stile, l'ambiente cosmopolita che dovette contraddistinguere i laboratori metallurgici tardoantichi e più tardi gli ateliers di corte come quelli di Eligio orafo e monetiere alla corte merovingia di Clotario II e Dagoberto (629-639) ... , autorizza ad immaginare anche in Lombardia la compresenza di artisti di formazione mediterranea classica, 'bizantini,' accanto a rappresentanti della componente germaico-orientale e nordica già parzialmente integrati nell'ambiente della Milano tardoantica ed ostrogota e affiancati poi

this process of integration, which ended sometime in the AD 8th century, Lombard Italy society had certainly abandoned the custom of social distinction based on ethnic affiliation in favor of material criteria based on the abilities and wealth of the individuals. This was made explicit by Lombard legislation of the period. This legislation, especially from the second half of AD 8th, never included ethnic identity as an intrinsic part of social organization. Romans were never mentioned in previous legislation – i.e. within the famous *Edictus Rothari regis* of AD 643 – because of its 'tribal' character and not because there was an intent to isolate particular ethnic entities. Society in this early period had no ethnic barriers.[338] Thus, it is possible to speak not of two separate peoples living next to each other within the same territory but of two cultural traditions moving towards a kind of osmosis related to common socio-economic conditions encountered by "Italians", Roman/Byzantines and Lombards living both in Lombardy and in Byzantine controlled areas. There is no question that the overall situation in Byzantine Italy shared many similarities with "ruralized cities of Lombard Italy," especially with regard to the general impoverishment of economic life, decreased commercial activities, a widening gap between living standards of the well-to-do class – members of the military and ecclesiastical hierarchy – and those below, who move from an urban to rural model of life slowly but in a greater degree after the AD 7th century. Reciprocal acculturation (i.e. integration), therefore, would have had to have taken place during, or, more precisely, from the AD 7th century. This integration was not as much and not only an eventual exchange of opposite practices and customs, such as different funerary rites. Thus, it was not a simplistic process implying that one side (the Lombard side) began to lose its traditions, including changes in the types of objects used as funerary goods – and acquired those of the other side (the Roman side). Rather, these changes took the form of elaboration of new practices, growing partly in parallel and in common, and which can also be traced to the evolution of the style of inhumation in necropolises. Despite the undoubted recession between the end of the AD 6th and the first half of the AD 7th centuries, the situation on the Italian peninsula during the Lombard period seems to be characterized by an extremely dynamic and flexible society where the Lombard side was open to influences, contacts and exchanges from and with Byzantine civilization, the so-called "guardian" of the Mediterranean heritage. At the same time, it is exactly from the AD 7th century that a strong standardization of Lombard craftsmanship production can be observed, typical of belts for hanging swords decorated in the zoomorphic *agemina* style, for bronze belts with triangular plates and multiple belts decorated with harmonious animal *agemina*. This type of products, manufactured in Lombard milieu, would slowly substitute objects of Byzantine origin within Lombard grave goods. All of these artifacts had a wide diffusion in Italian territory, including Byzantine areas, without any break in continuity all the way through the AD 7th century. At the end of that period, in fact, the practice of depositing grave goods was abandoned, making it quite complicated for present-day scholars to trace the path of further developments. Thus, the AD 7th century marks a crucial moment in the socio-cultural transformation of early medieval Italy in general and of the Lombard Kingdom in particular. It is at this stage that a profound cultural integration takes place which would be indispensable for the artistic heyday of the AD 8th century, a phenomenon that found its maximum expression in Lombard territories.[339] In light of this, the laws of King Astolf, of the sixties of the AD 8th century, prohibiting commercial exchanges with Byzantine areas, reflect the necessity of breaking with normal commercial relationships at the peripheral regions of the Kingdom.
Moreover, with their coming to Italy across the Alps, the Lombards had transferred not only their armies but almost an entire people, including a certain number of craftsmen with their technical heritage, something that was manifested by the tomb of a blacksmith in *Cividale*, a grave which dates to the time of the first immigrant generation.

As demonstrated in Chapter 3, this technical organization and apparatus of these smiths itself had a long and complex formation period. Its core may be traced back to knowledge inherited from the Late Iron Age La Tène culture, which was consolidated and further developed during the first centuries of the Christian era along the banks of the *Elbe*. Here, as far as the iron production and particularly its reduction is concerned, two key-stone features existed: the use of a particular type of a furnace – *the slag pit furnace* – and an extensive and decentralized production organization which was aimed at satisfying regional demands. Then, it acquired other diverse elements during the Lombard's stay in Central Europe, through contacts with provincial Roman culture, which was still quite lively during the AD 6th century as pointed out independently first by J. HENNIG and more recently by F. CURTA, who underlined how many Late Antique productive structures survived well into AD 6th century on the basis of the uniformity of both the distribution and the quality and shape of some working tools produced in the workshops of the region. However, the economic activities of these ateliers were no longer centered on cities and long-distance trade, but to the needs of great estates and to the *castra* along the limes. Other important connections were those with the Byzantines and with other oriental cultures, particularly through links with the

dalla componente pannonica di ultima generazione legata all'immigrazione longobarda; ed è fuori discussione il ruolo svolto dal bacino carpatico-danubiano come luogo di fusione, nella cultura materiale, di apporti orientali, germanici e tardoantichi/bizantini."

[338]On the historical development of Lombard laws and customs see, C. Azzara and S. Gasparri eds., *Le leggi dei Longobardi: storia, memoria e diritto di un popolo germanico* (Milan: Editrice La Storia, 1992); K. Fischer Drew trans. and ed., *The Lombard Laws* (Philadelphia (Pa): University of Pennsylvania Press, 1985).

[339]See M.Cl. Bournand, "Les artisans," 15-16; A. Doren, *Deutsche Handwerker und Handwerker-Bruderschaften im mittelalterlichen Italien* (Berlin, Prager, 1903), 49; D. Harrison, *The early state*, 38, 45, 200; S. Gasparri, *Prima delle Nazioni*, 148-49, 154; N. Zanini, *Le Italie Bizantine*, 207; P. Delogu, "Considerazioni conclusive," in L. Paroli ed., *L'Italia centro-settentrionale*, 430; Paroli, "La cultura materiale," 296-97.

Avars and other closely related Germanic groups, especially the Alamans and the Bavarians.[340] The resulting influences found their way into the Lombard technological milieu also thanks to the transfer and integration of portion of these populations, which carried such technological knowledge, within the body of Lombard society. The presence of *Norici*, Pannonians, proto-Bulgarians and members of other Germanic tribes in Italy from AD 568 as also attested in written sources, seems to be confirmed archaeologically. The appearance of previously unknown production technologies and instruments in Italy, as for the introduction of new types of agricultural tools and the diffusion of the stirrup within eastern Merovingian region, which do not belong to its Classical heritage and are clearly of North and Central European origin, must also be considered proof of their presence.[341]

From an archaeometallurgical point of view as well, the transition from Antiquity to the Middle Ages cannot be considered a simple mixture of late-Latin and Germanic elements, but rather as a confluence of multiple elements which resulted in a radical re-organization of metallurgical production on the peninsula. Within this context, one cannot deny the contribution of the so called Barbarian populations. The collapse of the Late Imperial economic system, which was not homogeneous either geographically or economically, had a profound effect on the organization of the highly specialized crafts, such as pottery production, and on iron making.

However, even if the disappearance of the Roman production organization created some discontinuity in the immediate aftermath of the collapse, it did not slow down or technically impoverish the processes involved in the reduction and processing of iron in any significant manner. In fact, based on analyses of iron slag found in Italy comparing ternary diagrams of FeO-CaO-SiO_2, it appears that the temperatures that could be produced in furnaces increased about 200 C° from the Etruscan period to the Roman while during the Early Middle Ages temperatures were increased by about another 50 C°, thus maintaining a continuous technological development.[342] This progressive trend was maintained notwithstanding the fact that the technology used in the Lombard period seems to have been remarkably different from that used in the Late Roman period, as shown by the example of *Bienno* where the emphasis seems to have been the achievement of higher temperatures rather than improving reducing conditions inside the furnace.

Moreover, as far as the use of mineral resources are considered, here again, one may observe the importance of Lombard metallurgical-technical traditions in restructuring iron production cycles, especially in Piedmont (particularly the area of *Canavese*), in Lombardy (near *Brescia* and *Bergamo*) and, to a certain degree, in Tuscany. This restructuring was characterized, on the one hand, by being generally collateral to agricultural activity, and, on the other hand, by the diffusion and spread of production centers within the territory, linked to the dynamics of the local economies.[343]

The restructuring of iron production systems and their focus on local markets is not only an Italian phenomenon. It also has parallels within the broader European context. The restructuring of production cycles which coincided with the integration of Germanic material culture and craftsmen's metal working techniques with iron-making technology inherited from Classical world certainly is not accidental. This re-organization took advantage of a tendency towards a non-centralized "schismatic" economic system, mainly centered on the demands of local markets, a situation endemic to Late Antique Western Europe.[344] The importance of local markets for iron making production in early Medieval Europe can be demonstrated either through archaeological research, such as those conducted in the Swiss *Jura*, or in written documents, such as one found in the Abbey of *Lorsch*, from AD 788, which indicates that the Abbey received 30 kg of iron annually from the village of *Weilnau*. All this attests that all of the so called old continent still possessed a regular iron production. However, this production was still characterized by an extensive system of exploitation and production of iron rather than an intensive one (as had been typical in Late Antiquity), with a number of production centers in locations throughout the territory/countryside. These workshops centers were not able to channel large quantities of half-finished material to a single and larger marketplace for further distribution and were, thus, tied to the regional with its particular needs.[345] Nonetheless, within early Medieval Europe, there was still a reliable metal supply functioning and able to fulfill market needs. Metal was still used as protective roof-revetment in the most important buildings as well as in the production of pipes and tools; precious metals were still used to mint coins. In fact, the old Roman mints in the western provinces continued to strike coins, initially in the name of the Emperor. Later limited issues with the names of the barbarian rulers were struck. Minting was still under the

[340]On the complex character of the Germanic smith, based on a core of technical knowledge and tools inherited from the Celtic culture of La Tene, see J. Henning, "Schmiedegräber," 73-77.
[341]F. Zagari and V. La Salvia, "Aspetti della produzione," 865; V. La Salvia and F. Zagari, "Cultura materiale e tradizione tecnica," 1005-06.
[342]About the consistent enhancement of heating inside iron smelting furnaces see R. Farinelli and R. Francovich, "Potere e attività mineraria," 445, especially note 6, and V. La Salvia, *Archaeometallurgy*, 23.

[343]V. La Salvia, "L'artigianato matallurgico," 22-24; V. La Salvia and F. Zagari, "Cultura materiale e tradizione tecnica," 958-68.
[344]As far as the definition of "schismatic economy" is concerned see, S. Mazzarino, *L'Impero romano*, 812-15. On the restructuring of the Italian economic and social system, starting from the Lombard period, and its Late Anitique roots see, S. Gelichi, "L'insediamento nella penisola italica durante il periodo longobardo," in J. Arce and P. Delogu eds., *Visigoti e Longobardi*, 234-35.
[345]L. Eschenlor and V. Serneels, *Les bas fourneaux Mérovingiens*, 12, 13-15, 117, 120-21; M. Cima, *Archeologia del Ferro*, 40; on the same topic see U. Zimmermann, "Mittelalterlicher Bergbau auf Eisen, Blei und Silber - Begrenzte Mittel und zahlreiche Veränderungen," in G. Magnusson ed., *The Importance of Ironmaking*, 275; V. La Salvia and F. Zagari, "Cultura materiale e tradizione tecnica," 1007.

control of royal authority as indicated by charters issued by *Cassiodorus* and the Lombard King *Rothari* where the royal privilege on minting is evident.

There is also some archaeological evidence for continuity in mining activities within Europe. Some late sixth-century Merovingian coins have been found in the Roman tin mines of *Abbaretz-Nozay* (Loire Atlantique). The silver mines at *Sotiel Coronada* in the Iberian Pyrites Belt also yielded some Visigoth coins. Moreover, the location of the Suevian, Visigoth, and Merovingian mints in the Iberian Peninsula and France indicates that silver and especially gold mining continued even after the collapse of the Roman productive organization. In fact, Sueves and Visigoths occupied the north western region of Spain which during Late Antiquity was already a flourishing mining district. The Visigoths placed 38 out of 79 of their mints in this area. In France, during the AD 7th century, 69 Frankish mints operated in the Limousin region, already known as a mining area from the Roman period. Altogether, 22 of these mints were located in proximity to old Roman mines. Thus, exploitation of metal ore resources did not come to an end with the "fall" of the Roman Empire, especially in those regions which had established and, therefore, preserved a long lasting mining tradition often reaching back to pre-Roman times.[346]

In addition to this, during the Merovingian period, metal workers were often buried as fully equipped warriors. This pattern of deposition indicates the rank of the inhumated as that of a freeman. Such a custom was also evident among the 'Lombards' as shown by graves from necropolises in *Brno*, *Poysdorf* and *Cividale*. Therefore, among the 'Lombards,' these craftsmen were considered freemen and were regarded as men of great importance. In fact, many written documents from the AD 8th century of the Lombard kingdom in Italy, clearly indicate that smiths and blacksmiths were often called *vires honesti* and were involved in the acquisition and sale of real estate.[347] Within Lombard kingdom, metalworkers were the only ones to achieve such an important social status among craftsmen relatively early. This situation, documented from the AD 8th century, suggests that within the society of the Lombard period in Italy there was a firmly established group of specialized artisans that possessed a sound technical tradition as shown by finds in the graves of these metalworkers (*Brno*, *Poysdorf*, *Cividale*) and in workshops (*Scharmbeck*, *Göhlen*, *Březno*). That such a tradition existed is confirmed by the homogeneity and the high quality of the metal artifacts that were produced by these craftsmen in these workshops and by the consistent use of standardized production processes.[348] This consistency in the *chaine opératoire* is evident in sword making. In fact, as far as sword production is concerned, the evidence from the metallographic analyses indicates a clear trend of continuity in the methods used in sword making, at least between the Pannonian and the Italian phases. The core of the sword was created by welding together various metal sheets with different carbon contents and edges were attached by hammering. Moreover, apparently, the edges were worked separately as half-finished products with completely different structural and, therefore, functional qualities. In fact, these were the very parts of the sword where enhancement of the carbon content was crucial. Nonetheless, an increase in carbon content of iron would have been useless without uniform distribution of the carbon inside the metal's structure. This method is called pattern welding and became a standard technique for blade-making throughout the Middle Ages. Such a conclusion is also confirmed by the analyses of other Central European artifacts, the knives of *Březno*. The analyses of the set of Migration Period objects from *Březno* and their comparison with the artifacts from the nearby site of *Opočno*, dated to the Late Roman period (AD 3rd-4th century), showed that the iron-making technology diverged strikingly for the two different periods especially concerning steel-making and iron hardening techniques. The knives from *Březno* were also produced throughout the composition of their bodies by hammering together bars of iron with different carbon contents. Therefore, these techniques represent the basic common features of 'Lombard' sword production. These analyses also indicate the substantial affinity of many products of Lombard artisans with those produced by craftsmen within the eastern Merovingian area. Thus, comparison and evaluation of the technical features of these swords and knives, certainly manufactured in a 'Lombard' milieu, make it possible to confirm the existence of a Germanic *koinè* with regard to some artifacts at least.

The aim of this investigation was to contribute to the study of interactions between Roman and Germanic cultures between Late Antiquity and the early Middle Ages, especially referring to the case of the Lombards. This was done from a particular point of view and with a particular methodology: through a complex investigation of Lombard iron production, its technology and its role in contemporary society. The sources utilised were archaeological and, particularly, archeometric and

[346]On agricultural tools, see M. Baruzzi, "I reperti in ferro dello scavo di Villa Clelia. Note sull'attrezzatura agricola nell'Alto Medioevo," *Studi Romagnoli* 29 (1978): 423-46. For a general survey, see R.F. Tylecote, *A history of Metallurgy*, 68; moreover, in 761 AD the piping at the monastery of S.Giulia of Brescia was made in lead, with regard to this, see G.P. Brogiolo, "Trasformazioni urbanistiche nella Brescia longobarda, 205; on coins, and mints finds and location and on archaeological evidence and written documents concerning metal ores exploitation, see P. Spufford, *Money and its Use in Medieval Europe* (Cambridge: University Press, 1988), 9-11, 23, 24; J.C. Edmondson, "Mining in the later Roman empire," 92, 99-101; Th. Zotz, "Schriftquellen zum Bergbau im frühen Mittelalter." in H. Steuer, and U. Zimmermann eds., *Montanarchaeologie in Europa. Berichte zum Internationalen Kolloquium 'Frühe Erzgewinnung und Verhüttung in Europa' in Freiburg im Breisgau vom 4. bis 7. Oktober 1990* (Sigmaringen: Jan Thorbecke Verlag, 1993), 184, 185-88.

[347]V. La Salvia, *Archaeometallurgy*, 30-31; F. Zagari and V. La Salvia, "Aspetti della produzione" 870-71.

[348]The presence of specialized craftsmen underlined the existence of a stratified society and of technically and culturally homogeneous workshops. On this topic see, J. de Baye, *Étude Archaelogique*, 111: V. La Salvia, "L'artigianato metallurgico," 15; F. Zagari and V. La Salvia, "Aspetti della produzione" 863-65, 866.

metallographical. This data, often neglected in traditional historiography, proved to be fundamental in expanding our understanding of the subject under investigation. Even if combined with a rather limited amount of available written sources, such a methodology enables to establish the 'original' technical background of each culture, the stages of interactions, their impact on the development of technological skills and production strategies as well as on their integration into a new economic and social context. This methodology, chiefly archeometric analysis, helps to reconstruct virtually beyond a shadow of doubt, that there was a complex and two-way interaction and influences between Roman and Germanic, for the Italian case the Lombard, cultures. Thus, through the lens of archaeology of production it becomes evident how the material base on which the general socio-economic environment of the so called Roman-Barbarian kingdoms was finally formed in the passage between Late Antiquity and the early Middle Ages.

BIBLIOGRAPHY

PRIMARY SOURCES

Azzara, C. and S. Gasparri eds. *Le leggi dei Longobardi: storia, memoria e diritto di un popolo germanico*. Milan: Editrice La Storia, 1992.

Caesar, Gaius Iulius. *The Gallic war*. H.J. Edwards, trans. London/ Cambridge, Mass.: Heinemann/Harvard University Press, 1958.

Capo, L. ed. *Paolo Diacono. Storia dei Longobardi*. Vicenza: Lorenzo Valla, 1992.

Cassius Dio. *Dio's Roman history*. E. Cary, trans. London/New York: Heinemann/Macmillan, 1914-1927.

Cipolla, C. ed., *Codice Diplomatico del Monastero di S. Colombano di Bobbio*, I. Roma 1918.

Codex Diplomaticus Langobardiae 1873.

Codice diplomatico longobardo. Vols. 1-2. Ed. Luigi Schiaparelli. Vol. 5. Ed. Herbert Zielinski. *Fonti per la storia d'Italia*. Rome: Istituto Storico Italiano per il Medioevo, 1929-1933, 1986.

Edictus Rothari regis. In *MGH, Legum*, tomus 4.

Fischer Drew, K. trans. and ed., *The Lombard Laws*. Philadelphia (Pa): University of Pennsylvania Press, 1985.

Fontes Iuri Romani Anteiustiniani. Florence 1940-43.

"La Saga di Egil." In M. Meli ed., *Antiche saghe nordiche*, vol. II. Milan: Mondadori, 1997, 459-714.

Maurice's Strategikon. G.T. Dennis, trans. Philadelphia: University of Pennsylvania Press, 1984.

Momsen, T. and P.M. Meyer eds. *Codex Theodosianus*. Berlin: 1905.

Muratori, L.A ed. *Antiquitates Italicae Mediaevi* t. II, *dissertatio* XXIV coll. pp.373-374.

Notitia Dignitatum, Notitia Occidentis, VIII, Insigna viri illustris magistri officiorum. Bonn: Böcking, 1839-1853.

Pauli Historia Langobardorum. Eds. Ludwig Conrad Bethmann and Georg Waitz, 12-187. In *MGH. SRL*.

Pliny the Elder. *Natural history*. Volume V, bks. 17-19. H. Rackham ed. London/Cambridge, (Mass.): Heinemann/Harvard University Press, 1950.

Procopius Caesariensis. *De Bello Gothico*. In *Procopius Caesariensis Opera Omnia*. Ed. J. Haury. Leipzig: Teubner, 1963.

Tacite. *La Germanie*. J. Perret, ed. Paris: Les Belles Lettres, 1949.

Tacite. *Histoires*. H. Le Bonniec, trans. and ed. Paris : Les Belles Lettres, 1989.

Tacite. *Annales*. P. Wuilleumier, trans. and ed. Paris: Les Belles Lettres, 1974.

Vegetius (Flavius) Renatus. *P. Flavi Vegeti Renati Epitoma Rei Militaris*. A. Önnerfors, ed. Stuttgart/Leipzig: Teubner, 1995.

Versum de Mediolano civitate. E. Duemmler, ed. In *MGH, Poetae Latini aevi carolini* vol. I, 24-26.

Lotharii I et Lotharii II diplomata, T. Schieffer, ed. In *M.G.H.*, *Diplomata karolinorum*, 3, n. 23, 93 and n. 35, 112-15.

SECONDARY LITERATURE

Åberg, N. *Die Goten und Longobarden in Italien.* Uppsala, Almqvist & Wiksell, 1923.

Ahumada Silva, I. "Cividale del Friuli. Necropoli di S. Mauro. Tomba 43 di cavallo e cavaliere." In E. Arslan and M. Buora eds. *L'oro degli Avari.* Milan: In-Form, 2000, 198-205.

Ahumada Silva, I. "Necropoli longobarde a Cividale ed in Friuli." In *Paolo Diacono ed il Friuli altomedievale (sec. VI-X), Atti del XIV congresso internazionale di studi sull'Altomedioevo. 24-29 settembre 1999.* Spoleto: CISAM, 2001, 321-56.

Ahumada Silva, I. "Testimonianze archeologiche avare a Cividale." *Forum Julii* 14 (1990): 63-67.

Aitchinson, L. *A History of Metals*, 2 vols. London: Macdonald & Evans, 1960.

Ajbabin, A.I. "Pogrebenie hazarskogo voina" [The grave of Khazar warrior]. *Sovjetskaja Arkheologija* 3 (1985):191-205.

Akrich, M. "Comment decrier les objects techniques?" *Techniques et Culture* 5 (1987): 49-63.

Akrich, M. "La construction d'un système socio-technique." *Anthropologie et Sociétés* 13 (1989): 31-54.

Ambroz, A.K. "Stremena i sedla rannego srednevekovja kak xronologičeskij pokazatel' (IV-VIIIvv.)" [Early medieval Stirrups and Saddles as chronological indicators, AD 4th-8th cents]. *Sovjetskaja Arkheologija* 4 (1973): 81-98

Angioni, G. *Il sapere della mano.* Palermo: Sellerio, 1986.

Ankner, D. "Zur Damaszierung der Spathen aus Alterding." In H. Helmuth, D. Ankner and H.-J. Hundt eds. *Das Reihengräberfeld von Alterding in Oberbayern*, vol II. Mainz: P. Von Zabern Verlag, 1996, 144-53.

Anthony, D. "Prehistoric Migration as social Process." In J. Chapman and H. Hamerow eds. *Migrations and invasions in archaeological explanation.* Oxford: BAR, I.S. 664, 1997, 21-32.

Arnold, B. "The past as propaganda: Totalitarian Archaeology in Nazi Germany." *Antiquity* 64 (1990): 464-78.

Arnoux, M. *Mineurs, férons et maitres de forge.* Paris: Editions du CTHS, 1993.

Azzara, C. *Le invasioni barbariche.* Bologna: Il Mulino, 1999.

Azzaroli, A. *Il cavallo nella Storia antica.* Milan: L.L. Edizioni equestri, distributore Dielle, 1975.

Babucke, V. "Nach Osten bis an den Lech." In *Die Alamannen.* Stuttgart: Theiss, 1997, 249-60.

Bailly-Maitre, M.-Chr. "Les methodes de l'Archéologie Miniere." In R. Francovich ed. *Archeologia delle Attività Estrattive e Metallurgiche.* Florence: All'Insegna del Giglio, 1993, 237-61.

Bálint, Cs. " Byzantinisches zur Herkunfsfrage des vielteigen Gürtels." In Cs. Bálint ed. *Kontakte zwischen Iran, Bysanz und der Steppe im 6.-7. Jahrhundert, Varia Archaeologica Hungarica* 10 (2000): 99-162.

Bálint, Cs. "Les tombes a ensevelissement de cheval chez les Hongrois aux IXe-XIe siecles." *Archivum Eurasiae Medii Aevi* 2 (1982): 5-32.

Bálint, Cs. "Über die Datierung der osteuropäischen Steppenfunde des frühen Mittelalters (Schwierigkeiten und Möglichkeiten)." *Mitteilungen des Archäologischen Instituts der Ungarischen Akademie der Wissenschaften* 14 (1985): 137-47.

Bálint, Cs. "Zur Geschichte und Archäologie der osteuropäischen Reiterhirten im Frühmittelalter." In *Reitervölker aus dem Osten. Hunnen + Awaren.* Bad Vöslau: Grasl, 1996, 202-04.

Bálint, Cs. *Die Archäologie der Steppe. Steppenvölker zwischen Wolga und Donau von 6. bis 9. Jahrhunderts.* Vienna: Böhlau, 1989.

Bank-Burgess, J. "An Websthul und Webrahmen." In *Die Alamannen*. Stuttgart: Theiss, 371-78.

Banks, M. *Ethnicity: Anthropological Constructions*. London: Routledge, 1996.

Banti, A.M. "Le invasioni barbariche e le origini delle nazioni." In A.M. Banti and R. Bizzochi eds. *Immagini della nazione nell'Italia del Risorgimento*. Roma: Carocci, 2002, 21-44.

Bárdos, E. "La necropoli di Zamárdi." In *Gli Avari. Un popolo d'Europa*. Tavaganacco (UD): Arti Grafiche Friulane, 1995, 151-64.

Barker, G. and J. Llyod eds. *Roman Landscapes. Archeological Survey in the Mediterranean Region*. London: British School at Rome, 1991.

Barkóczi, L. "A 6th century Cemetery from Keszthely-Fenékpuszta." *Acta Archeologica Academiae Scientiarum Hungaricae* 20 (1968): 275-311.

Barni, G. "Alamanni nel territorio lombardo." *Archivio storico lombardo* n.s. 16 (1938): 137-62.

Bartlett Wells, H. "A contribution to the history of the heat treatment of steel." *Journal of Arms and Armour Society* 6.8 (dec. 1969): 217-38.

Baruzzi, M. "I reperti in ferro dello scavo di Villa Clelia (Imola). Note sull'attrezzatura agricola nell'Altomedioevo." In R. Francovich and G. Noyè eds. *Archeologia e storia del Medioevo italiano*. Rome: NIS, 1987, 151-70.

Baruzzi, M. "I reperti in ferro dello scavo di Villa Clelia. Note sull'attrezzatura agricola nell'Alto Medioevo." *Studi Romagnoli* 29 (1978): 423-46.

Behrens, G., "Eiserne Webschwerter der Merowingerzeit." *Mainzer Zeitschrift* 41-43 (1946-1948): 138-43.

Belaiev, N.T. "Sur le damas oriental et les lames damassées." *Métaux et Civilisation* 1 (1945): 10-15.

Beninger, E. and H. Mitscha-Märheim. "Die Langobardenfriedhof von Poysdorf." *Archaeologia Austriaca* 40 (1966): 167-87.

Berattino, G. "I reperti della necropoli a Reihengräber di Borgomasino." *Bollettino di storia e arte canavesana* 7 (1981): 81-130.

Bertolini, O. *I Germani. Migrazioni e Regni nell'Occidente già Romano*, part 1. vol. 3 of *Storia Universale. Nuove Formazioni Politiche nel Mondo Mediterraneo Medievale*. Milan: F. Vallardi, 1965.

Bianchi Bandinelli, R. *Roma. L'arte nel centro del potere. Dalle origine al II secolo D.C.* Milan: RCS Libri, 2005.

Bianchi Bandinelli, R. *Roma. La fine dell'arte antica*. Milano: Rizzoli, 1999.

Bielenin, K. "Frhügeschichtliche Eisenverhütung auf den Gebiet Polens." In J. Gomori ed. *Traditions and Innovations in the early Medieval Iron Production*. Sopron/Somogyfajsz: Dunaferr-Somogy Archaeometallurgical Foundation/ Soproni Muzeum, 1999, 57-62.

Bielenin, K. "Frühgeschichtliche Eisenverhüttung im Heiligenkreuz-Gebirge (Góry Świętokrzysjkie). Allgemeine Bemerkungen." In *Beiträge des Symposium ,Eisengewinnug und –verarbeitung in der östlichen Germania Magna.' Humboldt Universität Berlin 26-27 April 1996*, *Ethnographisch-Archäologische Zeitschrift* 37.3 (1996): 293-308.

Bietti, A. and A.M. Bietti Sestrieri. "Problemi di teoria e di metodo in Archeologia preistorica." In M. Liverani, A. Palmieri, R. Peroni eds. *Studi di paletnologia in onore di Salvatore M. Pugliesi*. Rome: Università di Roma "La Sapienza," Dipartimento di scienze storiche, archeologiche e antropologiche dell'antichità, 1985, 13-29.

Bintliff, J. and H. Hamerow. "Europe between Late Antiquity and the Middle Ages. Recent archaeological and historical research in Western and Southern Europe." In J. Bintliff and H. Hamerow eds. *Europe between Late Antiquity and the Middle Ages. Recent archaeological and historical research in Western and Southern Europe*. Oxford: BAR, I.S. 617, 1995, 1-7.

Bitenc, P. and T. Knific eds. *Od Rimiljanov do Slovanov*. Lubiana: Narodni muzej Slovenije, 2001.

Bivar, A.D.N. "The strirrup and its Origin." *Oriental Art*, n.s. 2 (1955): 61-5.

Blackmore, Ch., M. Braithwaite and I. Hodder. "Social and Cultural patterning in Late Iron Age England." In B.C. Burnham and J. Kingsbury eds. *Space, hierarchy and society: interdisciplinary studies in social area analysis*. Oxford: BAR I.S. 59, 1979, 93-111.

Bloemers, J.H.F. "Acculturation in the Rhine\Meuse basin in the Roman period: some demographical considerations." In J.C. Barrett, A.P. Fitzpatrick and L. Macinnes eds. *Barbarians and Romans in North-west Europe from the later Republic to Late Antiquity*. Oxford: BAR, I.S. 471, 1989, 175-97.

Böhne, C. and H. Dannheimer. "Studien an Wurmbuntkliegen des frühen Mittelalters." *Bayerische Vorgeschichtblätter* 26 (1961): 107-22.

Bökönyi, S. "Analisi archeozoologica dello scheletro del cavallo nella necropoli di Vicenne." *Conoscenze* 4 (1988): 69-75.

Bökönyi, S. "Two more horse graves from Vicenne." In S. Capini and A. De Niro eds. *Samnium. Archeologia del Molise*. Rome: Quasar, 1991, 342-43.

Bóna, I. "Avar lovassír Iváncsáról" [The avar horseman burial of Iváncsá]. *Archaeologiai Értesítő* 97 (1970): 243-61.

Bóna, I. "Das langobardische Gräberfeld von Hegykő, Komitat Györ-Sopron." In P. Anreiter, L. Bartosiewicz, E. Jerem and W. Meid eds. *Man and the Animal World. Studies in Archaeozoology, Archaeology, Anthropology, and Palaeolinguistic in memoriam Sándor Bökönyi*. Budapest: Archaeolingua Foundation, 1998, 109-20.

Bóna, I. "Die Langobarden in Pannonien." In R. Busch ed. *Die Langobarden von der Unterelbe nach Italien*. Neumünster: Veröffentlichung des Hamburger Museums für Archäologie und Geschichte Harburgs 54, (Helms-Museum), 1988, 256-87.

Bóna, I. "Die Langobarden in Ungarn." *Acta Archaeologica Academiae Scientiarum Hungarica* 7 (1956): 183-244.

Bóna, I. "Gli Avari. Un popolo dell'Oriente nell'Europa altomedievale." In *Gli Avari. Un popolo d'Europa*. Tavaganacco (UD): Arti Grafiche Friulane, 1995, 13-47.

Bóna, I. "Hatodik századi germán temető Hegykőn" [Sixth-century German Cemetery at Hegykő]. *Soproni Szemle* 14 (1960): 233-41.

Bóna, I. "Hatodik századi germán temető Hegykőn" [Sixth-century German Cemetery at Hegykő]. *Soproni Szemle* 15 (1961): 131-40.

Bóna, I. "Hatodik századi germán temető Hegykőn" [Sixth-century German Cemetery at Hegykő]. *Soproni Szemle* 17 (1963): 136-44.

Bóna, I. "I Longobardi e la Pannonia." In *Atti del Convegno Internazionale sul Tema: la Civiltà dei Longobardi in Europa, Roma-Cividale 1971*. Rome: Accademia Nazionale dei Lincei, 1974, 241-55.

Bóna, I. "I Longobardi in Pannonia." In G.C. Menis ed. *I Longobardi*. Milan: Electa, 1990, 14-19.

Bóna, I. "Longobarden in Ungarn," *Archeološki Vesnik* 20-21 (1970-1971): 45-74.

Bóna, I. "Neue Nachbarn im Osten. Die Awaren." In H. Dannaheimer, H. Dopsch eds. *Die Bajuvaren. Von Severin zu Tassilo 488-788*. Munich/Salzsburg: Freistaat Bayern/Amt der Salzburger Landesregierung, 1988, 108-17.

Bóna, I. *The Dawn of the Dark Ages. The Gepids and the Lombards in the Carpathian Basin*. Budapest, Corvina Press, 1976.

Bóna, I., J. Gömöri, L. Mihok, and V. La Salvia. "Metallographic Analyses of Lombard Swords." *Východoslovenský Pravek* (special issue, 1999):180-86.

Bonafini, G. "Due reperti altomedievali in Valle canonica." In *Miscellanea di Studi bresciani sull'Alto Medioevo*.

Brescia: F. Apollonio e C., 1959, 117-26.

Borhy, L. "Romani e Pannoni. Aspetti dell'acculturazione in una provincia di frontiera." In G. Hajnóczi ed. *La Pannonia e l'Impero Romano. Atti del Convegno Internazionale. Accademia d'Ungheria e l'Istituto Austriaco di Cultura (Roma 13-16 Gennaio 1994)*. Milan: Electa, 1994, 71-82.

Borile, F. and C. Dondolato. "Il microscopio elettronico a scansione e le sue applicazioni in metallurgia." *La metallurgia italiana* 7/8 (1973): 425-34.

Bournand, M. C. "Les artisans dans le royaume Lombard." *Arte Lombarda* 4 (1959): 13-16.

Breccairoli Taborelli, L. "Tomba longobarda da Borgo d'Ale." *Quaderni della Soprintendenza Archeologica del Piemonte* 1 (1982): 103-23.

Brocchi, G.B. *Trattato mineralogico e chimico sulle miniere di ferro del dipartimento del Mella*, 2 vols. Brescia: 1808.

Brogan, O. "Trade between the Roman Empire and the free Germans." *Journal of Roman Studies* 26 (1936): 195-222.

Brogiolo, G.P. "Considerazioni sulle sequenze altomedievali nella zona monumentale della città romana." In F. Rossi ed. *Carta Archeologica della Lombardia. V. Brescia. La città*. Modena: F.C. Panini 1996, 257-63.

Brogiolo, G.P. "Trasformazioni urbanistiche nella Brescia longobarda:dalle capanne in legno al monastero regio di S. Salvatore." In Carla Stella and Gerardo Brentegani eds. *S.Giulia di Brescia. Arte storia di un monastero dai Longobardi al Barbarossa*. Brescia: Grafo, 1992, 179-206.

Brogiolo, G.P. and L. Castelletti. *Il territorio tra tardo antico e altomedioevo. Metodi di indagine e risultati*. Florence: All'Insegna del Giglio, 1992.

Brogiolo, G.P. and S. Gelichi. *Le ceramiche altomedievali (fine VI-X sec.) in Italia settentrionale: produzione e commerci*. Mantova, Padus, 1996.

Brogiolo, G.P. ed. *S. Giulia di Brescia: gli scavi dal 1980 al 1992: reperti preromani, romani e alto medievali*. Florence: All'Insegna del Giglio, 1999.

Brogiolo, G.P., S. Lusuardi Siena and P. Sesino. *Ricerche su Sirmione longobarda* (Ricerche di Archeologia altomedievale e medievale, 16). Florence: All'Insegna del Giglio, 1989.

Bromlej, J.V. *Etnos e Etnografia*. Rome: Editori Riuniti, 1975.

Bronowski, J. *L'ascesa dell'uomo*. Milan: Fratelli Fabbri Editori, 1976.

Brozzi, M. "Avari e Longobardi friulani." In *Gli Avari. Un Popolo d'Europa*. Tavaganacco (UD): Arti Grafiche Friulane, 1995, 57-67.

Brozzi, M. "La più antica necropoli longobarda in Italia — Cronologia relativa e osservazioni economiche." In A. Tagliaferri ed. *Problemi della civiltà e della economia longobarda Scritti in onore di G.P. Bognetti, Biblioteca della Rivista Economia e Storia*, 12 (1964): 117-24.

Brozzi, M. "Strumenti di orafo longobardo." *Quaderni ticinesi* 1 *(NAC)* (1972): 167-74.

Bruhn De Hoffmeyer, A. "Introduction to the History of European Swords." *Gladius* 1 (1961): 30-75.

Brulet, R. "La sépulture du roi Childéric à Turnoi et le site funéraire." In F. Vallet and M. Kazanski eds. *La noblesse romaine et les chefes barbares du 3e au 7e siécle*. Saint-Germain-en-Laye: AFAM, 1995, 309-26.

Brunner, K. "Continuity and Discontinuity of Roman Agricultural Knowledge in the early Middle Ages." In D. Sweeney ed. *Agriculture in the Middle Ages*. Philadelphia: University of Pennsylvania Press, 1995, 21-40.

Bryony, O. *Anthropology for archaeologists: an introduction*. London: Duckworth, 1981.

Burgarella, F. "Bizantini e Longobardi nell'Italia meridionale." In *I longobardi dei Ducati di Spoleto e Benevento. Atti del 16 Congresso internazionale di Studi sull'alto medioevo. Spoleto-Benevento 20-27 Ottobre 2002*. Spoleto: CISAM, 2003, 181-204.

Burnham B.C. and J. Kingsbury eds. *Space, hierarchy and society: interdisciplinary studies in social area analysis*. Oxford: BAR I.S. 59, 1979.

Burnham B.C. and J. Kingsbury. "Introduction." In B.C. Burnham and J. Kingsbury eds. *Space, hierarchy and society: interdisciplinary studies in social area analysis*. Oxford: BAR, I.S. 59, 1979, 1-11.

Callu, J.P. "I commerci oltre i confini dell'Impero." In *Storia di Roma. L'età Tardoantica I. Crisi e trasformazioni*. Turin: Einaudi, 1992, 487-524.

Cameron, A. *Procopius and the sixth century*. London: Duckworth, 1985.

Cameron, A. *The Late Roman Empire*. London: Fontana Press, 1993.

Cameron, A. *The Mediterranean world in late antiquity. AD 395-600*. London, New York: Routledge, 1993.

Canard, M. "La relation du voyage d'Ibn Fadlan chez les Bulgares de la Volga." *Annales de l'Institut d'etudes Orientales* 16 (1958): 41-145.

Capelle, T. "Bemerkungen zu einer Untersuchung der goldenen Miniaturenkette von Szilágysomlyó." *Communicationes Archaeologicae Hungariae* (1992): 75-79.

Cardini, F. *Alle origini della cavalleria medievale*. Florence: La Nuova Italia, 1991.

Carli, F. *Il mercato nell'età del Comune*. In Id. *Storia del commercio italiano*. Padova: CEDAM, 1934.

Castagnetti, A., M. Luzzati, G. Pasquali, and A. Vasina. *Inventari altomedievali di terre, coloni e redditi*, F.S.I., 104. Rome: Istituto Storico per il Medioevo, 1979.

Castellani, A. "Capitoli d'una introduzione alla grammatica storica italiana, II: L'elemento germanico." *Studi linguistici italiani* 11 (1985): 1-26 and 151-81.

Catarsi Dall'Aglio, M. "Materiali longobardi di provenienza ignota presso il Museo Archeologico di Parma." In M. Catarsi Dall'Aglio ed. *I Longobardi in Emilia occidentale*. Parma: Editoria Tipolitotecnica, 1993, 74.

Ceglia, V. "Campochiaro (CB) località Vicenne. La Necropoli altomedeievale." *Bollettino di Archeologia* 5-6 (1990): 213-17.

Ceglia, V. and B. Genito, "La necropoli altomedievale di Vicenne a Campochiaro." In S. Capini and A. De Niro eds. *Samnium. Archeologia del Molise*. Rome:Quasar, 1991, 329-34.

Chadwick, S.E. "The Anglo-Saxon Cemetery at Finglesham/Kent." *Medieval Archaeology* 2 (1958): 1-71.

Chang, K.C. *Shang Civilisation*. New-Haven: Yale University Press, 1980.

Cherubini, G. "Le campagne italiane dall'XI al XV secolo. L'intensificato sfruttamento dei terreni coltivati." In *Storia d'Italia*, IV, *Comuni e Signorie: istituzioni, società e lotte per l'egemonia*. Turin: UTET, 1981, 265-448 .

Chi, L. *Anyang*. Washington: University of Washington Press, 1977.

Christie, N. "Longobard weaponry and warfare, AD 1-800." *Journal of Roman Military Equipment Studies* 2 (1991): 1-26.

Christie, N. *From Constantine to Charlemagne. An Archaeology of Italy, AD 300-800*. Aldershot (UK): Ashagate, 2006.

Christie, N. *The Lombards*. Oxford: Blackwell, 1995.

Christlein, R. Die Alamannen. Archäologie eines lebendigen Volkes. Stuttgart: Theiss, 1991.

Ciampoltrini, G. "La falce del guerriero e altri appunti per la Tuscia fra VI e VII secolo." *Archeologia Medievale,* 20 (1993): 595-606.

Ciglenečki, S. "Die Eisenwerkzeug aus den befestigten Höhensiedlungen Slowenien aus der Völkerwanderungszeit." *Balcanoslavica* 10 (1983): 45-54.

Ciglenečki, S. "Poznorimski depo z Rudne pri Rudnici" [Late Roman finds from Rudne near Rudnici]. *Arheoloski Veštnik* 42 (1991): 225-32.

Ciglenečki, S. "Romani e Longobardi in Slovenia nel VI secolo." In *Paolo Diacono e il Friuli Altomedievale. Atti del XIV Congresso Internazionale di Studi sull'Alto Medioevo (24-29 Settembre, Cividale, Bottenicco di Moimacco, 1999).* Spoleto: CISAM, 2001, 179-200.

Cima, M. "Il cannecchio bresciano tra forno a manica ed altoforno." In *Dal basso fuoco all'altoforno. Atti del 1 Simposio Valle Camonica 1988: La siderurgia nell'antichità. Sibrium* 20 (1989): 275-94.

Cima, M. "Le Origini della Metallurgia del Ferro nel Canadese." *Rivista di Archeologia* 11 (1987): 113-23.

Cima, M. "Metallurgia in ambiente rurale al sito alto-medievale di Misobolo." *Archeologia Medievale* 13 (1986): 173-89.

Cima, M. *Archeologia del ferro.* Brescia-Turin: Edizioni Grafo-Edizioni Nautilus, 1991.

Cima, M. *Archeologia e Storia dell'industria di una valle.* Florence: All'Insegna del Giglio, 1981.

Cirese, A.M. "Homo Faber, Homo loquens. Testimonianza per Andrè Leroi-Gourhan." In A. Leroi-Gourhan, *Ambiente e Tecniche.* Milan: Jaca Book, 1994, 295-98.

Cirese, A.M. *Segnicità, Fabrilità, Procreazione.* Roma: CISU, 1984.

Citter, C. "I corredi funebri nella Toscana longobarda nel quadro delle vicende storico-archeologiche del popolamento." In L. Paroli ed. *L'Italia centro-settentrionale in età longobarda.* Florence: All'Insegna del Giglio, 1997, 185-211.

Citter, C. "I corredi nella Tuscia longobarda: produzione locale, dono o commercio? Note per una storia delle attività produttive nella Toscana altomedievale." In G.P. Brogiolo ed. *Documenti di Archeologia. Sepolture tra VI e VIII secolo. 7° Seminario di Monte Barro, Gardone Rivera 1996.* Mantova: SAP, 1998, 179-95.

Claude, D. "Die Handwerker der Merowingerzeit nach den erzählenden und urkundichen Quellen." In H. Jankuhn ed. *Das Handwerk in vor-und frühgeschichtlicher Zeit.* Göttingen: Vandenhoeck & Ruprecht, 1980, 204-66.

Cleere, H. and D. Crossley. *The Iron industry of the Weald.* Leicester: University Press, 1985.

Costin, C. L. "Craft Specialization: Issues in Defining, Documenting and Explaining the Organization of Production." In M.B. Schiffer ed. *Archaeological Method and Theory.* Tucson: The University of Arizona Press, 1991, 1-56.

Craddock, P.T. *Early Metal Mining and Production.* Edinburgh: University Press, 1995.

Csallány, D. *Archäologische Denkmäler der Awarenzeit in Mitteleuropa: Schriftum und Fundorte.* Budapest: Hungarian Academy of Sciences, 1956.

Cucini Tizzoni C. "Un forno da ferro longobardo nelle Alpi italiane: Ponte di Val Gabbia - Val Camonica (Bienno, Brescia)." *Notiziario di Archeologia Medievale* 65 (1995): 8-9.

Cucini Tizzoni, C. and M. Tizzoni. "Early Ironworking in Northern and Central Italy. A critical Review of the Evidence." In *Proceedings of the International Conference Early Ironworking in Europe. Abstracts 19-25, Settembre 1997.* Plas Tan y Bwlch (Galles, U.K.): Snowdonia National Park, 1997, 40-42.

Cucini Tizzoni, C. and M. Tizzoni. *La miniera perduta.* Bienno: Comune di Bienno, 1999.

Curta, F. *TheMaking of the Slavs.* Cambridge: University Press, 2001.

Daim, F. "Archäologie der Awaren." In *Reitervölker aus dem Osten. Hunnen + Awaren*. Bad Vöslau: Grasl, 1996, 199-201.

Daim, F. "Die Bayern, die Nachbarn der Awaren westlich der Enns." In *Reitervölker aus dem Osten. Hunnen+ Awaren*. Bad Vöslau: Grasl, 1996, 308-15.

Daim, F. "The Avars. Steppe people of Central Europe." *Archaeology* 37.2 (1984): 33-39.

Dark, K.R. *Theoretical Archaeology*. London: Duckworth, 1995.

David, J. *L'outil*. In *Typologie des sources du Moyen Age occidental* 78. Turnhout: Brepols, 1997.

de Baye, J. *Étude Archaelogique. Epoque des Invasions Barbares. Industrie Langobarde*. Paris: Librarie Nisson, 1883.

De Marchi, P. M. "Note su produzione e scambi nella Lombardia di età longobarda: l'esempio degli scudi da parata." In G. P. Brogiolo ed. *II Congresso Nazionale di Archeologia Medievale. Musei Civici, Chiesa di Santa Giulia (Brescia, 28 settembre-1 ottobre 2000)*. Florence: All'Insegna del Giglio, 2000, 284-91.

De Marchi, P.M. "Calvisano e la necropoli d'ambito longobardo in località Santi di sopra. La pianura tra Oglio, Mella e Chiese nell'altomedioevo." In L. Paroli ed. *L'Italia centro-settentrionale in età longobarda*. Florence: All'insegna del Giglio 1997, 377-411.

De Marchi, P.M. "Catalogo dei materiali altomedievali delle civiche raccolte archeologiche di Milano." *Notizie dal Chiostro del Monastero Maggiore*, 4 (1988), suppl. IV.

De Marchi, P.M. "Gli scudi da parata longobardi in Lombardia: luoghi e centri di potere." In P. Baj ed. *Studi in memoria di Carlo Mastorgio*. Gavirate: Nicolini, 2002, 61-84.

De Marchi, P.M. "I ritrovamenti di Boffalora d'Adda." In *Nuovi Contributi agli Studi Longobardi in Lombardia. Atti del convegno (Arsago Seprio 29 Settembre 1984)*. Busto Arsizio: Gualdoni, 1986, 21-30.

De Marchi, P.M. "La ceramica longobarda. Osservazioni." In R. Fiorillo and P. Peduto eds. *III Congresso Nazionale di Archeologia Medievale, Castello di Salerno, Complesso di Santa Sofia (Salerno, 2-5 ottobre 2003)*. Florence: All'Insegna del Giglio, 2003, 14-20.

De Marchi, P.M. and A. Breda. "Il territorio bresciano in età longobarda e la necropoli di Leno." In C. Bertelli and G.P. Brogiolo eds., *Il futuro dei Longobardi. L'Italia e la costruzione dell'Europa di Carlo Magno*, Ginevra\Milano: Skira, 2000, 472-95.

de Rijk, P. "Eiseverhüttung und Eisenverarbeitung im nordwestlichen Elbe-Weser Raum." *Ethnographisch-Archäologische Zeitschrift* 37 (1996): 325-33.

de Rijk, P. "Examinations on iron slags in the north-western German coastal region." *Early Iron* 1 (1996): 49-52.

de Rijk, P. "Iron production and manifacturing in the Elbe-Weser-area." In G. Magnusson ed. *The Importance of Ironmaking: Technical Innovation and Social Changes. Papers presented at the Norgberg Conference on May 8-13, 1995*. Stockholm: Jernkontorets Bergshistorika Utskott, 1995, 81-86.

De Vingo, P. "Metal handicraft among Western Germanic population in the 6th-7th centuries: working, manufacturing and decoration techniques." In *Proceeding of the International Conference Archaeometallurgy in Europe (Milan 24-26 September 2003)*, vol. 1. Milan: AIM, 2003, 607-16

Decaens J. "Un nouveau cimetière du haut moye age en Normandie, Hérouvillette (Calvados)." *Archéologie Medievale* 1 (1971): 1-126

Delogu, P. "Considerazioni conclusive." In L. Paroli ed. *L'Italia centro-settentrionale in età longobarda*. Florence: All'insegna del Giglio 1997, 425-30.

Delogu, P. "L'editto di Rotari e la società del VII secolo." In J. Arce and P. Delogu eds. *Visigoti e Longobardi*. Florence: All'Insegna del Giglio, 2001, 329-55.

Delort, R. *La vita quotidiana nel Medioevo*. Rome/Bari: Laterza 1997.

Demougeot, E. *De l'Unitè á la division de l'Empire Romain (395-410). Essai sur le gouvernament imperial.* Paris: Maisonneuve, 1951.

Ditten, H. "Protobulgaren und Germanen im 5.-7. Jahrhundert." *Bulgarian Historical Review* 3 (1980): 51-77.

Dopsch, A. "Istituzioni agrarie dei regni germanici dal V al IX secolo." In *Storia Economica Cambridge. L'agricoltura e la società rurale nel Medioevo*, vol. 1. Turin: Einaudi, 1976, 223-53.

Doren, A. *Deutsche Handwerker und Handwerker-Bruderschaften im mittelalterlichen Italien.* Berlin, Prager, 1903.

Driehaus, J. "Fürstergräber und Eisenerze zwischen Mittelrhein, Mosel und Saar." *Germania* 43 (1965): 32-49.

Edmondson, J.C. "Mining in the later Roman Empire and beyond: Continuity or Disruption ?" *The Journal of Roman Studies* 79 (1989): 84-102.

Edward Stevens, C. "Agricoltura e vita rurale nel Tardo Impero Romano." In *Storia Economica Cambridge. L'agricoltura e la società rurale nel Medioevo*, vol. 1. Turin: Einaudi, 1976, 116-55.

Eibl-Eibesfeldt, I. *Etologia Umana. Le basi biologiche e culturali del comportamento.* Turin: Bollati Boringhieri, 1993.

Emmerling, J. "Die metallografische Untersuchungen der Schwertklingen aus Münsingen." *Jahrbuch des Bernischen Historischen Museums in Bern* 47-48 (1967-1968): 147-90.

Emmerling, J. "Technologische Untersuchungen an eisernen Bodenfunden." *Alt-Thüringen* 12 (1972): 267-320.

Eschenlor, L. and V. Serneels. *Les bas fourneaux Mérovingiens de Boécourt, les Boulies.* Porrentruy: Office du Patrimonie Historique Société Jurassienne d'émulation, 1991.

Evans, J.A.S. *The Age of Justinian.* London/New York: Routledge, 1996.

Farinelli, R. and R. Francovich. "Potere e attività mineraria nella Toscana altomedievale." In R. Francovich and G. Noyè eds. *La storia dell'Alto Medioevo italiano alla luce dell'archeologia.* Florence: All'Insegna del Giglio, 1994, 443-65.

Fasoli, G. "Aspetti di Vita economica e sociale nell'Italia del secolo VII." In *Caratteri del secolo VII in Occidente, Settimane di Studio del Centro Italiano di Studi sull'Alto Medioevo, 5. Spoleto 23-29-IV-1957.* Spoleto: CISAM, 1958, 103-59.

Fennert, M. "Metallurgiche Aspekte zur Eisengewinnung im Rennofen unter direkter Bezugnahme auf eine spätrömerzeitliche Verhüttungsstelle bei Zethlingen, Kr. Salzwedel." *Archäologische Informationen aus der Altmark* 3 (1992): 36-40.

Ferrero, E. "Borgomasino. Sepolcreto barbarico scoperto nell'abitato." *Notizie degli Scavi di Antichità* (1893): 189.

Fettich, N. *Das awarenzeitliche Gräberfeld von Pilismarót-Basaharc.* Budapest: Akadémiai kiadó, 1965.

Finó, J.F. "Notes sur la production du fer et la fabrication des Armes en France au Moyen Âge." *Gladius* 1 (1964): 47-66.

Fiorini, M. "Postfazione." In Leroi-Gourhan, A. *Ambiente e Tecniche.* Milan: Jaca Book, 1994, 299-304.

Fitz, J. "Economic life." In A Lengyel and G.T.B. Randan eds. *The Archaeology of Roman Pannonia.* Budapest/Lexington: Academia/University Press of Kentucky, 1980, 323-35.

Forni, G. "Gli aratri dell'Europa nord-occidentale dalla preistoria al Mille." In R. Comba, F. Panero eds. *Il seme, l'aratro, la messe. Le coltivazioni frumentarie in Piemonte dalla preistoria alla meccanizzazione agricola.* Cuneo: Società per gli Studi Storici, Archeologici ed Artistici della Provincia di Cuneo, 1996, 37-114.

Forni, G. "Strumenti aratori in Aquileia Romana. Loro origine, evoluzione, tipologia. Il contesto socio-economico." *Antichità Altoadriatiche*, 35 (1989): 313-34.

Forrer, R. *Die Schwerter und Schwertknaufe der Sammlung Carl von Scwerzenbach-Bregenz.* Leipzig: K.W. Hiersmann, 1905.

Francovich, R. "Introduzione." In R. Francovich ed. *Archeologia delle Attività Estrattive e Metallurgiche.* Florence: All'Insegna del Giglio, 1993, 3-9.

Francovich, R. "Premessa." In T. Mannoni and A. Molinari eds. *Scienze in Archeologia.* Florence: All'Insegna del Giglio, 1990, 1-10.

Francovich, R. "Villaggi dell'altomedioevo: invisibilità sociale e labilità archeologica." In M. Valenti. *L'insediamento altomedievale nelle campagne toscane. Paesaggi, popolamento e villaggi fra VI e X secolo.* Florence: All'Insegna del Giglio, 2004, IX-XXII.

Francovich, R. and G. Noyè eds. *La storia dell'Alto Medioevo italiano alla luce dell'archeologia.* Florence: All'Insegna del Giglio, 1994.

Francovich, R. and R. Hodges. *From Villa to Village. The Transformation of the Roman Countryside in Italy, c. 400-1000.* London: Duckworth, 2003

Friesinger, H. and Adler, H. *Die Zeit der Völkerwanderung in Niederösterreich.* In *Wissenschaftliche Schriftenreihe Niederösterreich*, Bd. 41/42. St. Pölten, Wien: Nö. Pressehaus, 1979.

Fulford, M.G. "Roman material in barbarian society c. 200 B.C.- c. A.D. 400." In T.C. Champion and J.V.S. Magaw eds. *Settlement and Society: aspects of West European prehistory in the first millenium B.C.* Leicester : Leicester University Press, 1985, 91-108.

Fumagalli, V. "Strutture materiali e funzioni dell'azienda curtense. Italia del Nord: sec. VIII e IX." *Archeologia medievale*, 7 (1980): 21-29.

Gabotto, F. "Un millennio di storia eporediese." *Eporediensia, Biblioteca della società storica subalpina* 4 (1900): 1-262.

Galloni, P. *Il sacro artefice.* Roma/Bari: Laterza, 1998.

Garam, E. "Sepolture di cavalli." In *Gli Avari. Un popolo d'Europa*, Tavaganacco (UD): Arti Grafiche Friulane, 1995, 143-49.

Gaspari A., M. Guštin, I. Lazar, and Ž. Trkman. "Late Roman tool finds from Celje Gradišče at Zbelovska gora nad Sv. Pavel above Vrtovin." In M. Feugère, M. Guštin eds. *Iron, Blacksmith and Tools. Ancient European Craft. Acts of the Instrumentum 12 Conference at Podjreda (Slovenia). April 1999.* Montagnac: Éditions Monique Mergoil, 2000, 187-203.

Gasparri, S. "I Germani immaginari e la realtà del regno. Cinquanta anni di studi sui Longobardi." In *I Longobardi di Spoleto e Benevento. Atti del XVI Congresso Internazionale di Studi sull'Alto Medioevo (Spoleto/Benevento 20-27 Ottobre 2002).* Spoleto: CISAM, 2003, 3-28.

Gasparri, S. "Tra antichità e Medioevo: i modelli sociali ed economici dei popoli barbarici e il loro impatto con il mondo mediterraneo." In V. Castronovo ed. *Storia dell'economia mondiale*, 1. *Permanenze e mutamenti dall'antichità al Medioevo.* Rome-Bari: Laterza, 1996, 317-34.

Gasparri, S. *Prima delle Nazioni.* Rome: Carocci, 2000.

Gaudenti, A. "Le società delle Arti in Bologna nel sec. XIII. I loro statuti e le loro matricole." *Bollettino dell'Istituto storico italiano* 21 (1899): 7-126.

Geary, P.J. *Before France and Germany: the creation and transformation of the Merovingian world.* New York: Oxford University Press, 1988.

Geary, P.J. *The myth of nations: the medieval origins of Europe.* Princeton: Princeton University Press, 2003.

Gelichi, S. "L'insediamento nella penisola italica durante il periodo longobardo." In J. Arce and P. Delogu eds. *Visigoti e Longobardi.* Florence: All'Insegna del Giglio, 2001, 219-38.

Genito, B. "Archaeology of the early medieval nomads in Italy: the horse-burials in Molise (7th century) south-central Italy." In Cs. Bálint ed. *Kontakte zwischen Iran, Byzanz und der Steppe in 6.-7. Jharhundert. Varia Archeologica Hungarica* 9 (2000): 229-47.

Genito, B. "Materiali e problemi." *Conoscenze* 4 (1988): 49-67.

Genito, B. "Sepolture con cavallo da Vicenne (CB): un rituale nomadico di origine centroasiatica." In S. Gelichi ed. *I Congresso Nazionale di Archeologia Medievale. Auditorium del Centro Studi della Cassa di Risparmio di Pisa (Pisa, 29-31 maggio 1997)*. Florence: All'Insegna del Giglio, 1997, 286-89.

Gibbon, G. *Anthropological Archaeology*. NewYork: Columbia University Press, 1984.

Giostra, C. "Catalogo." In L. Pejrani Baricco. *Presenze Longobarde. Collegno nell'Alto medioevo*. Turin: Soprintendenza per i Beni Archeologici del Piemonte, 2004, 73-152.

Giostra, C. *L'arte del metallo in età longobarda*. Spoleto: CISAM, 2000.

Giovannini, A. "La necropoli di Romans d'Isonzo." In *Paolo Diacono ed il Friuli altomedievale (sec. VI-X), Atti del XIV congresso internazionale di studi sull'Altomedioevo 24-29 settembre 1999*. Spoleto: CISAM, 2001, 595-653.

Giuntella, A.M. "Il ducato di Spoleto: persistenze e trasformazioni nell'asseto territoriale (Abruzzo e Sabina)." In *I longobardi dei Ducati di Spoleto e Benevento. Atti del 16 Congresso internazionale di Studi sull'alto medioevo. Spoleto-Benevento 20-27 Ottobre 2002*. Spoleto: CISAM, 2003, 763-99.

Giuntella, A.M. et alii. "Recenti indagini nella catacomba di Castelvecchio Subequo (Aq)." *Rivista di Archeologia Cristiana* 2.67 (1991): 249-321.

Godelier M. *L'Idéel et le Matériel*. Paris: Fayard, 1984.

Godłowski K."The Chronology of the Late Roman and Early Migration Periods in Central Europe." *Universitas Jagellonica Acta Scientiarum Litterarumque CCXVII Schedae Archaeologicae*, fasc. 11 (1970).

Godłowski, K. "Der römische Handel in die Germania Libera aufgrund der archäologischen Quellen." In K. Düwell, H, Jankuhn, H. Siems, and D. Timpe eds. *Untersuchungen zu Handel und Verkehr der vor- und frühgeschichtlichen Zeit in Mittel-und Nordeuropa*. Göttingen: Vandenhoeck & Ruprecht, 1985, 337-66.

Goffart, W. *Barbarians and Romans*. Princeton (N.J.): Princeton University Press, 1980.

Goldstein, J.I. and H.Yakowitz. *Practical scanning electron microscopy, electron and ion microprobe analysis*. New York: Plenum Press, 1975.

Gömöri, J. "Az avar kori és X-XI. századi vaskohászat régészeti emlékei Somogy megyében" [*Achaeometallurgical site of Somogy county during Avaric and early Arpadian period*]. *Somogy Megyei Múzeumok Közleményei* 14 (2000):163-218.

Gömöri, J. "Nemeskér-type iron smelting workshops from the time of Onogur colonization of Pannonia. Excavations in Zámardi." In J. Gomori ed. *Traditions and Innovations in the early Medieval Iron Production*. Sopron/Somogyfajsz: Dunaferr-Somogy Archaeometallurgical Foundation/ Soproni Muzeum, 1999, 149-59.

Gömöri, J. and P. Kisházi. "Iron ore utilization in the Carpathian Basin up to the Middle Ages with special regard to bloomeries in western Transdanubia." In *Neogen Mineral Resources in the Carpathian Basin*. Budapest: Hungarian Academy of Sciences, 1985, 323-55.

Gömöri, J. *Az Avar kori és Árpád-kori vaskohászat régészeti emlékei Pannóniában* [*The Archaeometallurgical sites of Pannonia from the Avar and early Arpadian period*]. Sopron: Hillebrand, 2000.

Graenert, G. "Langobardinnen in Alamannien. Zur Interpretation mediterranen Sachgutes in südwestdeutschen Frauengräbern." *Germania* 78 (2000): 417-47.

Grünewald, Ch. "Das alamannische Graeberfeld von Unterthuerhein, Bayerisch-Schwaben." In *Materialhefte der Bayerische Vorgeschichtsblätter* 59. Munich: Michael Lassleben Kallmuenz, 1988.

Guglielmetti, A. "Ceramica di età longobarda dall'area del Capitolium: analisi di una struttura produttiva." In F. Rossi ed. *Carta Archeologica della Lombardia. V. Brescia. La città.* Modena: F.C. Panini, 1996, 265-83.

Hachmann, R. *Les Germains*. Geneve: Les Èditions Nagel, 1971.

Hamerow, H. "Shaping settlements: Early Medieval communities in Northwest Europe." In J. Bintliff and H. Hamerow eds. *Europe between Late Antiquity and the Middle Ages. Recent archaeological and historical research in Western and Southern Europe*. Oxford: BAR, I.S. 617, 1995, 8-18.

Hammer, P. and H.-U. Voß. "Metallkundliche Untersuchungen an römischen und germanischen Funden des Elbegebietes." In C. Bridger and C. von Carnap-Bornheim eds. *Römer und Germanen – Nachbarn über Jahrhunderte*. Oxford: BAR, I.S. 678, 1997, 25-28.

Hampel, J. *Alterthümer des frühen Mittelalters in Ungarn*, 3 vols. Braunschweig: F. Vieweg und Sohn, 1905.

Harris, M. *Cultural materialism: The struggle for a science of culture*. New York: Random House, 1979.

Harrison, D. *The early state and towns*. Lund: University Press, 1993.

Haseloff, G. *Die germanische Tierornamentik der Völkerwanderungszeit*. Berlin-New York: 1981.

Haudricourt, A.G. "Contribution à la géographie et à l'ethnologie de la voiture." In Id. *La technologie science humaine*. Paris: MSH, 1987, 141-55.

Healy, J.F. *Mining and Metallurgy in the Greek and Roman World*. London: Thames and Hudson, 1978.

Hedeager, L. "Empire, frontier and the barbarian hinterland: Rome and Northern Europe from AD 1-400." In M. Rowlands, M. Larsen and K. Kristiansen eds. *Centre and Periphery in the Ancient world*. Cambridge:University Press, 1990, 125-40.

Hedeager, L. "The creation of Germanic identity. A European origin myth." In P. Brun, S. van der Leeuw and C.R. Whittaker, *Frontieres d'empire. Nature et signification des frontières romaines. Actes de la Table Ronde Internationle de Nemours, 21-22-23 mai 1992*. Nemours: A.P.R.A.I.F, 1992, 121-31.

Heggers, H.J. *Der römische Import im freien Germanien*. Hamburg: Hamburgisches Museum für Völkerkunde und Vorgeschichte, 1951.

Henning, J. "Eisenverarbeitungswerkstätten im unteren Donaugebiet zwischen Spätantike und Frühmittelalter." *Zeitschrift für Archäologie* 21 (1987): 59-73.

Henning, J. "Frühgeschichtliche Landwirtschaft Südosteuropas: Vom Großgrundbesitz zur Großgrundwirtschaft." In F. Horst and B. Krüger eds. *Produktivkräfte und Produktionsverhältnisse in ur- und frühgeschichtlicher Zeit*. Berlin: Akademie-Verlag, 1985, 301-08.

Henning, J. "Schmiedegräber nördlich der Alpen. Germanisches Handwerk zwischen keltischer Tradition und römischem Einfluß." *Saalburg-Jahrbuch* 46 (1991): 65-82.

Henning, J. "Untersuchungen zur Entwicklung der Landwirtschaft in Südeosteuropa im Übergang von der Spätantike zum frühen Mittelalter." *Ethnographisch-Archäologische Zeitschrift* 25 (1984): 123-30.

Henning, J. "Zum Problem der Entwicklung materieller Produktivkräfte bei den germanischen Staatbildungen." *Klio*, 68 (1986): 128-38.

Henning, J. "Zur Datierung von Werkzeug- und Agrargerätenfunden im germanischen Landnahmegebiet zwischen Rhein und oberer Donau (Der Hortfund von Osterburken)." *Jahrbuch des römisch-germanischen Zentralmuseums Mainz* 32 (1985): 570-94.

Henning, J. *Südosteuropa Zwischen Antike und Mittelalter*. Berlin: Akademie Verlag, 1987.

Hinde, R.A. *Le basi biologiche del comportamento umano*. Bologna: Zanichelli, 1979.

Hinton, D. A. "A smith's hoard from Tattershall Thorpe, Linconshire: a synopsis." *Anglo Saxon England*, 22 (1993): 147-66.

Hodder, I. "Economic and Social Stress in Material Culture Patterning." *American Antiquity* 44 (1979): 446-54.

Hodder, I. ed. *Symbolic and Contextual Archaeology*. Cambridge: Cambridge University Press, 1982.

Hodder, I. *Symbols in Action*. Cambridge: Cambridge University Press, 1982.

Hodder, I. *The Present Past*. London: B. T. Batsford, 1982.

Hodges R. and W. Bowden eds. *The sixth century: production, distribution and demand*. Leiden/Boston: Brill, 1998.

Hodges, R. "Rewriting the Rural History of Early Medieval Italy: Twenty-five Years of Medieval Archaeology Reviewed." *Rural History* 1 (1991): 17-36.

Hodges, R. *Early Medieval Archaeology*. Gwynedd: Headstart History, 1991.

Hodges, R. *Wall-to-Wall History: The Story of Roystone Grange*. London: Duckworth, 1991.

Hodson, F.R. "Inferring status from burial in Iron Age Europe. Some recent attempts." In B.C. Burnham and J. Kingsbury eds. *Space, hierarchy and society: interdisciplinary studies in social area analysis*. Oxford: BAR, I.S. 59, 1979, 23-30.

Hoffman, M. "The warp-weighted loom." *Studies in the History of Thechnology. Studia Norvegica*, 14. Oslo: Universitetforlaget, 1964.

Incitti M. "La necropoli longobarda della Selvicciola." In E. Herring, R. Whitehouse, J. Wilkins eds. *Papers of the Fourth Conference of Italian Archaeology, 4*. London: Accordia Research Centre, 1992, 213-17.

Incitti, M. "La necropoli altomedievale della Selvicciola ad Ischia di Castro (VT) ed il territorio castrense in età longobarda." In L. Paroli ed. *L'Italia centro-settentrionale in età longobarda*. Florence: All'Insegna del Giglio, 1997, 213-38.

Jones, A.H.M. *The Later Roman Empire: A Social, Economic and Administrative Survey*, 3 vols. Oxford: Blackwell, 1964.

Jones, S. *The Archaeology of Ethnicity*. London/New York: Routledge, 1997.

Jöns, H. "Iron production in Northern Germany during the Iron Age." In C. Fabech and J. Ringtved eds. *Settlement and Landscape. Proceedings of a Conference in Århus, Denmark. May 4-7 1998*. Århus: University Press, 1999, 249-60.

Jørgensen L. "*Castel Trosino* and *Nocera Umbra*. A chronological and social analysis of family burial practices in Lombard Italy (6th-8th cent AD)." *Acta Archaeologica* 62 (1991): 1-58.

Kenk, R. *Früh- und hochmittelalterliche Gräber von Kudyrge im Altai. Frühmittelalterliche Gräber aus West-Tuva*. In *Materialen zur allgemeinen und vergleichenden Archäologie,* Bd. 3 u. 4. Munich: Beck, 1982.

Kilbride-Jones, H. *Celtic craftsmanship in bronze*. London: Trowbridge & Esher, 1980.

Kiss, A. "Germanen im awarenzeitlichen Karpatenbecken." In F. Daim ed., *Awarenforschungen. Studien zur Archäologie der Awaren* 4. Vienna: 1992, 36-38.

Kiszely, I. *The Anthropology of the Lombards*. Oxford: BAR, I.S. 61, 1979.

Knific, T. "Vojščaki iz mesta Karnija" [Worriors from Karnij]. *Kranjski Zbornik* (1995): 23-40.

Knific, T. and D. Svoljšak. "Grobovi Langobardskih vojščacov iz Solkana (Nove Gorice)" [Graves of Lombard warriors from Solkan (Nova Gorica)]. *Arheološki Vestnik* 35 (1984): 277-90.

Knight J.K. *The End of Antiquity. Archaeology, Society and Religion in early medieval Western Europe AD 235-700.* Stroud: Tempus, 1999.

Koch, U. "Alamannische Gräber der ersten Hälfte des 6. Jahrhunderts in Südbayern." *Bayerische Vorgeschichtsblätter* 34 (1969): 162-93.

Koch, U. *Das fränkische Gräberfeld von Klepsau im Hohenloherkreis.* In *Forschungen und Berichte zur Vor- und Frühgeschichte in Baden-Würtemberg* Band 38. Stuttgart: Theiss, 1990.

Koch, U. *Das Reihegräberfeld bei Schretzheim.* In *Germanische Denkmäler der Völkerwanderungszeit*, 2 vols., ser. A, Band 13. Berlin: Gebr. Mann Verlag, 1977.

Koebner, R. "Popolamento stanziale e colonizzazione dell'Europa." In *Storia Economica Cambridge. L'agricoltura e la società rurale nel Medioevo*, vol. 1. Turin: Einaudi, 1976, 3-115.

Kolendo, J. "Les influences de Rome sur les peuples de l'Europe central habitant loin des frontieres de l'Empire." *Klio* 63 (1981):453-72.

Kollautz, A. "Awaren, Langobarden und Slaven in Noricum und Istrien." *Karinthia* 1 (1965): 619-45.

Koščević, R. and R. Makjanić. *Siscia. Pannonia Superior.* Oxford: BAR, I.S. 621, 1995.

Kovrig, I "Contribution au probleme de l'occupation de la Hongrie par les Avars." *Acta Archeologica Academiae Scientiarum Hungaricae*, 6 (1955): 163-92.

Kovrig, I. "Deux tombes avares de Törokbálint." *Acta Archeologica Accademiae Scientiarum Hungaricae* 9 (1957): 119-33.

Krempuš, R. "Krvavica beu Vransko in Slowenien, Höhensiedlung des 3. bis 6. Jahrhunderts." In M. Feugère and M. Guštin eds. *Iron, Blacksmith and Tools. Ancient European craft. Acts of the Instrumentum 12 Conference at Podjreda (Slovenia). April 1999.* Montagnac: Éditions Monique Mergoil, 2000, 209-31.

Kunow, J. *Negotiator et vectura. Händler und Transport im freien Germanien.* Marburg: Philipps-Universität Marburg, 1980.

Kyzlassov, I.L. "O proisxoženii stremjan" [On the origin of Stirrup]. *Sovjetskaja Arkheologija* 3 (1973): 25-36.

La Salvia, V. "Archaeometallurgy as a Source for the History of Changes and Developments within Medieval Daily Life." In G. Jaritz ed. *History of Medieval Life and the Sciences.* Vienna: Austrian Academy of Sciences, 2000, 105-15.

La Salvia, V. "Archeometallurgia." In R. Francovich, D. Manacorda eds. *Dizionario di archeologia.* Rome-Bari: Laterza, 18-24.

La Salvia, V. "Aspetti dell'economia dell'Italia alto medievale. Artigianato e commercio fra Longobardi e Bizantini." In O. Merisalo and P. Pahta eds. *Frontiers in the Middle Ages.* Louvain la neuve: Brepols, 2006, 348-72.

La Salvia, V. "*De la Pirotechnia* of Biringuccio. A Key to the Craftsmen Production. Technical Questions and Aspects of Organisation of Production." In M. Lette and M. Oris eds. *Technology and Engineering. Proceedings of the XX International Congress of History of Science (Liège, 20-26 July, 1997)*, Vol. 7, of *De Diversis Artibus, Collection of Studies from the International Academy of the History of Science,* by E. Poulle and R. Halleux (Turnhout: Brepols, 2000), 13-25.

La Salvia, V. "Gap or Continuity? Mining in Early Medieval Italy." In G. Magnusson ed. *The Importance of Ironmaking: Technical Innovation and Social Changes. Papers presented at the Norgberg Conference on May 8-13, 1995.* Stockholm: Jernkontorets Bergshistorika Utskott, 1995, 262-71.

La Salvia, V. "L'artigianato metallurgico dei Longobardi alla luce delle fonti archeologiche con particolare riferimento alla lavorazione del ferro. Suggerimenti e problemi." *Archeologia medievale* 25 (1998): 7-26.

La Salvia, V. "La fabbricazione delle spade delle grandi invasioni. Per la storia del 'processo diretto' nella lavorazione del ferro." *Quaderni Medievali* 44 (dicembre 1997): 30-55.

La Salvia, V. "Notes on Early Medieval Ironmaking in Italy." In J. Gömöri ed. *Traditions and Innovations in the early Medieval Iron Production*. Sopron/Somogyfajsz: Dunaferr-Somogy Archaeometallurgical Foundation/ Soproni Muzeum, 1999, 83-87.

La Salvia, V. *Archaeometallurgy of Lombard Swords*. Florence: All'Insegna del Giglio, 1998.

La Salvia, V. and F. Zagari. "Cultura materiale e tradizione tecnica: la metallurgia del ferro dei Longobardi in Italia." In I Longobardi di Spoleto e Benevento. In *I Longobardi di Spoleto e Benevento. Atti del XVI Congresso Internazionale di Studi sull'Alto Medioevo (Spoleto/Benevento 20-27 Ottobre 2002)*. Spoleto: CISAM 2003, 945-1008.

László, Gy. *Steppenvölker und Germanen. Kunst der Völkerwanderungszeit*. Vienna/Munich: Schroll, 1970.

László, Gy. *The Art of the Migration Period*. Miami: University Press, 1974.

Latour, B. and P. Lemonnier. "Introduction: genèse sociale des techniques, genèse technique des humains." In B. Latour and P. Lemonnier eds. *De la Prehistoire aux Missiles ballistiques*. Paris: La Découvert, 1994, 9-24.

Lavan L. and W. Bowden eds. *Theory and Practice in Late Antique Archaeology*. Leiden/Boston: Brill, 2003.

Lavan, L. "Late Antique Archaeology: an Introduction." In Lavan L. and W. Bowden eds. *Theory and Practice in Late Antique Archaeology*. Leiden/Boston: Brill, 2003, VII-XVI.

Leger, A. *Les travaux public, les mines et la metallurgie aux temps des romains*. Paris: 1875.

Leineweber R. and F. Kirsch. "Ein römerzeitlicher Werksattkomplex zur örtlichen Eisengewinnung von Zethlingen, Kr. Salzwedel." *Ausgrabungen und Funde* 34 (1989): 180-86.

Leineweber, R. "Ein germanischer Siedlings- und Bestattungsplatze bei Zethlingen in der Altmark." *Magdeburger Blätter* (1990): 39-48.

Leineweber, R. "Ein spätrömerzeitlicher Verhüttungsplatz Bereich einer zeitgleichen Brandgräberfeldes von Zethlingen, Kr. Salzwedel." *Jahresschrift für mitteldeutsche Vorgeschichte* 72 (1989): 97-120.

Leineweber, R. "Langobardenwerkstatt Zethlingen – lebendiges Museum mit archäologischen Experimenten." *Archäologische Mitteilungen aus Nordwestdeutschland* 6 (1991): 119-29.

Leineweber, R. "Römerzeitliche Eisenverhüttung in der Altmark. Archäologischer Befund und Rekonstruktion." In Espelund, A. ed. *Bloomery Ironmaking during 2000 Years*. Trondheim: Metallurgisk Institutt Universitet i Trondheim, 1993, 41-50.

Lemonnier, P. "Choix techniques et representation de l'enferment chez le Anya de Novelle-Guinée." In B. Latour and P. Lemonnier eds. *De la Prehistoire aux Missiles ballistiques*. Paris: La Découvert, 1994, 253-72.

Lengyel A and G.T.B. Randan eds. *The Archaeology of Roman Pannonia*. Budapest/Lexington: Academia/University Press of Kentucky, 1980.

Leroi-Gourhan, A. "Notes pour une histoire des aciers." In A. Leroi-Gourhan, *Le fil du temps*. Paris: Fayard, 1983, 33-47.

Leroi-Gourhan, A. *Il gesto e la parola*. Turin: Einaudi, 1977.

Leroi-Gourhan, A. *Milieu et Techniques*. Paris: Éditions Albin Michel, 1973.

Leube, A. "Die Eisengewinnung und –verarbeitung während der römischen Kaiser- und Völkerwanderungszeit im Gebiet zwischen Elbe und Oder." *Ethnographisch-Archäologische Zeitschrift* 33.4 (1992): 471-98.

Leube, A. "Studien zur Wirtschaft und Siedlung bei den germanischen Stämmen im nördlichen Mitteleuropa während des 1.-5./6. jh. u.Z." *Ethnographisch-Archäologische Zeitschrift* 33 (1992): 132-36.

Liborio, C., T. Marinig and A. Storti. "*Regiones augusteae X et XI*: ritrovamenti di manufatti in ferro databili dall'età augustea all'età altomedievale." In *Dal basso fuoco all'altoforno. Atti del 1 Simposio Valle Camonica 1988: La siderurgia nell'antichità. Sibrium* 20 (1989): 23-37.

Liebeschuez, J.H.W.G. "The uses and abuses of the concept of "decline" in later Roman history, or was Gibbon politically incorrect?" In L. Lavan ed. *Recent Research in Late-Antique Urbanism. Journal of Roman Archaeology Supplementary Series* 42 (2001): 233-45.

Liborio, C., T. Marinig and A. Storti. "*Regiones augusteae X et XI*: ritrovamenti di manufatti in ferro databili dall'età augustea all'età altomedievale." In *Dal basso fuoco all'altoforno. Atti del 1 Simposio Valle Camonica 1988: La siderurgia nell'antichità. Sibrium* 20 (1989): 23-37.

Littauer, M.A., "Early stirrups," *Antiquity* 55 (1981): 99-105.

Lloyd, K. E. "Behavioural anthropology: A review of Marvin Harris' Cultural Materialism." *Journal of Experimental Analysis of Behaviour* 43(2) (March 1985): 279–87.

Lopez, R.S. "Discorso inaugurale." In *Artigianato e tecnica nella società dell'altomedioevo occidentale. Atti della Settimana di Studio del Centro Italiano di Studi sull'Alto Medioevo,18. Spoleto 2-8-IV-1970*. Spoleto: CISAM, 1971, 15-39.

Lopreato, P. "La necropoli di S. Mauro-Cividale" In E. Arslan and M. Buora eds. *L'oro degli Avari*. Milan: In-Form, 2000, 190-93.

Lot, F. et al. *Les Destinees de l'Empire en Occident (395-888)*. Paris: P.U.F., 1940-41.

Lucchelli, T.M. *La Moneta nei rapporti tra Roma e l'Europa barbarica: aspetti e problemi*. Florence: La Nuova Italia, 1998.

Lusuardi Siena, S. and C. Giostra. "L'artigianato metallurgico longobardo attraverso la documentazione materiale: dall'analisi formale all'organizzazione produttiva." In *I Longobardi di Spoleto e Benevento. Atti del XVI Congresso Internazionale di Studi sull'Alto Medioevo (Spoleto/Benevento 20-27 Ottobre 2002)*. Spoleto: CISAM, 2003, 916-58.

Lusuardi-Siena, S. ed. *Ad Mensam. Manufatti d'uso da contesti archeologici fra tarda antichità e medioevo*. Udine: Del Bianco, 1994.

Maddin, R., A. Hauptmann and D. Baatz. "A metalographic examination of some iron tools from the Saalburgmuseum." *Saalburg Jahrbuch* 46 (1991): 5-23.

Mahias, M.-C. "Façonage de l'argile et de la société en Inde." In B. Latour and P. Lemonnier eds. *De la Prehistoire aux Missiles ballistiques*. Paris: La Découvert, 1994, 187-201.

Mannoni, T. "Introduzione all'Archeometria." In T. Mannoni and A. Molinari eds. *Scienze in Archeologia*. Florence: All'Insegna del Giglio, 1990, 27-39.

Mannoni, T., A. Cucchiara, A. Rabbi. "Scorie e forni di S.Giulia e la metallurgia nel medioevo." In C. Stella and G. Brentegani eds. *S.Giulia di Brescia. Arte storia di un monastero dai Longobardi al Barbarossa*. Brescia: 1992, 212-14.

Manoni, T. and E. Giannichedda. *Archeologia della Produzione*. Turin: Einaudi, 1996.

Marchis, V. "*Homo faber*. Le tecniche della produzione nel mondo antico." In V. Castronovo ed. *Storia dell'economia mondiale*, 1. *Permanenze e mutamenti dall'antichità al Medioevo*. Rome-Bari: Laterza, 1996, 27-42.

Markey, T.L. "Germanic in the Mediterranean: Lombards, Vandals, and Visigoths." In F.M. Clover and R.S. Humphreys eds. *Tradition and innovation in Late Antiquity*. Madison(USA)/London: The University of Wisconsin Press, 1989, 51-71.

Martin, M. "Die goldene Kette von Szilágysomlyó und das frühmerowingische Amuletettgehänge der westgermanischen Frauentracht." In U. von Freeden and A. Wieczorek eds. *Perlen. Archäologie, Techniken,*

Analysen. Akten des Internationalen Perlensymposiums in Mannheim vom 11. bis 14. November 1994. Bonn: Dr. Rudolf Habelt GmbH, 1997, 349-72.

Martin, M. "Zur Interpretation des langobardischen Gräberfeldes von Várpalota, Komitat Veszprém." *Basler Beiträge zur Ur- und Frühgeschichte* (1976): 194-99.

Mastrelli, C.A. "La terminologia longobarda dei manufatti." In *Atti del Convegno Internazionale sul Tema: la Civiltà dei Longobardi in Europa, Roma-Cividale 1971*. Rome: Accademia Nazionale dei Lincei, 1974, 275-69.

Mauss, M. "Le Tecniche del corpo." In *Teoria generale della magia efd altri Saggi*. Torino: Einaudi, 1950, 383-409.

Mauss, M. "Les Technique et la Technologie." In *Œuvres*. Paris: Minuit, 1941, 250-56.

Mazzarino, S. *L'Impero romano*, 3 vols. Rome/Bari: Laterza, 1980.

Mazzarino, S. *La Fine del Mondo Antico*. Milano: BUR, 1999.

Mazzarino, S. *Stilicone*. Milan: Rizzoli, 1990.

Mazzi M.S. and S. Raveggi eds. *Gli uomini e le cose nelle campagne fiorentine del Quattrocento*. Florence: Olschki, 1983.

Mazzocchi, C.M. " ΚΑΤΑΓΡΑΦΑΙ dello Strategikon di Maurizio." *Aevum* 55 (1981): 111-38.

Melucco Vaccaro, A. "Agere de arte, agere per arte: la trasmissione dei saperi tecnici fra tradizione colta e fonti materiche." In *Morfologie sociali e culturali in Europa fra Tarda Antichità e Alto Medioevo. Atti della XXV Settimana di Studio del CISAM (Spoleto 3-9 Aprile1997)*. Spoleto: CISAM, 1998, 343-76.

Melucco Vaccaro, A. *I Longobardi in Italia*. Milan: Longanesi, 1988.

Mengarelli, R. "La necropoli barbarica di Castel Trosino presso Ascoli Piceno." *Monumenti Antichi, pubblicati a cura dell'Accademia dei Lincei*, 12 (1902): coll. 145-380.

Menghin, W. *Das Schwert im frühen Mittelalter*. Stuttgart: Theiss, 1988.

Menghin, W., T. Springer and E. Wamers eds. *Germanen, Hunnen und Awaren. Schätze der Völkerwanderungszeit*. Nuremberg: Verlag Germanisches Nationalmuseum, 1987.

Menis, G.C., ed. *I Longobardi*. Milan: Electa, 1990.

Menke, M. "Archeologia Longobarda tra la bassa Elba e l'Ungheria." In S. Gasparri and P. Cammarosano eds. *Langobardia*. Udine: Casamassima, 1990, 35-106.

Micheletto, E. and L. Pejrani Baricco. "Archeologia funeraria ed insediativa in Piemonte tra V e VII secolo. In L. Paroli ed. *L'Italia centro-settentrionale in età longobarda*. Florence: All'insegna del Giglio, 1997, 295-344.

Mitscha-Märheim, H. *Die frühmittelalterlichen Gräberfunde von Mistelbach, Katzelsdorf, Münchendorf und Schwechat*. Wien/Leipzig: Kühne, 1941.

Mócsy, A. *Pannonia and Upper Moesia. History of the middle Danube provinces of the Roman Empire*. London: Routledge & Keagan, 1974.

Molinari, A. "Il ciclo di lezioni sulla ricerca applicata in Archeologia ed un progetto di laboratorio." In T. Mannoni and A. Molinari eds. *Scienze in Archeologia*. Florence: All'Insegna del Giglio, 1990, 11-20.

Monneret De Villard, U. "L'organizzazione industriale nell'Italia longobarda durante l'Alto medioevo." *Archivio Storico Lombardo* 5 ser., 46 (1919): 1-83.

Monti, F. *Le corporazioni nell'evo antico e nell'alto medioevo*. Bari: Laterza, 1934.

Moosleiter, F. "Handwerk und Handel." In H. Dannaheimer and H. Dopsch eds. *Die Bajuvaren. Von Severin zu Tassilo 488-788*. Munich/Salzburg: Freistaat Bayern/Amt der Salzburger Landesregierung, 1988, 208-19.

Morin, D. "Sauvegarde et protection des sites archeologique miniers. L'exemple de la France de l'est." In G. Magnusson ed. *The Importance of Ironmaking: Technical Innovation and Social Changes. Papers presented at the Norgberg Conference on May 8-13, 1995.* Stockholm: Jernkontorets Bergshistorika Utskott, 353-59.

Motykovà, K. and R. Pleiner. "Die römerzeitliche Siedlung mit Eisenhütten in Orech bei Prag." *Památky Archeologické* 78, (1987): 371-448.

Müller Wille, M. "Pferdegrab und Pferdeopfer im frühen Mittelalter." *Berichten van de Rijksdients voor het Oudheikundig Bodenmonderzoek* (1970-71): 119-248.

Müller, R. "Die Keszthely-Kultur." In *Reitervölker aus dem Osten. Hunnen + Awaren.* Bad Vöslau: Grasl, 1996, 265-74.

Müller, R. "La cultura di Keszthely." In *Gli Avari. Un popolo d'Europa.* Tavaganacco (UD) 1995, 165-72.

Müller-Wille, M. "Settlement and non-agrarian production from the high mountain region to the shoreline." In C. Fabech and J. Ringtved eds. *Settlement and Landscape. Proceedings of a Conference in Århus, Denmark. May 4-7 1998.* Århus: University Press, 1999, 205-11.

Murialdo, G., E. Bonora, C. Falcetti, F. Ferretti, A. Fossati, T. Mannoni, G. Vicino and G. Imperiale. "Il Castrum Tardo-Antico di S. Antonino di Perti, Finale Ligure (Savona): fasi stratigrafiche e reperti dell'area D. Seconde notizie preliminari sulle campagne di scavo." *Archeologia Medievale* 15 (1988): 355-96.

Naess, J.R. ed. *Arkeologi og Etnisitet* [*Archaeology and Ethnicity*]. Stavanger: Arkeologisk museum i Stavanger, 1985.

Narroll, R. "On Ethnic Unit Classification." *Current Anthropology* 5.4 (1964): 283-312.

Nelson, M.C. "The study of technological organization." In M.B. Schiffer ed. *Archaeological Method and Theory.* Tucson: The University of Arizona Press, 1991, 57-100.

Neuffer-Müller, Chr. *Das frankische Gräberfeld von Iversheim.* In *Germanische Denkmäler der Völkerwanderungszeit*, Band 6. Berlin: Gebr. Mann Verlag, 1977.

Norbach, L. "Organising iron production and settlement in Northwestern Europe during the Iron Age." In C. Fabech and J. Ringtved eds. *Settlement and Landscape. Proceedings of a Conference in Århus, Denmark. May 4-7 1998.* Århus: University Press, 1999, 237-47.

Obcinnokova, B.B. "Pogrebenie drevnetjurskogo voina v central'noj Tuve."[The Grave of an ancient Turkic warrior in central Tuva]. *Sovjetskaja Arkheologija*, 3 (1982): 210-18.

Odorici, F. *Storie bresciane dai primi tempi fino all'età nostra.* Brescia: 1854.

Oexle, J. "Merowingerzeitliche Pferdbestattungen-Opfer oder Beigaben? Frühmittelalterliche Studien." *Jahrbuch des Instituts für Frühmittelalterforschung der Universität Müster* 18 (1984): 122-72.

Oexle, J. *Studien zu merowingerzeitlichem Pferdegeschirr am Beispiel der Trensen.* In *Germanische Denkmäler der Völkerwanderungszeit*, ser. A, Band 16. Mainz am Rhein: von Zabern, 1992.

Ohlhaver, H. *Der germanische Schmied und sein Werkzeug.* Raumburg (Saale): Lippert & C., 1939.

Okely J. "An Anthropological contribution to the History and Archaeology." In B.C. Burnham and J. Kingsbury eds. *Space, hierarchy and society: interdisciplinary studies in social area analysis.* Oxford: BAR, I.S. 59, 1979, 81-92.

Olsen, B. "Norwegian Archaeology and People without (Pre-)History: Or How to Create a Myth of a Uniform Past." *Archaeological Rewiev from Cambridge* (1986) 5: 25-42

Ong, W. *Oralità e scrittura.* Bologna: Il Mulino, 1986.

Osmuk, N. "Nove Antične najdbe Povir" [New Roman finds from Povir]. *Goriški Letnik* 3 (1976): 71-87.

Panazza, G. "Note sul materiale barbarico trovato nel bresciano." In A. Tagliaferri ed. *Problemi della civiltà e dell'economia longobarda. Scritti in onore di G.P. Bognetti, Biblioteca della Rivista Economia e Storia* 12 (1964): 137-70.

Pani Ermini, L. "Città fortificate e fortificazione delle città italiane fra V e VI secolo." In L, Pani Ermini, *'Forma' e Cultura della città altomedievale*, eds. A.M. Giuntella and M. Salvatore. Spoleto: CISAM, 2001, 45-58.

Pani Ermini, L. "Il recupero dell'altura nell'Alto Medioevo," in L. Pani Ermini, *'Forma' e Cultura della città altomedievale*, eds. A.M. Giuntella and M. Salvatore. Spoleto: CISAM, 2001, 59-110

Panseri, C. *La tecnica di fabbricazione delle lame d'acciaio presso gli antichi. Quaderno 2.* Milan: Centro per la Storia della Metallurgia, 1957.

Panseri, C. *Ricerche metallografiche sopra una spada da guerra del XII sec. Quaderno 1.* Milan: Centro per la storia della metallurgia, 1954.

Parain, Ch. "L'evoluzione delle tecniche agricole." In *Storia Economica Cambridge. L'agricoltura e la società rurale nel Medioevo*, vol. 1. Turin: Einaudi, 1976, 155-222.

Parker Pearson, M. "Beyond the pale: Barbarian social dynamics in western Europe." In J.C. Barrett, A.P. Fitzpatrick and L. Macinnes eds. *Barbarians and Romans in North-west Europe from the later Republic to Late Antiquity.* Oxford: BAR, I.S. 471, 1989, 198-221.

Paroli, L. "Aspetti e problemi dell'archeologia della produzione in età longobarda." In M. S. Arena and L. Paroli eds. *Le arti del fuoco in età longobarda.* Rome: Museo dell'Alto Medioevo, 1994, 11-18.

Paroli, L. "Introduzione." In L. Paroli ed. *La necropoli altomedievale di Castel Trosino. Bizantini e Longobardi nelle Marche.* Milan: Silvana Eitoriale, 1995, 17-25.

Paroli, L. "La cultura materiale nella prima età longobarda." In J. Arce and P. Delogu eds. *Visigoti e Longobardi.* Florence: All'Insegna del Giglio, 2001, 257-304.

Paroli, L. "La Necropoli di Castel Trosino: un laboratorio archeologico per lo studio dell'età longobarda." In L. Paroli ed. *L'Italia centro-settentrionale in età longobarda.* Florence: All'insegna del Giglio 1997, 91-111.

Paroli, L. "La Necropoli di Castel Trosino: un riesame critico." In L. Paroli ed. *La necropoli altomedievale di Castel Trosino. Bizantini e Longobardi nelle Marche.* Milan: Silvana Eitoriale, 1995, 197-212.

Paroli, L. "The langobardic finds and the archaeology of central Italy." In K. Reynold Brown, D. Kidd and C.T. Little eds. *From Attila to Charlemagne.* New York: Yale University Press, 2000, 140-63.

Pasquali G. "Santa Giulia di Brescia." In Castagnetti, A., M. Luzzati, G. Pasquali and A. Vasina. *Inventari altomedievali di terre, coloni e redditi*, F.S.I., 104, Rome: Istituto Storico per il Medioevo, 1979, 41-94.

Pasquali, G. "Gestione economica e controllo sociale di S.Salvatore-S.Giula dall'epoca longobarda all'età comunale." In C. Stella and G. Brentegani eds. *S.Giulia di Brescia. Arte storia di un monastero dai Longobardi al Barbarossa.* Brescia: Grafo, 131-45.

Pasqui, A. and R. Paribeni. "Necropoli barbarica di Nocera Umbra." *Monumenti Antichi, pubblicati a cura dell'Accademia dei Lincei*, 25 (1918): coll. 137-352.

Pejrani Baricco L. "Schede." In G.C. Menis ed., *I Longobardi.* Milan: Electa, 1990, 344-48.

Pernicka E. "Förderschwerpunkt Archäometallurgie – Eine Bilanz." *Archäologisches Nachrihtenblatt* 3.1 (1998): 79-85.

Petru P. "Zgodn jesredn jeveška naselbina na Polhograjski Gori nad Polhovin Gradcem" [Early Medieval settlement of Polhograjska Gora ob Polhovo Gradec. *Arheološki Vestnik* 18 (1967): 455-60.

Piaskowski J. "Studia nad lokalizacja starożytnych Kotynów" [Study of the localization of the ancient Cotins tribe]. *Acta Archaeologica Carpathica* 3/1-2 (1981): 77-102.

Piaskowsky, J. "Metallkundliche Untersuchungen an Eisengegenständen aus dem Gräberfeld von Környe." *Acta Archaeologica Academiae Scientiarum Hungaricae* 26 (1974): 117-30.

Pieri, I. "Da Totila ad Autari." *Atti dell'Accademia di Scienze, lettere ed arte di Palermo* s. 4, 12 (1951-52): 5-107.

Pierpoint, S.J. "Who's who in the Northern British Bronze Age." In B.C. Burnham and J. Kingsbury eds. *Space, hierarchy and society: interdisciplinary studies in social area analysis*. Oxford: BAR, I.S. 59, 1979, 31-45.
Piggott, S. *Wagon, Charriot and Carriage*. London: Thames and Hudson, 1992.

Pleiner, R. "Der Handel mit eisen im östlichen Mitteleuropa im 4. bis 9. Jahrhundert." *Early Medieval Studies* 3 (1971): 13-21.

Pleiner, R. "Extensive Eisenverhüttungsgebiete im freien Germanien." In *Symposium. Ausklang der Laténe-Zivilisation und Anfänge der germanischen Besiedlung im mittleren Donaugebiet*. Bratislava: Slovak Academy of Sciences, 1977, 297-305.

Pleiner, R. "Small-scale Rural Ironmaking in Migration Period Bohemia." In Gert Magnusson ed. *The Importance of Ironmaking: Technical Innovation and Social Changes. Papers presented at the Norgberg Conference on May 8-13, 1995*. Stockholm: Jernkontorets Bergshistorika Utskott, 1995, 115-23.

Pleiner, R. "Spätkaiserzeitliche und Völkerwanderungszeitliche Stahlklinge aus Nordwestböhmen im Lichte der Metallographie." *Jahrbuch der Römisch-Germanischen Zentralmuseums Mainz* 35.2 (1991): 479-704.

Pleiner, R. "The Technology of Iron in the Bloomery Period: A Brief Survey of the Archaeological Evidence." In R. Francovich ed. *Archeologia delle Attività Estrattive e Metallurgiche*. Florence: All'Insegna del Giglio, 1993, 533-60.

Pleiner, R. *Iron in Archaeology*. Prague: Institute of Archaeology of the Czech Academy of Sciences, 2000.

Pleinerová, I. "Germanische und slawische Komponenten in der altslawische Siedlung Březno bei Louny." *Germania* 43 (1965): 121-38.

Pohanka, R. *Die eisernen Agrargeräte der römischen Kaiserzeit in Österreich – Studien zur römischen Agrartechnologie in Rätien, Noricum und Pannonien*. Oxford: BAR, I.S. 298, 1986.

Pohl, W. "Introduzione." In W. Pohl. *Le origini etniche dell'Europa*. Rome: Viella 2000, 1-38.

Pohl, W. "Le identità etniche nei Ducati di Spoleto e Benevento." In *I Longobardi di Spoleto e Benevento. Atti del XVI Congresso Internazionale di Studi sull'Alto Medioevo (Spoleto/Benevento 20-27 Ottobre 2002)*. Spoleto: CISAM, 2003, 79-103.

Pohl, W. *Die Awaren. Ein Steppenvolk in Mitteleuropa*. Munich: C.H.Beck 1988.

Power, E. *La vita nel Medioevo*. Turin: Einaudi 1966.

Profumo M.C. "Le marche in età longobarda: aspetti storico-archeologici." In L. Paroli ed. *La necropoli altomedievale di Castel Trosino. Bizantini e Longobardi nelle Marche*. Milan: Silvana Eitoriale, 1995, 125-83.

Prown J.D. "The truth of material culture: history or fiction?" In S. Lubar and W.D. Kingery eds. *History from things. Essays on material culture*. Washington/London: Smithsonian Books, 1993, 1-19.

Randsborg, K. "The Migration Period: Model History and Treasure." In Hodges R. and W. Bowden eds. *The sixth century: production, distribution and demand*. Leiden/Boston: Brill, 1998, 61-88.

Rhé, G. and N. Fettich. *Jutas und Öskü. Zwei Gräberfelder aus der Völkerwanderungszeit in Ungarn*. In *ΣΚΥΘΙΚΑ* 4. Paraha: Seminarium Kondakovianum, 1931.

Ricci, M. "La produzione di merci di lusso e di prestigio a Roma da Giustiniano a Carlo Magno." In M.S. Arena ed. *Roma dall'antichità al Medioevo. Archeologia e storia*. Milan: Electa, 2001, 79-87.

Ricci, M. "Relazioni culturali e scambi commerciali nell'Italia centrale romano-longobarda alla luce della Crypta Balbi in Roma." In L. Paroli ed. *L'Italia centro-settentrionale in età longobarda*. Florence: All'insegna del Giglio, 1997, 239-73.

Richter, M. *The oral tradition in the early middle ages*. In *Typologie des sources du Moyen Age occidental*, 71. Turnhout: Brepols, 1994.

Romanini, A.M. "Il concetto di classico e l'altomedioevo." In *Magistra Barbaritas*. Milan: Garzanti/Scheiwiller, 1984, 665-78.

Rosa, G. "Metallurgia storica bresciana." *Commentari dell'Ateneo di Brescia* (1877): 90-104.

Rosa, G. "Sulle miniere di ferro della Lombardia." *Commentari dell'Ateneo di Brescia* (1843): 161-70.

Rosner, G. "Das awarenzeitliche Gräberfeld in Szekszárd-Bogyiszlói Straße." *Monumenta Avarorum Archaeologica* 3, 1999.

Rosner, G. A "Szekszárd-Bogyiszlói úti avar temető ló és lovas temetkezései" [Burials of horses and horsemen in the Avar necropolis of Szekszárd at Bogyiszló street]. *Szekszárdi Béri Balogh Ádám Múzeum Évkönyve* 6-7 (1975-76): 79-109.

Rossi, F. ed. *Nuove ricerche sul 'Capitolium' di Brescia. Scavi, studi e restauri*. Milan: Et, 2002, 449-558.

Roth, H. *Die Ornamentik der Langobarden in Italien*. Bonn: Theiss, 1973.

Roth, H. *Kunst und Handwerk im frühen Mittelalter*. Stuttgart, Theiss, 1986.

Rotili, M. "Le necropoli di tradizione germanica." *Archeologia Medievale* 10 (1983): 143-74.

Rotili, M. *La Necropoli longobarda di Benevento*. Napoli: Università di Napoli. Istituto di Storia Medioevale e Moderna, 1977.

Rouse, I. "The place of 'peoples' in prehistoric research." *Journal of the Royal Anthropological Institute of Great Britain and Ireland*, 95.1 (1965): 1-15.

Roymans, N. "Romanisation and the transformation of a martial elite-ideology in a frontier province." In P. Brun, S. van der Leeuw and C.R. Whittaker eds. *Frontieres d'empire. Nature et signification des frontières romaines. Actes de la Table Ronde Internationle de Nemours, 21-22-23 mai 1992*. Nemours: A.P.R.A.I.F, 1992, 33-50,

Rupp, C. "Catalogo." In L. Paroli ed. *Umbria Longobarda*. Rome: Edizioni De Luca, 1997, 89-130.

Rupp, C. "La necropoli longobarda di Nocera Umbra (loc. Il Portone): l'analisi archeologica." In L. Paroli ed. *Umbria Longobarda*. Rome: Edizioni De Luca, 1997, 23-40.

Rupp, C. *Das langobardische Gräberfeld von Nocera Umbra*. Unpublished Dissertation, University of Bonn, 1994.

Sabatini F. *Riflessi linguistici della dominazione longobarda nell'Italia mediana e meridionale*. Florence: Olschki, 1965.

Sackett, J.R. "Style and Ethnicity in the Kalahari: A Reply to Wiessner." *American Antiquity* 50.1 (1985): 154-59.

Sagadin, M. "Late Antique wood working tools from Grdavov hrib near Kamnik (Slovenia)." In M. Feugère, M. Guštin eds. *Iron, Blacksmith and Tools. Ancient European craft. Acts of the Instrumentum 12 Conference at Podjreda (Slovenia). April 1999*. Montagnac: Éditions Monique Mergoil, 2000, 205-08.

Salač, V. "Die Bedeutung der Elbe für die böhmisch-sächsischen Kontakte in der Laténezeit." *Germania* 76 (1998):573-617.

Salamon A. and A.Cs. Sós, "Pannonia –Fifth to Ninth Centuries." In A Lengyel and G.T.B. Randan eds. *The Archaeology of Roman Pannonia*. Budapest/Lexington: Academia/University Press of Kentucky, 1980, 397-425.

Salamon, Á. and I. Erdélyi. *Das völkerwanderungszeitliche Gräberfeld von Környe*. Budapest: Akadémiai Kiadó, 1971.

Salin, E. "Les techniques metallurgiques après les grandes invasions." In *Les fer atravers les ages, Actes du Colloque International. Nancy 3-6.10.1955*. Nancy: Université, 1956, 45-56.

Salin, E. *La Civilisation merovingienne d'apres les Sepoltures, les Textes, les Laboratoires*, 4 vols. Paris: Edition A. et J. Picard et C., 1949-1959.

Salvi, L. "La fusione del ferro all'uso catalano in Campania." In *Dal basso fuoco all'altoforno. Atti del 1 Simposio Valle Camonica 1988: La siderurgia nell'antichità. Sibrium* 20 (1989): 317-20.

Scafile, F. "Di alcuni oggetti in ferro rinvenuti a Belmonte." *Ad Quintum*, 3 (1972): 28-32.

Schmid, H. "Die montangeologischen Voraussetzungen des ur- und frühgeschichtlichen Eisenhüttenwesens im Gebiet des mittleren Burgenlands (Becken v. O. Pullendorf)." *Wissenschaftliche Arbeiten Burgenland* 59 (1977): 11-23.

Schmidt, B. *Die späte Völkerwanderungszeit in Mitteldeutschland*. Halle (Saale): Max Niemeyer Verlag, 1961.

Schmiedt L. *Allgemeine Geschichte der Germanische Völker*. Munich/Berlin: Oldenbourg, 1909.

Schmiedt L. *Geschichte der deutschen Stämme*. Munich: C.H. Beck'sche Verlagbuchhandlung, 1938.

Schneeweiss, J. "Die ur- und frühgeschichtliche Eisenverhüttung und –verarbeitung im westlichen Odergebiet." In *Beiträge des Symposium, Eisengewinnug und –verarbeitung in der östlichen Germania Magna. Humboldt Universität Berlin 26-27 April 1996. Ethnographisch-Archäologische Zeitschrift* 37.3 (1996): 335-63.

Schutz, H. *Tools, Weapons and Ornaments*. Leiden/Boston/Köln: Brill, 2001.

Scott, B. "Archaeology and National Identity: The Norwegian Example." *Scandinavian Studies* 68.3 (1996): 321-42.

Scott, D.A. *Metallography and Microstructure of Ancient and Historic Metals*. Singapore: Tien Wah Press Ldt, 1991.

Shanks, M. and Tilley Chr. *Social Theory and Archaeology*. Cambridge: Polity Press, 1987.

Shephard, J. "The social identity of the individual isolated barrows and barrow cemeteries in Anglo-Saxon England." In B.C. Burnham and J. Kingsbury eds. *Space, hierarchy and society: interdisciplinary studies in social area analysis*. Oxford: BAR I.S. 59, 1979, 47-79.

Simon, M.L. "De l'explotation des mines et de la metallurgie en Toscane pendant l'antiquité et le Moyen âge." *Annales des Mines ou recueil de memoires sur l'explotation des mines*, 5th ser., 14 (1858): 557-615.

Slicher van Bath, B.A. *The agrarian history of Western Europe. AD 500-1850*. London: Edward Arnlod Ldt., 1963.

Smirnov, K.F. "Arheologiceskie stat'i v Trudah Saratovskogo oblastonogo Muzeja kraevedenija" [Archaeological papers in the works of Saratov regional museum]. *Sovjetskaja Arkheologija* 4 (1962): 270-73.

Soffia, E., ed."La necropoli longobarda di Trezzo sull'Adda." *Ricerche di Archeologia Altomedievale e Medievale* 12-13. Florence: All'Insegna del Giglio, 1986.

Solmi A. *Le associazioni in Italia avanti le origini del comune*. Modena: Società Tipografica Modenese, 1898.

Souchopová V. *Pocátky západoslovanského hutnictivi zeleza ve svetle prameni z Moravy* [*The Beginnings of the Metallurgy of Iron among Western Slavs in the Light of Sources from Moravia*]. Brno: Insitute of Archaology of the Czech Academy of Sciences, 1995.

Spazier, I. "The Germanic Iron Smelting Complex at Wolkenberg in Lower Lusacia, Southern Brandenburg." In L. Nørbach ed. *Prehistoric and Medieval Iron Smelting in Scandinavia and Europe*. Åarhus: University press, 2003, 37-42.

Sperl, G. "Il sentiero del ferro." In *Dal basso fuoco all'altoforno. Atti del 1 Simposio Valle Camonica 1988: La siderurgia nell'antichità. Sibrium* 20 (1989): 17-22.

Sprandel, R. "Bergbau und Verhüttung im frühmittelalterlichen Europa." In *Artigianato e tecnica nella società dell'altomedioevo occidentale. Settimana di Studio del Centro Italiano di Studi sull'Alto Medioevo,18. Spoleto 2-8-IV-1970*. Spoleto: CISAM, 1971, 583-601.

Springer, M. "Germanisches und anderes in der 'Kriegskunst' des Maurikios und anderswo." In Jh. Irmischer ed. *Die Literatur der Spätantike—polyethnisch und polyglottisch betrachtet*. Amsterdam: A.M. Hakkert Verlag, 1997, 183-86.

Spufford, P. *Money and its Use in Medieval Europe*. Cambridge: University Press, 1988.

Stabile, G. "La torre di Babele: confusione dei linguaggi e impotenza tecnica." In J.-C. Maire Vigueur and A. Paravicini Bagliani eds. *Ars et Ratio. Dalla torre di Babele al ponte di Rialto*. Palermo: Sellerio, 1990, 247-77.

Stadler, P. "Das germanisches Substrat: Langobarden, Gepiden und anderen germanische Völkerschaften im frühawarischen Material." In *Reitervölker aus dem Osten. Hunnen + Awaren*. Bad Vöslau: Grasl, 1996, 281-82.

Staffa, A.R. "La persistenza di logiche tardoantiche nella difesa dell'Abruzzo dai Longobardi. Reperti inedito da *Castrum Truentinum* e Crecchio." In M. Buora ed. *'Miles Romanus' dal Po al Danubio nel Tardoantico*. Pordenone: Lucaprint, 2002, 251-72.

Staffa, A.R. "Un quadro di riferimento per Castel Trosino: presenze longobarde fra Marche e Abruzzo." in In L. Paroli ed. *La necropoli altomedievale di Castel Trosino. Bizantini e Longobardi nelle Marche*. Milan: Silvana Eitoriale, 95-123.

Steuer, H. "Handel und Fernbeziehungen." In *Die Alamannen*. Stuttgard: Theiss, 1997, 403-15.

Stork, I. "Friedhof und Dorf, Herrendorf und Adelsgrab." In *Die Alamannen* Stuttgard: Theiss, 1997, 290-310.

Stupperich, R. "Export oder Technologietransfer? Beobachtungen zu römischen Metallarbeiten in Germanien." In C. Bridger and C. von Carnap-Bornheim eds. *Römer und Germanen – Nachbarn über Jahrhunderte*. Oxford: BAR, I.S. 678, 1997, 19-24.

Svoboda, B. "Zur Frage der Langobarden in Böhmen." In Tagliaferri A. ed. *Problemi della civiltà e dell'economia longobarda*. Milan: Giuffrè, 1964, 55-64.

Svoboda, B. *Cechy v dobe stehovani narodu* [*Bohemia during the Migration Period*]. Prague: Academia, 1965.

Szádeczky-Kardoss, S. "Der awarisch-türkische Einfluss auf die byzantinische Kriegskunst um 600 (Anmerkungen zum Strategikon des Maurikios)." In *Avarica. Über die Awarengeschichte und ihre Quellen. Acta Universitatis de Attila József Nominatae. Acta Antiqua et Archaeologica* 24 (Szeged, 1986): 203-13.

Szádeczky-Kardoss, S. "Histoire des Avars et leur heritage en Europe." In S. Csernus and K. Korompay eds. *Les Hongrois et l'Europe: conquete et integration*. Paris/Szeged: Publications de l'Institut Hongrois de Paris 1999, 149-70.

Tagiaferri, A. "Aspetti e Limiti dei Traffici Internazionali Longobardi (489-568 d.C.)." *Economia e Storia* 4 (1963): 517-43.

Tagliaferri, A. "Il Friuli e l'Istria nell'Altomedioevo." *Antichità Altoadriatiche* 2/2 (1972): 273-94.

Tagliaferri, A. "Industria artistica nell'Italia longobarda nel VII secolo." *Economia e storia* 3 (1964): 472-98.

Tagiaferri, A. "Note sull'economia longobarda dagli stanziamenti nordici al primo ducato italiano." *Economia e Storia* 4 (1972): 425-43.

Tagliaferri A. *Strutture sociali e sistemi economici precapitalistici*. Milano: Giuffrè, 1972.

Tanelli G., M. Benvenuti and I. Mascaro "Aspetti giacimentologici dei Minerali estratti in Età Preindustriale." In R. Francovich ed. *Archeologia delle Attività Estrative e Metallurgiche*. Florence: All'Insegna del Giglio, 1993, 263-76.

Teegen, W.R. “Zur Metallversorgung germanischer Buntmetallschmiede am Beispiel des Pyrmonter Brunnenfundes und des Moorfundes von Strückhausen.” In C. Bridger and C. von Carnap-Bornheim eds. *Römer und Germanen – Nachbarn über Jahrhunderte*. Oxford: BAR, I.S. 678, 1997, 29-35.

Tejral, J. “Die Langobarden nördlich der mittleren Donau.” In R. Busch ed. *Die Langobarden von der Unterelbe nach Italien*. Neumünster: Veröffentlichung des Hamburger Museums für Archäologie und Geschichte Harburgs 54 (Helms-Museum), 1988, 39-53.

Tejral, J. “Les fédérés de l’Empire et la formation des royaumes barbares dans la règion di Danube moyen à la lumière des données archéologiques.” *Antiquités Nationales* 29 (1997): 137-66.

Tejral, J. *Grundzüge der Völkerwanderungszeit in Mähren*, vol. 4.2. Prague: Studie Archeologické no ústavo Ceskoslovenske Ved v Brne, 1976.

Tholander, E. “Did the Blast Furnace occur before the 16th Century?” In *Dal basso fuoco all’altoforno. Atti del 1 Simposio Valle Camonica 1988: La siderurgia nell’antichità. Sibrium* 20 (1989): 105-14.

Tocík, A. *Slawisch-awarisches Gräberfeld in Štúrovo*. Bratislava/Bonn: Vydavatel’stvo Slovenskej akadémie vied/Habelt, 1968.

Todd, M. *Northern Barbarians, 100 BC- AD 300*. London: Hutchison, 1985.

Tomka, P. “Il Costume.” In *Gli Avari. Un popolo d’Europa*. Tavagnacco (UD): Arti Grafiche Friulane, 1995, 81-92.

Tondo, L. “Moneta longobarda di Montebamboli.” *Archeologia medievale* 17 (1990): 761.

Traina, G. “La fine del regno d’Armenia.” In *Atti dei Convegni Lincei, 201. La Persia e Bisanzio. Roma 14-18 ottobre 2002*. Roma: Accademia Nazionale dei Lincei, 2004, 353-72.

Trkman, Ž. “Sv. Pavel nad Vrtovinom” [St. Paul above Vrtovin]. *Varstvo Spomenikov* 31 (1989): 243.

Tylecote, R.F. *A History of Metallurgy*. London: The Metals Soc., 1976.

Vaglia, U. “Tracce di vita longobarda in Val Sabbia.” In *Miscellanea di Studi bresciani sull’Alto Medioevo*. Brescia: F. Apollonio e C., 1959, 127-28.

Valenti, M. *L’insediamento altomedievale nelle campagne toscane. Paesaggi, popolamento e villaggi fra VI e X secolo*. Florence: All’Insegna del Giglio, 2004.

Valič, A. “Ajdovska Luknja.” *Varstvo Spomenikov* 27 (1985): 272-74.

Vallet, F. “Une tombe de rich cavalier lombard découvert a Castel Trosino.” In F. Vallet and M. Kazanski eds. *La noblesse romaine et les chefs barbares du 3e au 7e siécle*. Saint-Germain-en-Laye: AFAM, 1995, 335-49.

Valsecchi, F. *Le corporazioni nell’organismo politico del medioevo*. Milan: Edizioni Alpes, 1931.

van Nie, M. “Three Iron Production Areas in the Netherlands: Contrasts and Similarities.” In G. Magnusson ed. *The Importance of Ironmaking: Technical Innovation and Social Changes. Papers presented at the Norgberg Conference on May 8-13, 1995*. Stockholm, Jernkontorets Bergshistorika Utskott, 1995, 100-06.

Veek, W. *Die Alamannen in Würtemberg*, 2 vols. Berlin-Lepzig: de Gryuter, 1931.

Vida, T. “Merovingische Spathagurte der Awarenzeit.” *Communicationes Archaeologicae Hungaricae* (2000): 161-75.

Vidale, M. *Che cos’è l’etnoarcheologia*. Rome: Carocci, 2004.

Vitali, M. “La ceramica longobarda.” In G.P. Brogiolo ed. *S. Giulia di Brescia : gli scavi dal 1980 al 1992: reperti preromani, romani e alto medievali*. Florence: All’Insegna del Giglio, 1999, 175-220.

Volpe, G. *Medioevo italiano*. Florence: Vallecchi, 1928.

von Freeden, U. "Das frühmittelalterliche Gräberfeld Von Moos-Burgstall." *Bericht des Römisch-germanischen Kommission* 68 (1987): 493-637.

von Freeden, U. "Das Grab eines awarischen Reiters von Moos-Burgstall, Niederbayern." *Bericht des Römisch-germanischen Kommission* 66 (1985): 5-24.

von Hessen, O. "A proposito della produzione di ceramica nel periodo delle migrazioni nell'Europa centrale e meridionale." In *Artigianato e tecnica nella società dell'alto medioevo occidentale. Atti Settimane di Studi del Centro Italiano di Studi sull'alto medioevo*, 17, vol. 2. Spoleto: CISAM, 1971, 749-64.

von Hessen, O. "Tecniche di Lavorazione." In G. C. Menis ed. *I Longobardi.* Milan: Electa, 1990, 208-09.

Voss, J. "Antiquity Imagined: Cultural Values in Archaeological Folklore." *Folklore* 98.1 (1987): 80-90.

Voss, O. "Eisenproduktion und Versorgung mit Eisen in Skandinavien vor der Vikingerzeit." *Early Medieval Studies* 3 (1971): 22-30.

Voss, O. "Snourp: An Iron Producing Settlement in west Jutland, 1st-7th century AD." In G. Magnusson ed. *The Importance of Ironmaking: Technical Innovation and Social Changes. Papers presented at the Norgberg Conference on May 8-13, 1995.* Stockholm, Jernkontorets Bergshistorika Utskott, 1995, 132-39

Wallace, A.F.C. *Culture and Personality.* New York: Random House, 1967.

Ward-Perkins, B. *The Fall Of Rome and the End of Civilization.* Oxford: University Press, 2005.

Wegewitz, W. "Ein Rennfeuerofen aus einer Siedlung der älteren Römerzeit in Scharmbeck (Kreis Harburg)." *Nachricthen aus Niedersachsens Urgeschichte* 26 (1957): 3-25.

Wells, P.S. "Identities, material culture and change: Celts and Germans in late Iron Age Europe." *Journal of European Archaeolgy* 3.2 (1995): 169-85.

Wells, P.S. *Culture contact and Culture change: Early Iron Age Central Europe and the Mediterranenan World.* Cambridge:University Press, 1980.

Wenskus, R. *Stammesbildung und Verfassung.* Köln-Graz: Bönlau Verlag, 1961.

Werner, J."Ein byzantinischer Steigbugel aus Caričin Grad." In N. Duval and V. Popovic eds. *Caričin Grad. I. Les basiliques B et J de Caričin Grad* (Rome: Ecole Française de Rome, 1984), 147-55.

Werner, J. "Eine langobardische Schild von Ischl an der Alz, ge. Seeon (Oberbayern)." *Bayerische Vorgeschichtsblätter*, 18-19 (1974): 45-58.

Werner, J. "Fernhandel und Naturalwirschaft im östlichen Merowingerreich nach archäologischen und numismatischen Zeugnisse." *Bericht des Römisch-germanischen Kommission* 42 (1961): 307-42.

Werner, J. "Italienisches und koptisches Bronzegeschirr des 6. und 7. Jahrhundert nordwärts der Alpen." In *Memnosymon Theodor Wiegand* (Munich: F. Bruckmann Verlag, 1938), 74-86.

Werner, J. "Langobardischer Einfluss in Süddeutschland während des 7. Jahrhundert im Lichte der archäologische Funde." In *Atti del I Congresso internazionale di Studi longobardi. Spoleto 27-30 settembre 1951.* Spoleto: CISAM, 1952, 521-24.

Werner, J. "Nomadischen Gürtel bei Persern, Byzantynern und Langobarden." In *Atti del Convegno Internazionale sul Tema: La Civiltà dei Longobardi in Europa. Roma-Cividale 1971* (Rome: Accademia Nazionale dei Lincei, 1974, 109-39.

Werner, J. "Zur Verbereitung frühgeschichtlicher Metallarbeiten (Werkstatt-Wanderhandwerk-Handel-Familien Verbindung)." *Early Medieval Studies* 1 (1970): 65-81.

Werner, J. *Das alamannische Fürstergrab von Wittischlinge*. In *Veröffentlichungen der Kommission zur vergleichenden Archäologie römischer Alpen- und Donauländer*, Band 2. Munich: Verlag C.H. Beck, 1950.

Werner, J. *Die Langobarden in Pannonien*. In *Abhandlungen, Bayerische Akademie der Wissenschaften philosophisch-historische Klasse*, n. F. H. 55 A/B. Munich: Bayerische Akademie der Wissenschaften, 1962.

Wewer, F. "Das Schwert im Mythos und Handwerk." *Arbeitsgemeinschaft fur Forschung des Landes Nordrhein Westfalen* 91 (1961): 7-71.

Wheeler, M. *La civiltà romana oltre i confini dell'impero*. Turin: Einaudi, 1963.

White, Jr. L. "Medieval roots of modern technology." In L. White Jr., *Medieval Religion and Technology. Collected Essays*. Berkely, Los Angeles, London: University of California Press, 1978, 75-91.

White, Jr. L. "Technology and Invention in the Middle Ages." In L. White Jr., *Medieval Religion and Technology. Collected Essays*. Berkely, Los Angeles, London: University of California Press, 1978, 1-22.

White, Jr. L. "The life of the silent majority." In L. White Jr., *Medieval Religion and Technology. Collected Essays*. Berkely, Los Angeles, London: University of California Press, 1978, 133-47.

White, Jr. L. *Medieval Technology and Social Change*. Oxford: University Press, 1964.

White, K.D. *Greek and Roman technology*. London: Clarendon Press, 1984.

Whittaker, C.R. "What happens when frontiers come to an end?" In P. Brun, S. van der Leeuw and C.R. Whittaker eds. *Frontieres d'empire. Nature et signification des frontières romaines. Actes de la Table Ronde Internationle de Nemours, 21-22-23 mai* 1992. Nemours: A.P.R.A.I.F, 1992, 133-41.

Whittaker, D. "The Economy of Roman Frontier." In E. Aerts, J. Andreau and P. Ørsted eds. *Models of Regional Economies in Antiquity and the Middle Ages to the 11th. Century. Proceedings of the 10th International Economic History Congres. Leuven, August 1990*. Leuven: University Press, 1990, 59-70.

Wickham, C. "L'Italia e l'Alto Medioevo." *Archeologia medievale* 15 (1988): 105-24.

Wiessner, P. "Style or Isochrestic Variation? A Reply to Sackett." *American Antiquity* 50.1. (1985): 160-66.

Williams, B.F. "Of Straightening Combs, Sodium Hydroxide, and Potassium Hydroxide in Archaeological and Cultural-Anthropological Analyses of Ethnogenesis." *American Antiquity* 57.4. (Oct., 1992): 608-12.

Wilson, A. J. *The Living Rock*. Cambridge: Woodhead Publishing Ldt., 1994.

Wolagiewicz, R. "Der Zufluß römischer Importe in das Gebiet nordlich der mittleren Donau in der alteren Kaiserzeit." *Zeitschrift für Archäologie* 4 (1970): 222-49.

Wolf, E.R. "Perilous Ideas: Race, Culture and People." *Current Anthropology* 35.1 (1994): 1-12.

Wood, I. "The frontiers of Western Europe: developments east of the Rhine in the 6th century." In Hodges R. and W. Bowden eds. *The sixth century: production, distribution and demand*. Leiden/Boston: Brill, 1998, 231-53.

Woolf, A. "Romancing the Celts. A segmentary approach to acculturation." In R. Laurence and J. Berry eds. *Cultural Identity in the Roman Empire*. London & New York: Routledge, 1998, 111-24.

Woolf, G. "European social development and roman imperialism." In P. Brun, S. van der Leeuw and C.R.Whittaker eds. *Frontieres d'empire. Nature et signification des frontières romaines. Actes de la Table Ronde Internationle de Nemours, 21-22-23 mai* 1992. Nemours: A.P.R.A.I.F, 1992, 13-20.

Yalçin, U. and B. Wollschläger. "Die Eisenverhüttung bei den Elb-Germanen in Südwestmecklenburg." *Das Altertum* 40, (1994): 5-12.

Ypey, J. "Au sujet des armes avec damas soudé en Europe." *Archéologie Medievale* 11 (1981): 147-65.

Ypey, J. "Drei römische Dolche mit tauschierten Scheiden aus niederlandischen Sammlungen." *Berichten van de Rijksdienst voor het Oudheidkundig Bodemonderzoek* 10-11 (1960-61): 347-62.

Ypey, J. "Vroeg-middeleeuwse zwaarden uit Nederlandse verzamelingen."*Berichten van de Rijksdienst voor het Oudheidkundig Bodemonderzoek* 14 (1964): 89-96.

Zagari, F. and V. La Salvia. "Aspetti della produzione metallurgica longobarda. Note sulla tomba del fabbro di Grupignano e sullo sviluppo dell'attrezzatura agricola." In *Paolo Diacono e il Friuli alto medievale (Secc. VI-X). Atti del XIV Congresso Internazionale di Studi sull'Alto Medioevo*. Spoleto: CISAM, 2001, 863-86.

Zanini, N. *Le Italie Bizantine. Territorio, insediamenti ed economia nella provincia bizantina d'Italia (VI-VIII secolo)*. Bari: Edipuglia, 1998.

Zaseckaja, I.P. "Osobennosti pogrebal'nogo obrjada gunnskoj epohi na territorii stepej niznego povolz'ja i severnogo pricernomr'ja" [The specific grave-goods of Hunnic period on the Steppes of the lower Volga basin and northern Black Sea area]. *Arheologiceskih Sbornik* 13 (1971): 64-69.

Zenetti, P. and H. Zeiß. "Alamannische Webschwerter im Museum Dillingen a.d. Donau." *Germania* 16 (1932): 307-08.

Zimmermann, U. "Mittelalterlicher Bergbau auf Eisen, Blei und Silber - Begrenzte Mittel und zahlreiche Veränderungen." In G. Magnusson ed. *The Importance of Ironmaking: Technical Innovation and Social Changes. Papers presented at the Norgberg Conference on May 8-13, 1995*. Stockholm, Jernkontorets Bergshistorika Utskott, 1995, 272-80.

Zotz, Th. "Schriftsquellen zum Bergbau im frühen Mittelalter." In H. Steuer, and U. Zimmermann eds. *Montanarchäologie in Europa. Berichte zum Internationalen Kolloquium 'Frühe Erzgewinnung und Verhüttung in Europa' in Freiburg im Breisgau vom 4. bis 7. Oktober 1990*. Sigmaringen: Jan Thorbecke Verlag, 1993, 183-99.

Figure 1
Iron reduction process through the direct method.
From R. Pleiner 2000

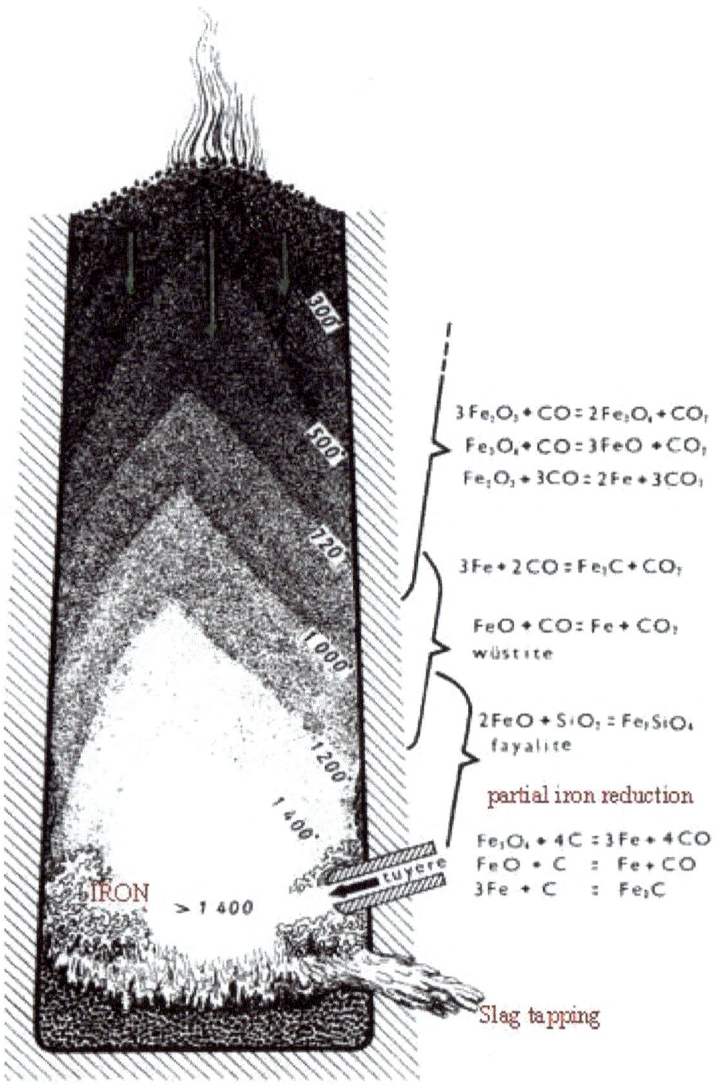

Figure 2
Most important European iron mining districts of the Iron Age.
From R. Pleiner 2000

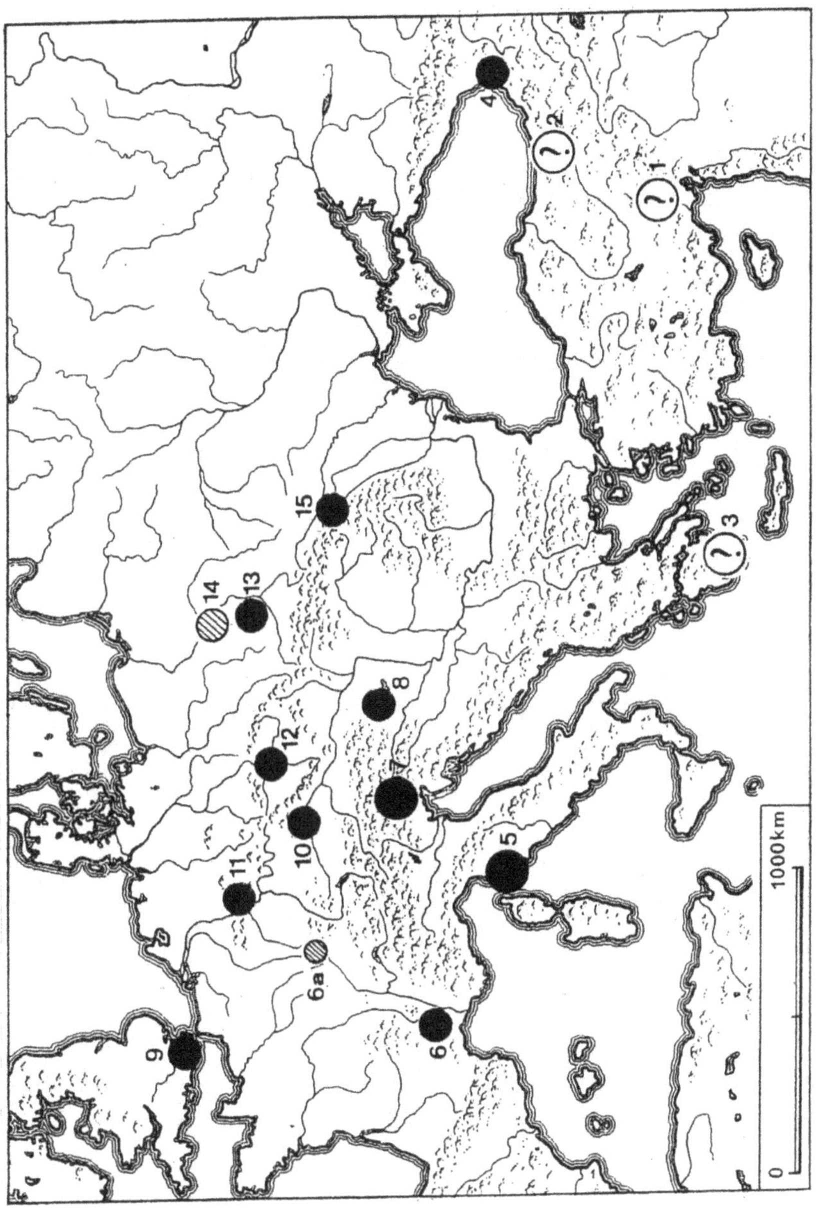

Figure 3
Profile rendition of slag pit furnace

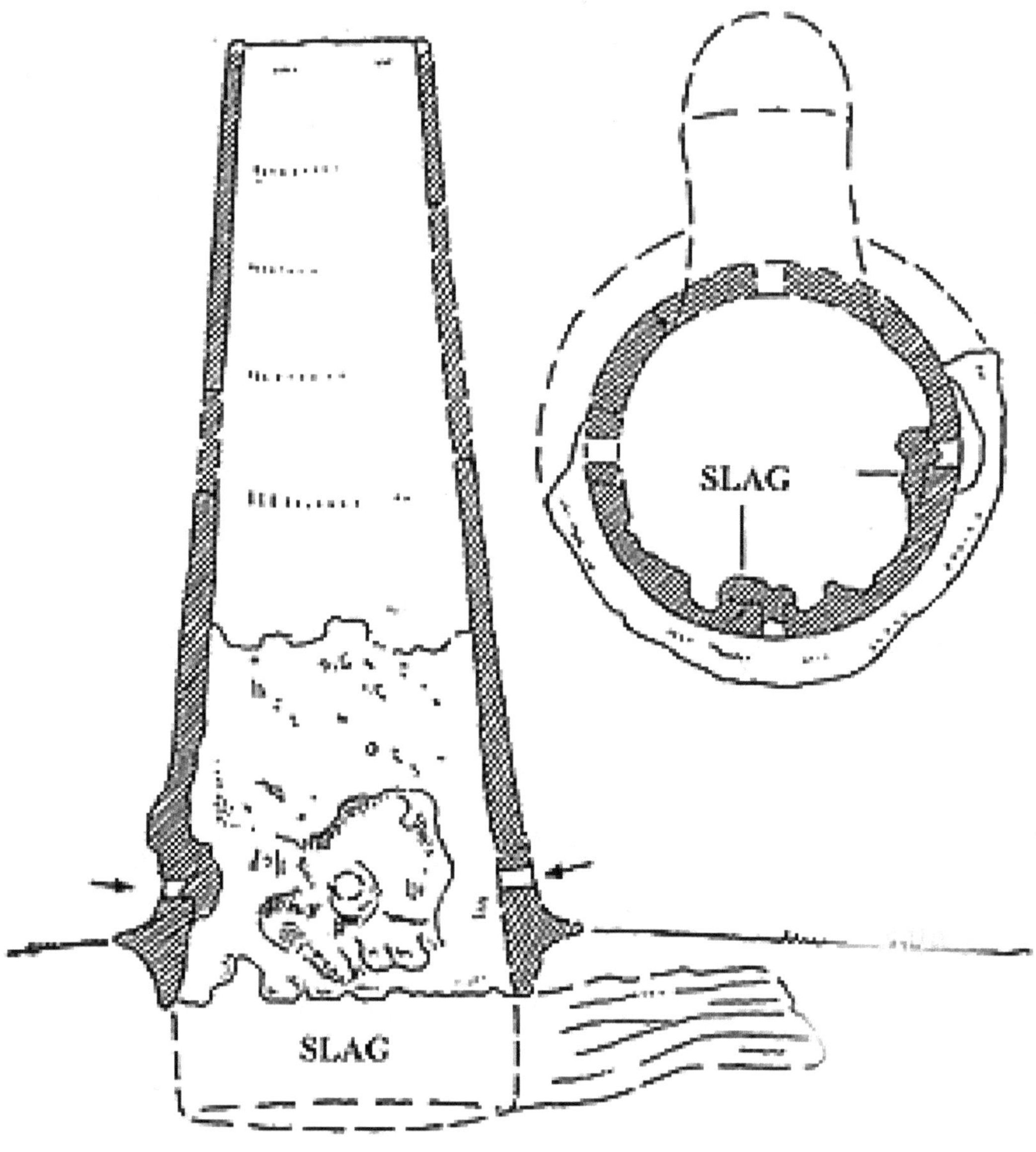

Figure 4
'Lombard' iron making sites in Northern and Central Europe

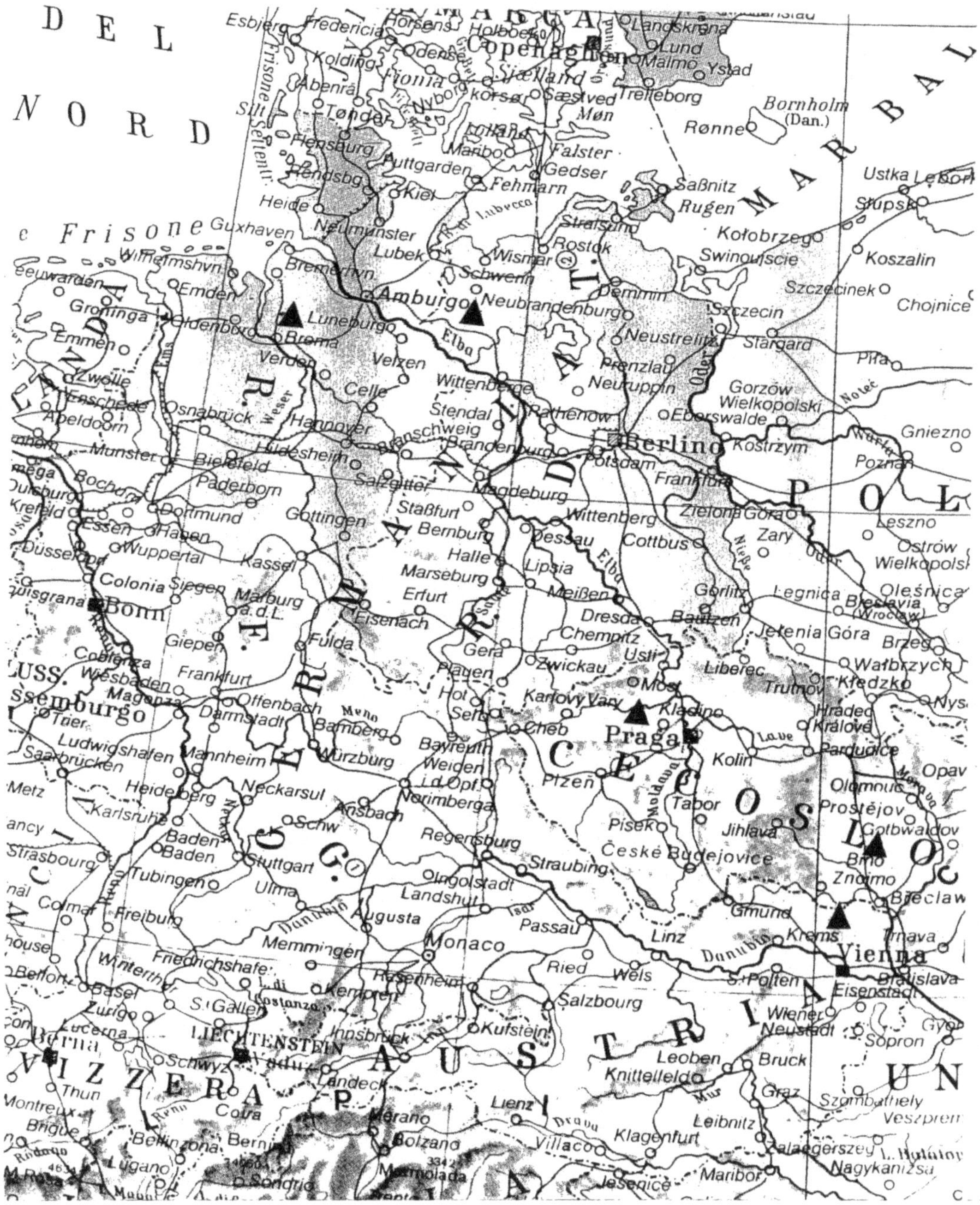

Figure 5
The locations of *Sharmbeck* and *Göhlen* (Germany) iron making sites

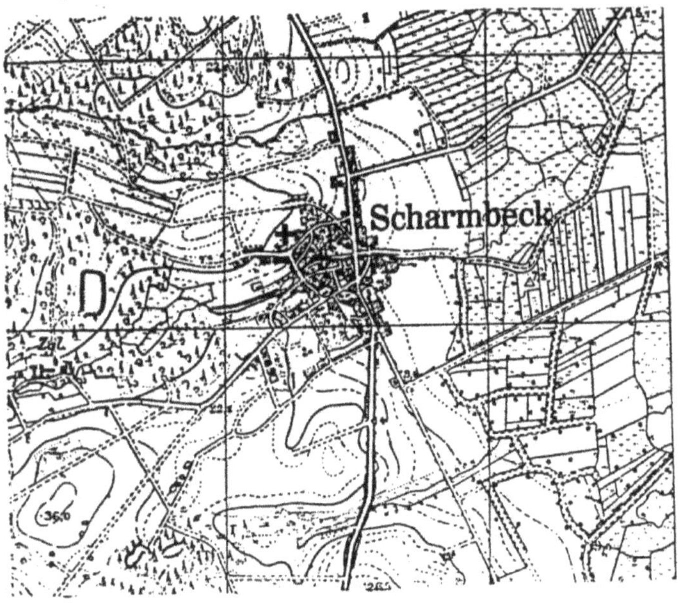

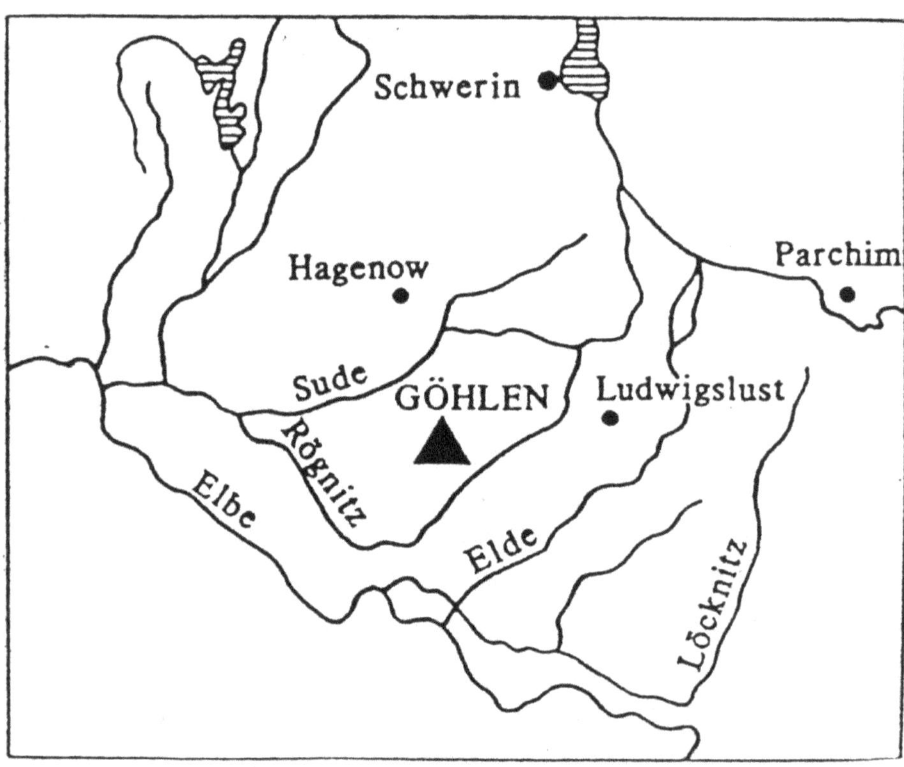

Figure 6
Location of the iron-making site of *Zethlingen* (Germany).
From R. Leineweber 1993

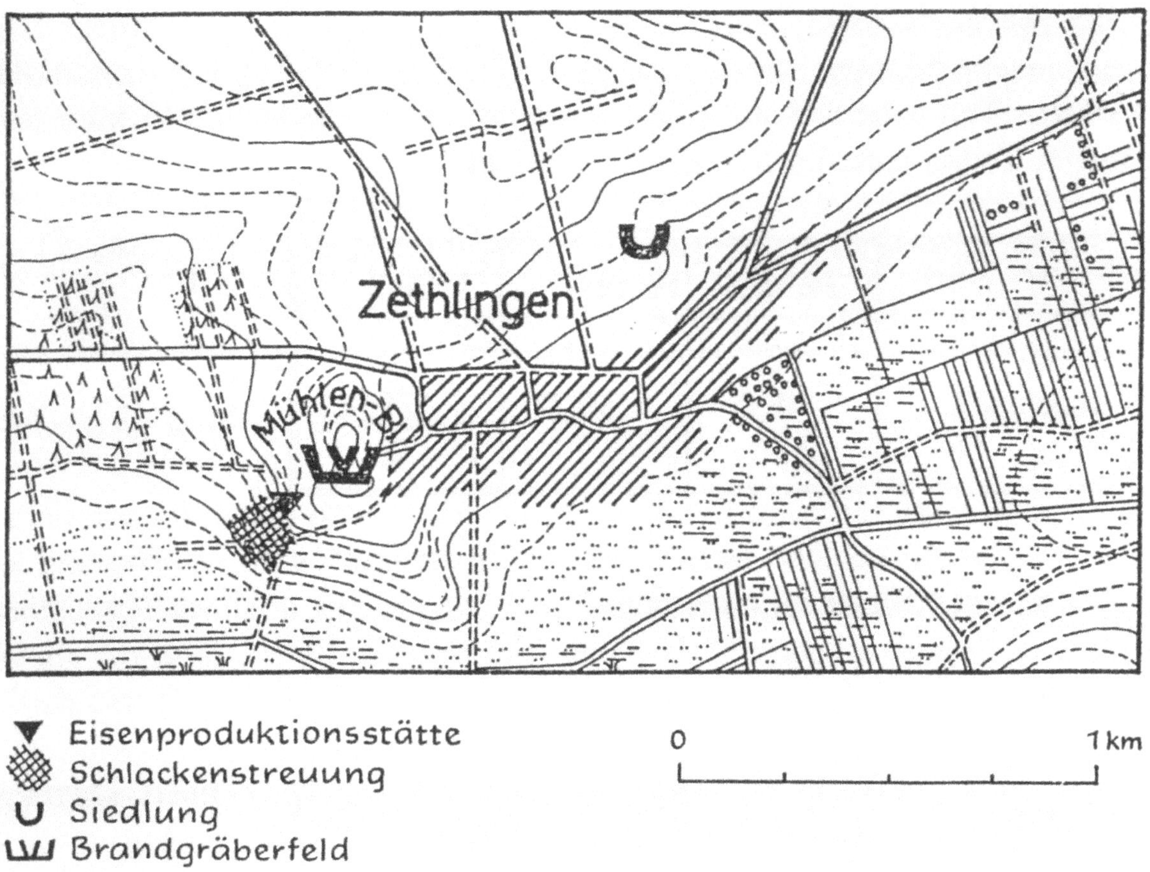

Figure 7
The sets of smithing tools of the graves of *Poysdorf* (Austria) (a) and *Brno* (Czech Republic) (b)

a

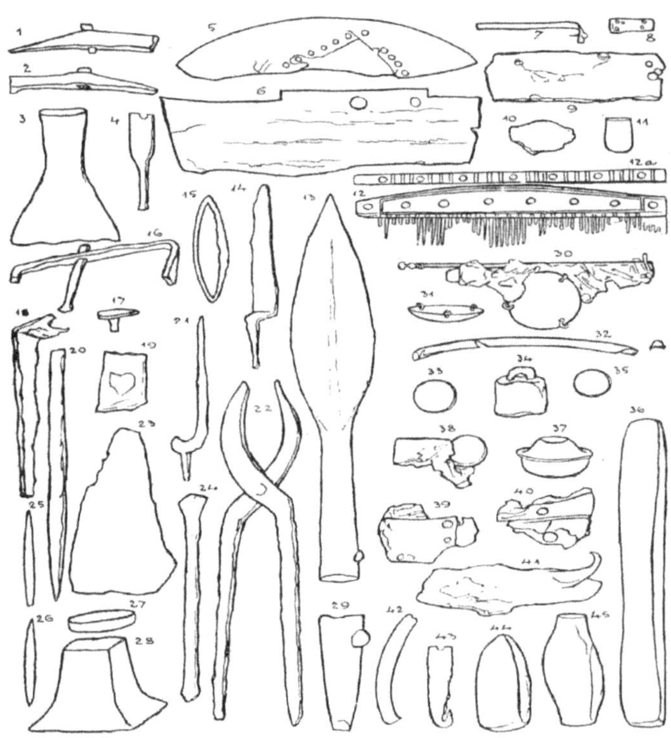

Figure 8
Rendition of the slag pit furnace of Březno
From Pleiner 1995.

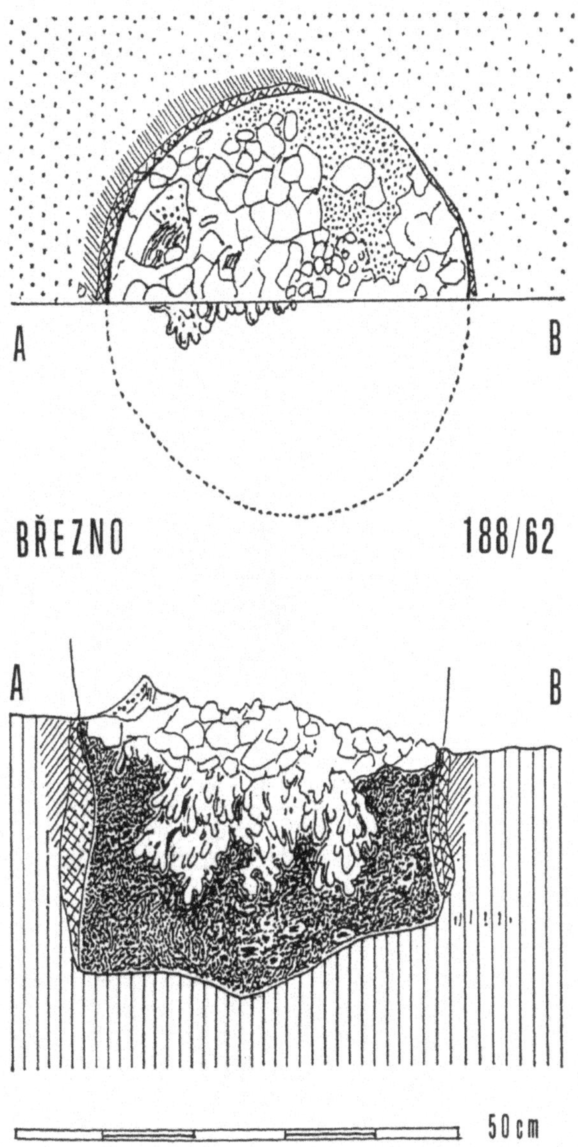

Fig. 2 Brezno. Plan and section of the slag-pit bloomery furnace, feature 188/62.

Figure 9
Finds of iron tools from *Noricum* and *Carinthia*.
From Bitenc P. and T. Knific 2001.

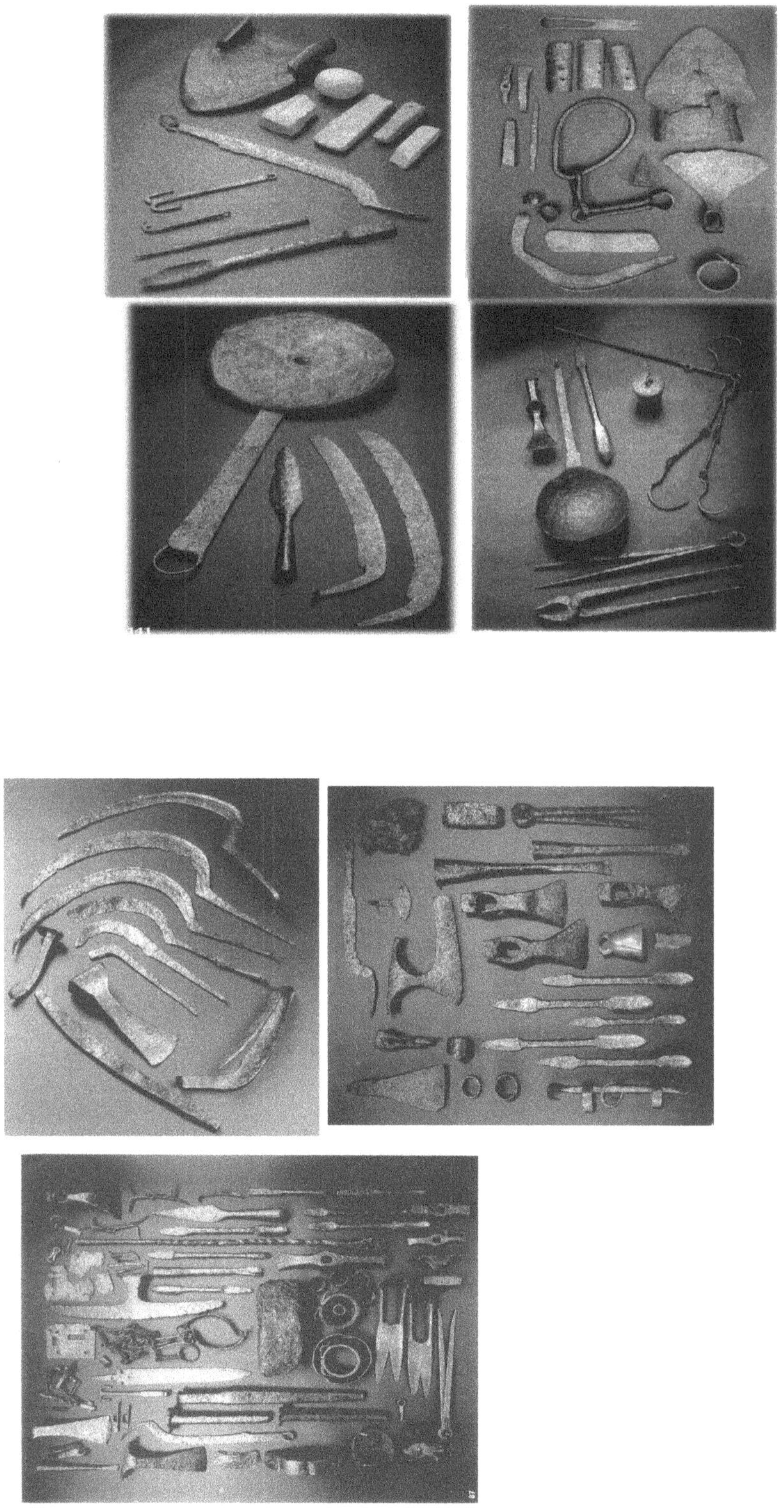

Figure 10
Lombards and Gepids in the Carpathian Basin

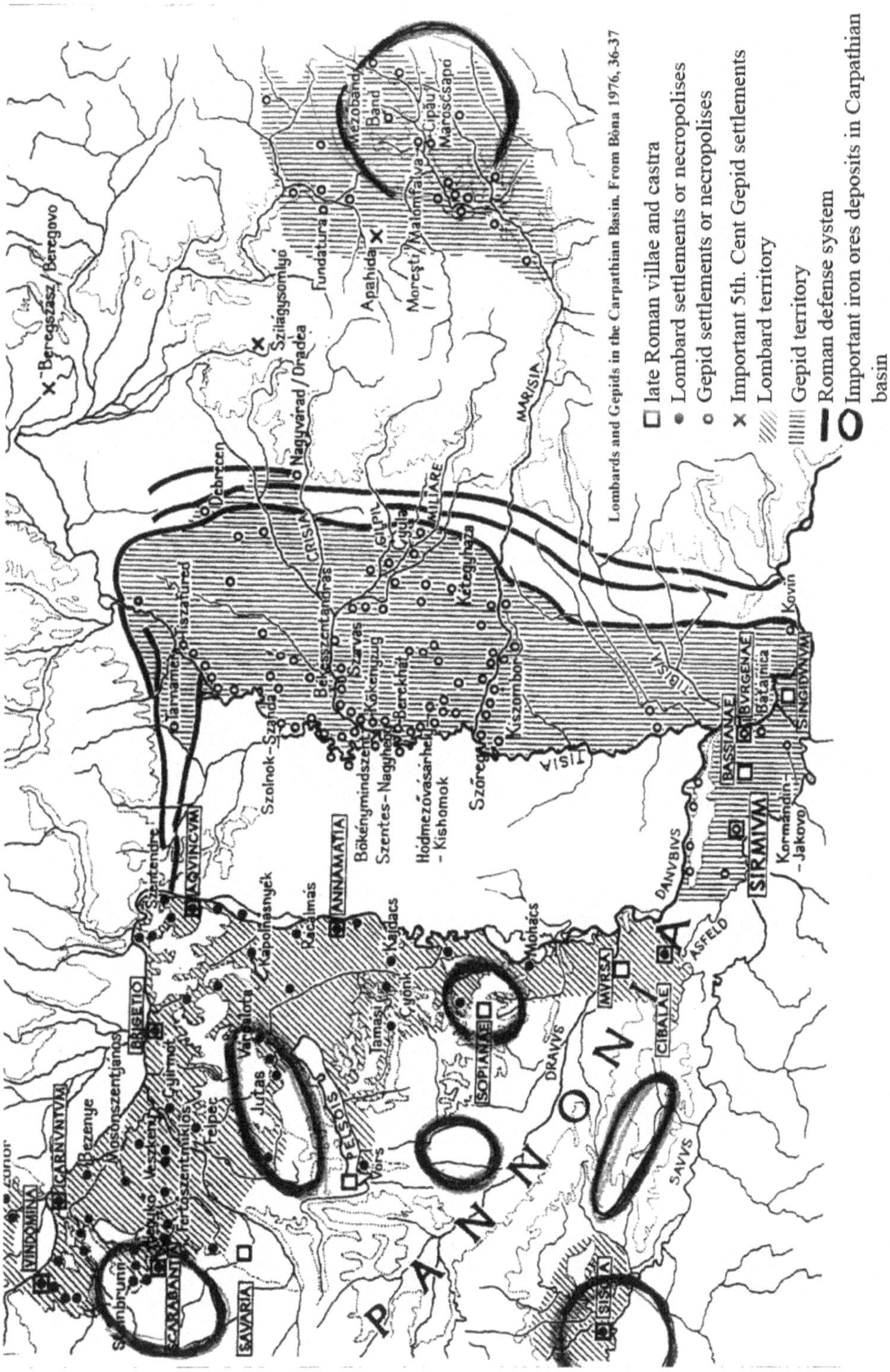

Figure 11
Metallographic microstructures of swords # 65.59.1 (A) and # 65.43.1 (B) from the necropolises of *Hegykő*(Hungary)

DA:
VÝCHODOSLOVENSKÝ PRAVEK, SPECIAL ISSUE, KOŠICE 1999.

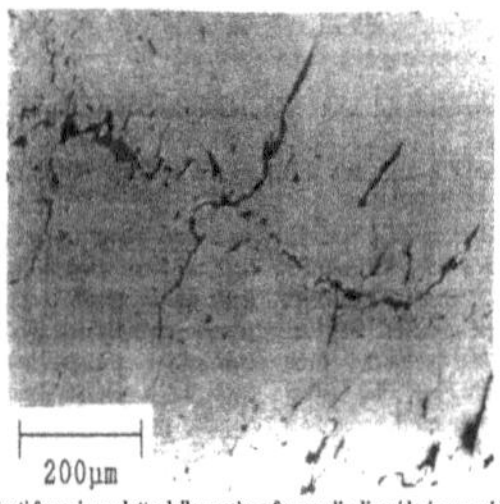

Silicati ferrosi, prodotto della reazione fra scaglie di ossidazione e sabbia nel corso della forgiatura; dalla spada 65.59.1

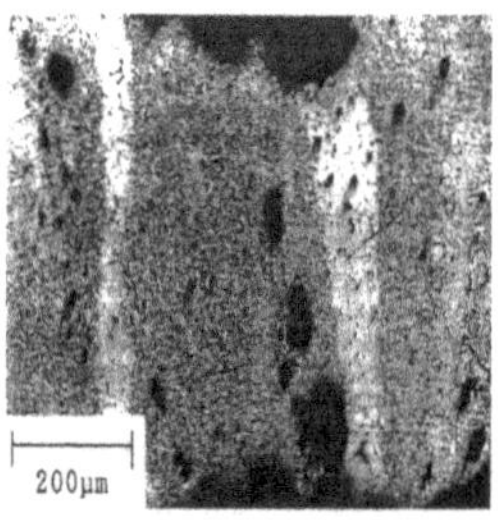

Pattern welding nella sezione dalla spada 65.59.1, dopo attacco col nital.

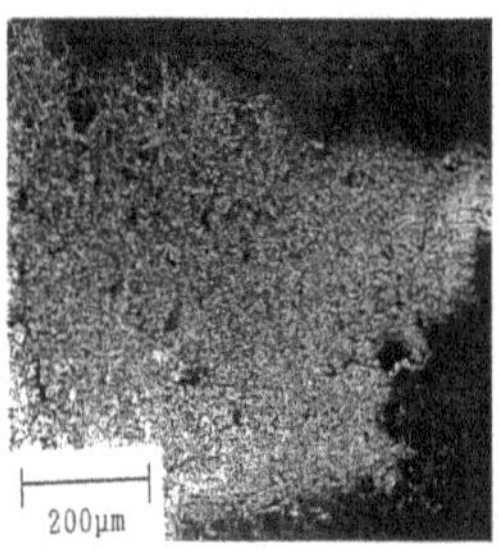

Struttua ferritico-perlitica poco omogenea sulla lama della spada 65.59.1, dopo attacco col nital.

A

Da:
VÝCHODOSLOVENSKÝ PRAVEC, SPECIAL ISSUE, KOSICE 1999

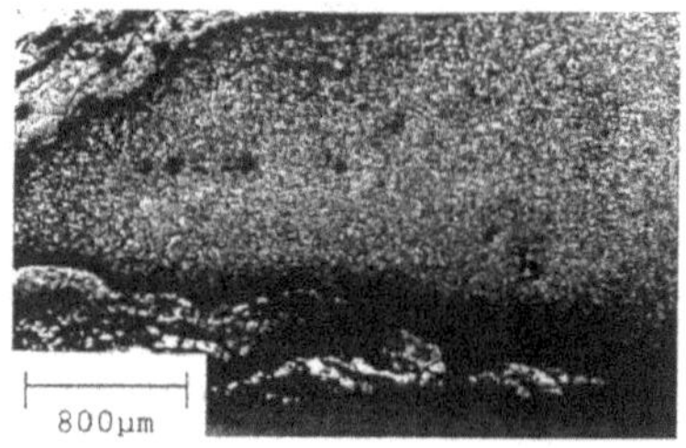

Struttura omogenea sulla lama della spada 65.34.1, dopo attacco col nital.

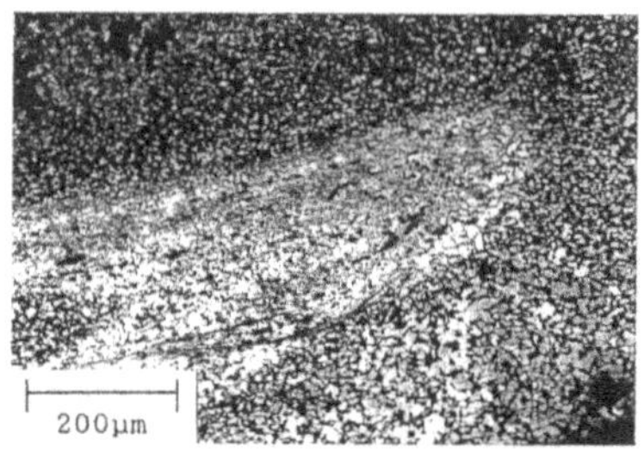

Pattern welding nella sezione dalla spada 65.59.1, dopo attacco col nital.

Dettaglio della omogeneità della struttura sorbitica in parte della spada 65.34.1, dopo attacco col nital.

B

Figure 12
Iron objects from the necropolis of
***Kajdacs-Homokbánya* and *Tamási-Csikólegelő* (Hungary).**
Weaving sword 73.148.1 indicating sampling places

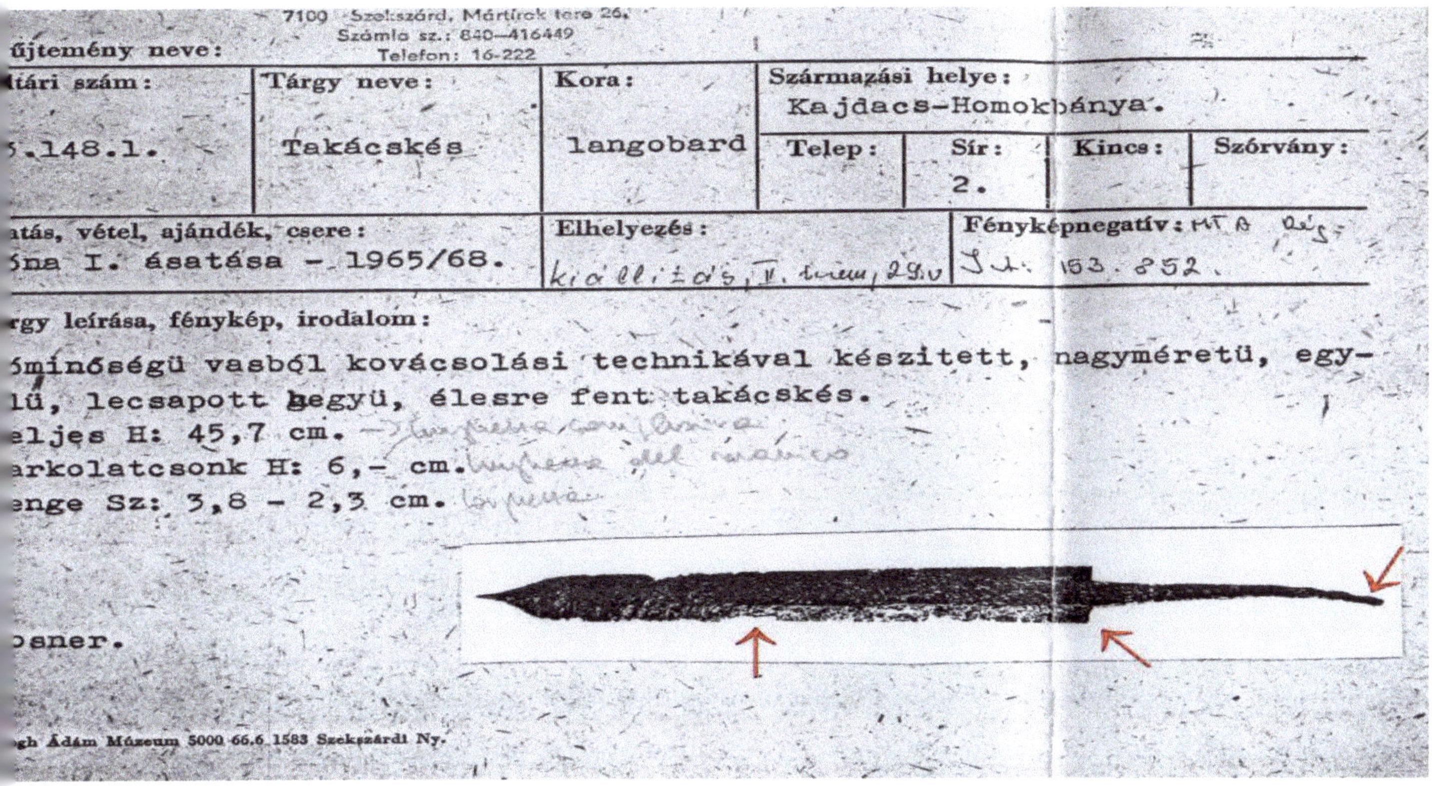
7100 Szekszárd, Mártírok tere 26.
Számla sz.: 840—416449
Telefon: 16-222

ijtemény neve:

tári szám:	Tárgy neve:	Kora:	Származási helye: Kajdacs-Homokbánya.			
5.148.1.	Takácskés	langobard	Telep:	Sír: 2.	Kincs:	Szórvány:

atás, vétel, ajándék, csere:	Elhelyezés:	Fényképnegatív:
óna I. ásatása – 1965/68.	kiállitás, V. terem, 29.v	153. 852.

rgy leírása, fénykép, irodalom:
ominőségü vasból kovácsolási technikával készitett, nagyméretü, egy-
lü, lecsapott hegyü, élesre fent takácskés.
eljes H: 45,7 cm.
arkolatcsonk H: 6,– cm.
enge Sz: 3,8 – 2,3 cm.

sner.

gh Ádám Múzeum 5000 66.6 1583 Szekszárdi Ny.

Figure 13
Iron objects from the necropolis of *Kajdacs-Homokbánya* and *Tamási-Csikólegelő* (Hungary). Metallographic anlyses of weaving sword 73.148.1. Pictures 1-6

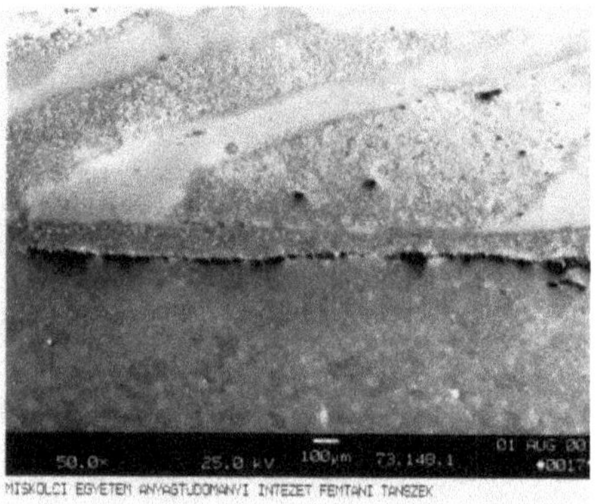

Picture 1

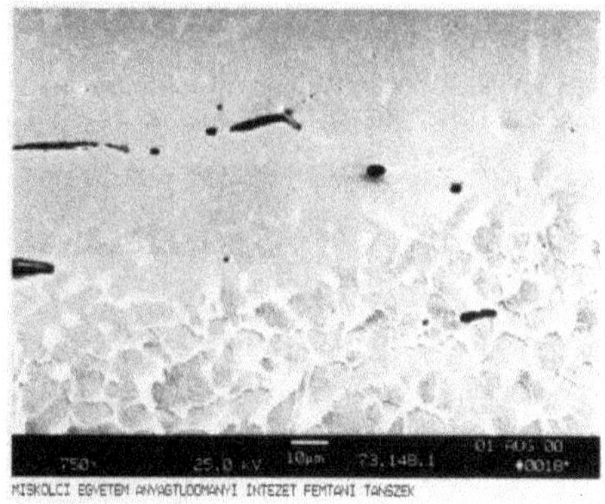

Picture 2

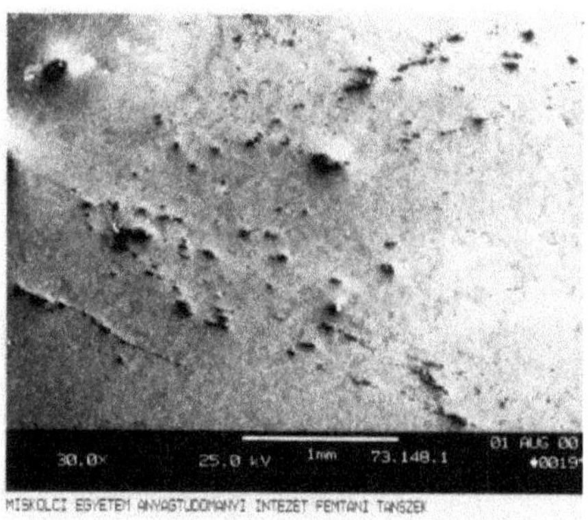

Picture 3

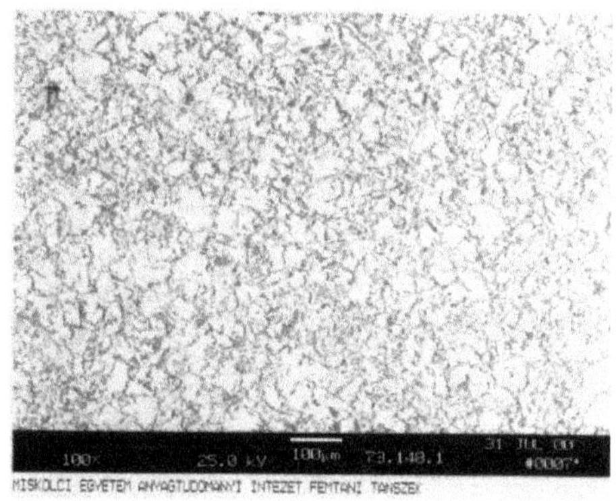

Picture 4

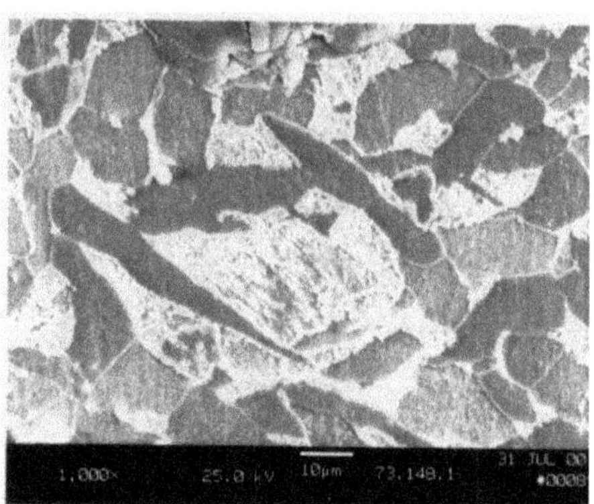

Picture 5

Picture 6

Figure 14
Iron objects from the necropolis of
***Kajdacs-Homokbánya* and *Tamási-Csikólegelő* (Hungary).**
Metallographic anlyses of weaving sword 73.148.1
Pictures 7-8

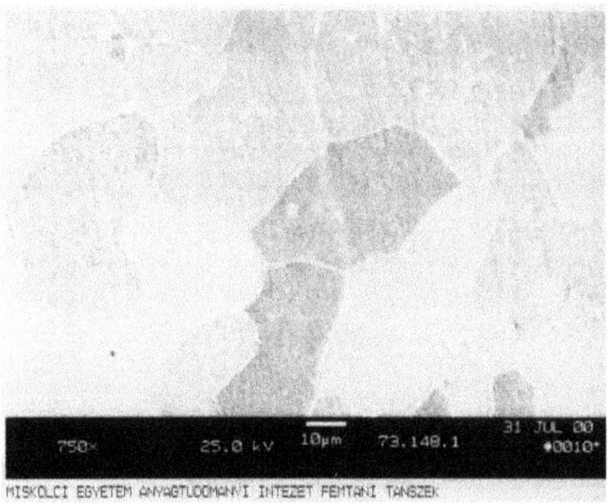

Picture 8

Picture 9

Figure 15
Small "weaving sword" from the Lombard necropolis of *Collegno* (Province of *Turin*, Piedmont). Grave 60 AD 660 ca. 13.8 cm long and 1.8 cm wide, the tang measuring 3.5 cm. Courtesy of Dr. C. Giostra

Figure 16
Iron objects from the necropolis of
Kajdacs-Homokbánya and *Tamási-Csikólegelő* (Hungary).
Iron knife 75.6.4

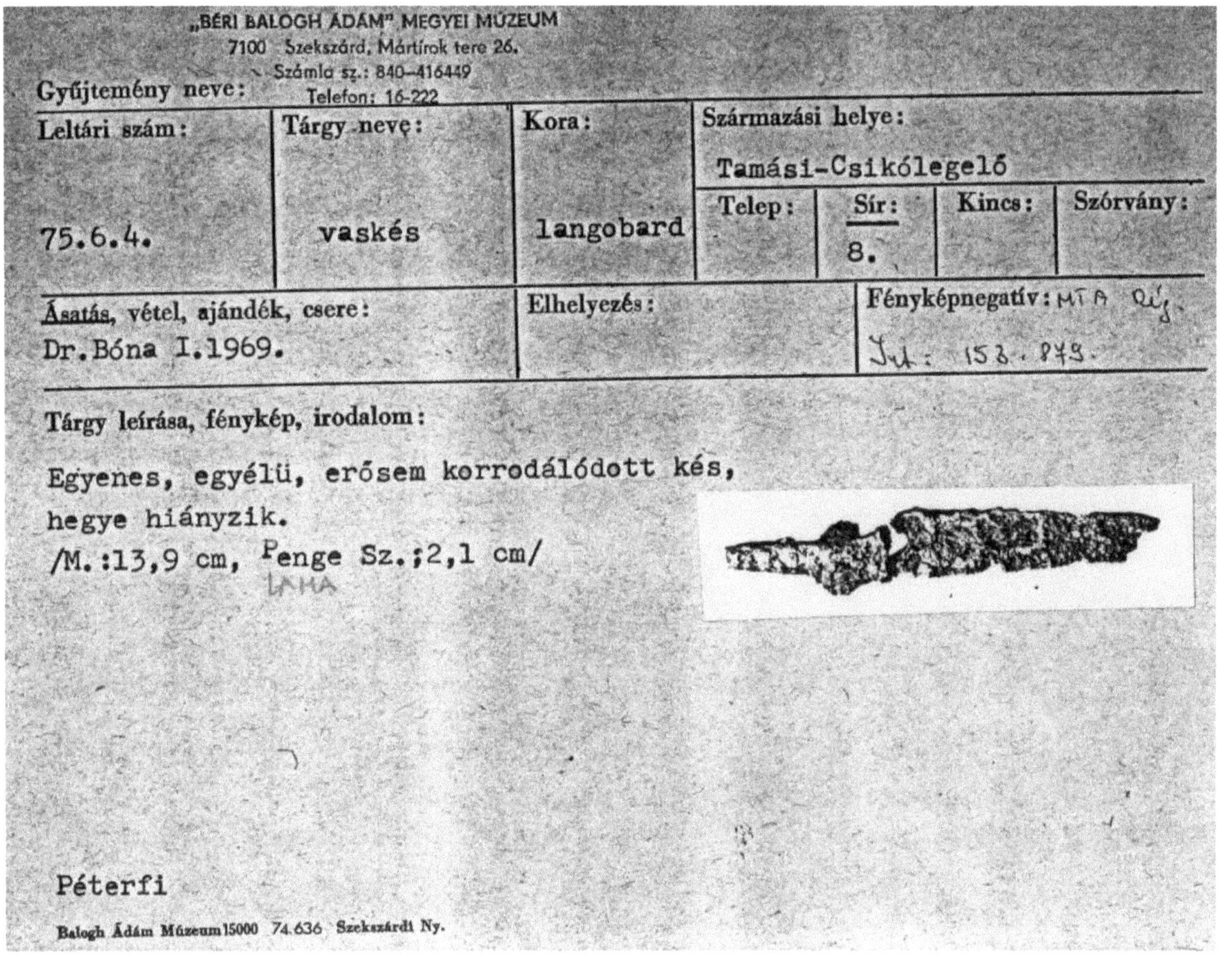
„BÉRI BALOGH ÁDÁM" MEGYEI MÚZEUM
7100 Szekszárd, Mártírok tere 26.
Számla sz.: 840–416449
Telefon: 16-222

Gyűjtemény neve:

Leltári szám:	Tárgy neve:	Kora:	Származási helye: Tamási-Csikólegelő			
75.6.4.	vaskés	langobard	Telep:	Sír: 8.	Kincs:	Szórvány:

Ásatás, vétel, ajándék, csere:	Elhelyezés:	Fényképnegatív: MTA Rég.
Dr.Bóna I.1969.		Ltsz: 153.879.

Tárgy leírása, fénykép, irodalom:

Egyenes, egyélü, erősem korrodálódott kés,
hegye hiányzik.
/M.:13,9 cm, Penge Sz.;2,1 cm/

Péterfi

Balogh Ádám Múzeum15000 74.636 Szekszárdi Ny.

Figure 17
Iron objects from the necropolis of
***Kajdacs-Homokbánya* and *Tamási-Csikólegelő* (Hungary).**
Metallographic analysis of iron knife 75.6.4
Picutres 9-11

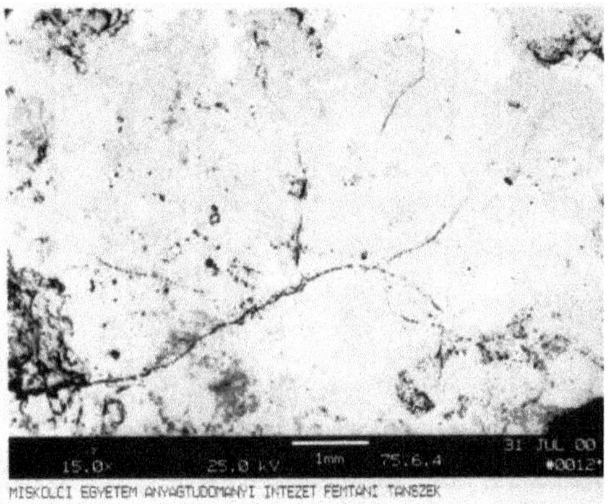

Picture 9

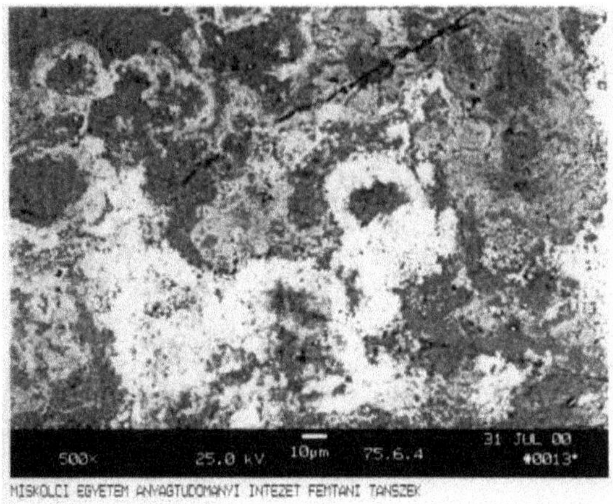

Picture 10

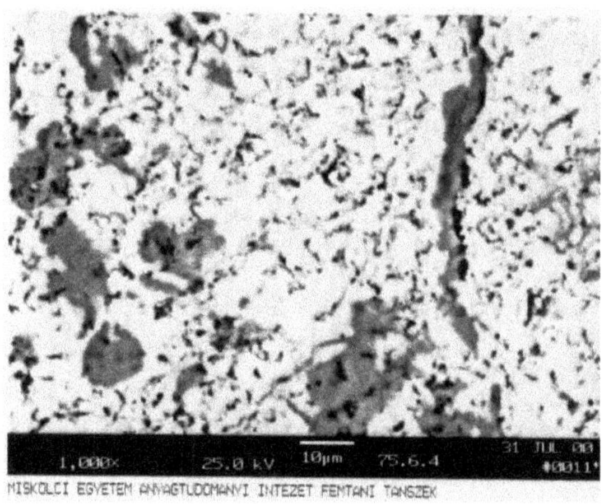

Picture 11

Figure 18
Iron objects from the necropolis of
***Kajdacs-Homokbánya* and *Tamási-Csikólegelő* (Hungary).**
Iron drill 75.18.4

„BÉRI BALOGH ÁDÁM" MEGYEI MÚZEUM
7100 Szekszárd, Mártírok tere 26.
Számla sz.: 840–416449
Telefon: 16-222

Gyűjtemény neve:

Leltári szám:	Tárgy neve:	Kora:	Származási helye:
75.18.4.	vas furó	langobard	Tamási-Csikólegelő

Telep:	Sír:	Kincs:	Szórvány:
	2o.		

Ásatás, vétel, ajándék, csere: Dr.Bóna I. 1969.

Elhelyezés:

Fényképnegatív: MTA Rég. Ltsz: 153.877.

Tárgy leírása, fénykép, irodalom:

Nagyméretü, kör átmetszetü, kihegyezett vasrud. Ellentett végét kissé ellapitották s hurokba visszakalapálták.
/H.:2o,7 cm, V.:1,3 cm/

Figure 19
Iron objects from the necropolis of *Kajdacs-Homokbánya* and *Tamási-Csikólegelő* (Hungary) .
Metallographic analysis of iron drill 75.18.4. Pictures 12-17

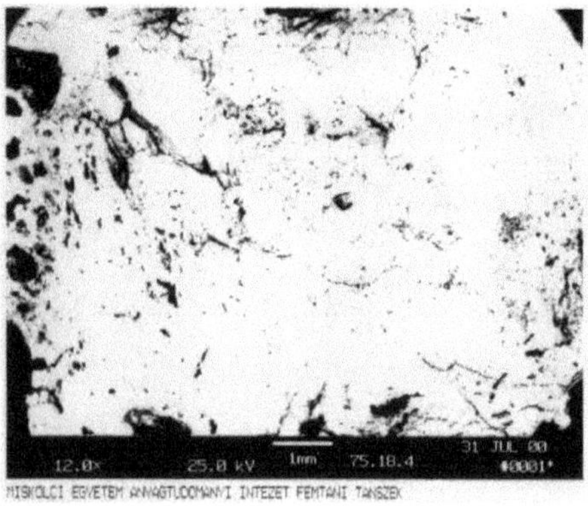

Picture 12

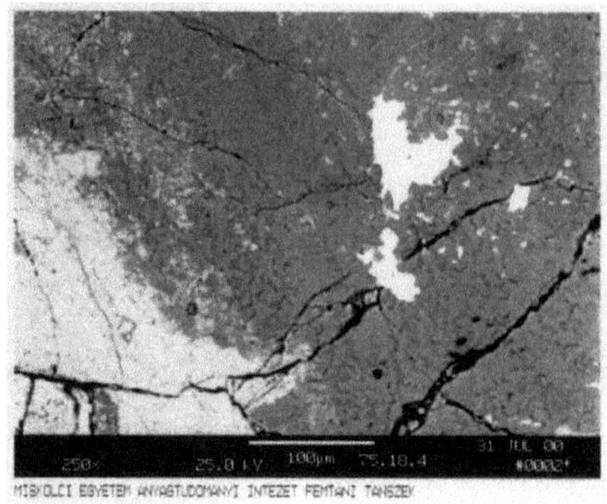

Picture 13

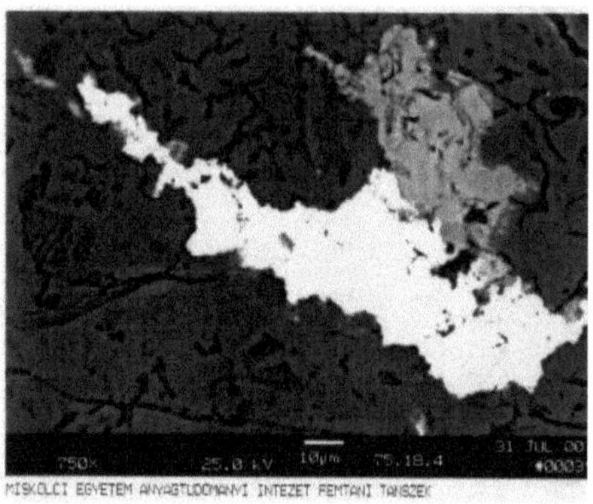

Picture 14

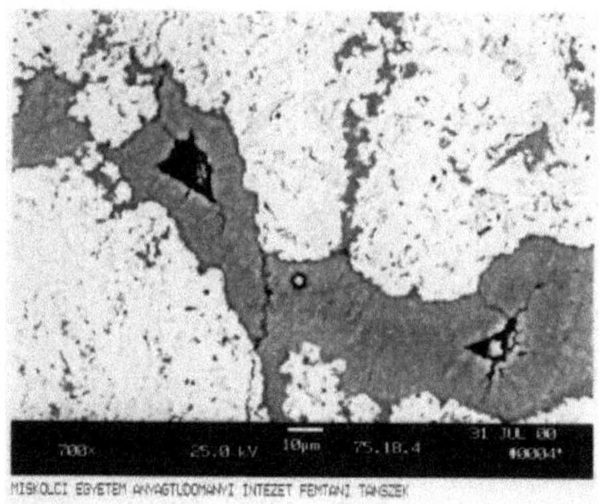

Picture 15

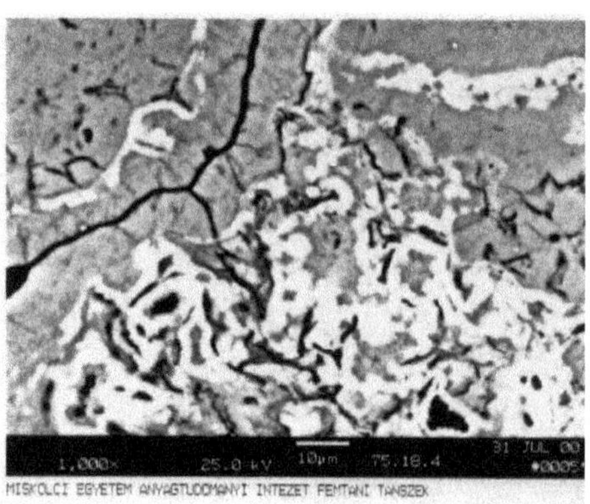

Picture 16

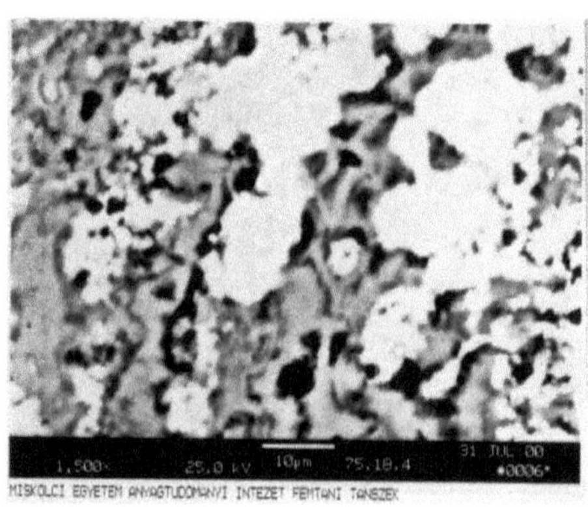

Picture17

Figure 20
Iron objects from the necropolis of
***Kajdacs-Homokbánya* and *Tamási-Csikólegelő* (Hungary).**
Iron shield fragments 75.22.2

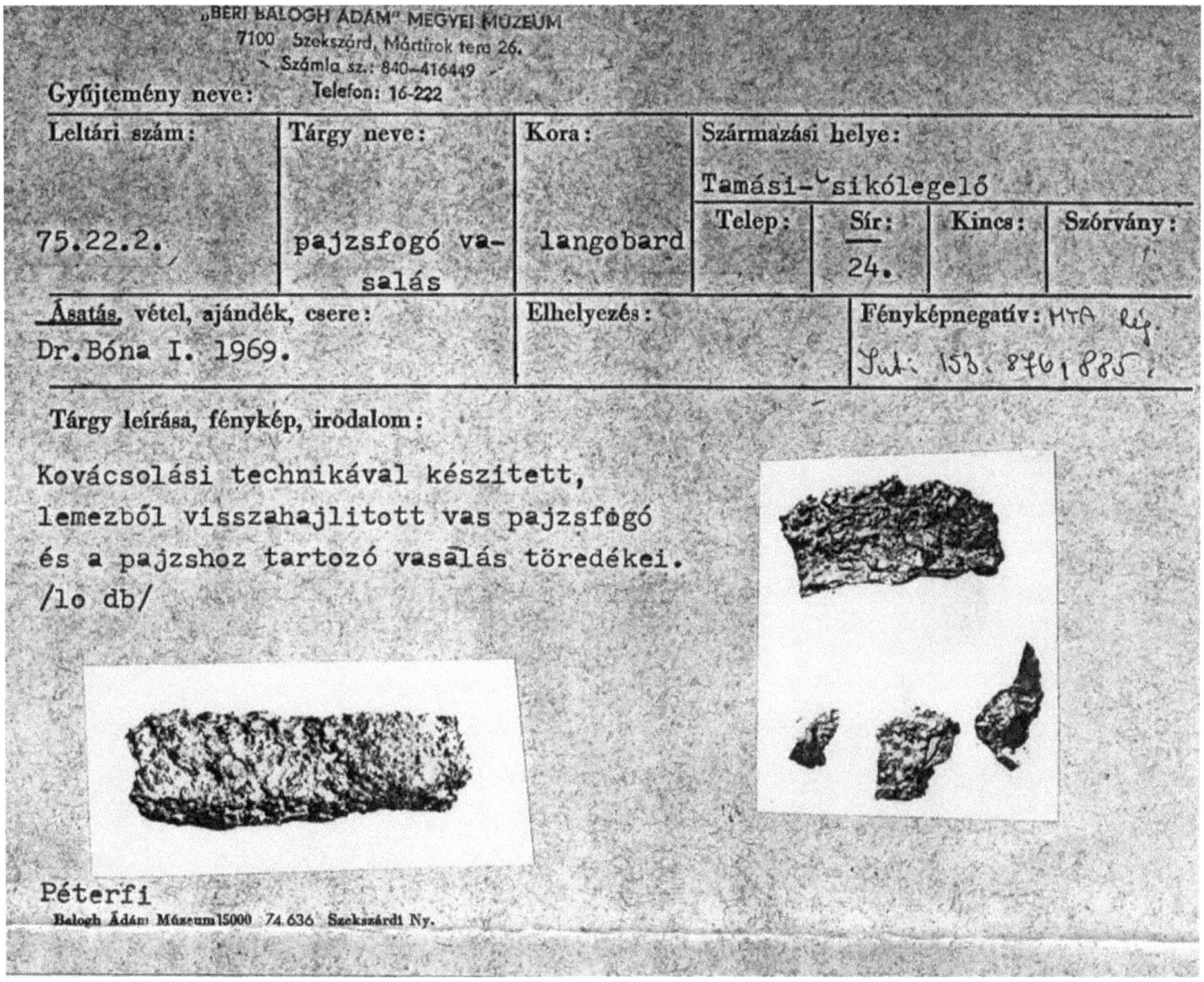

„BÉRI BALOGH ÁDÁM" MEGYEI MÚZEUM
7100 Szekszárd, Mártírok tere 26.
Számla sz.: 840–416449
Telefon: 16-222

Gyűjtemény neve:

Leltári szám:	Tárgy neve:	Kora:	Származási helye: Tamási-Csikólegelő			
			Telep:	Sír:	Kincs:	Szórvány:
75.22.2.	pajzsfogó vasalás	langobard		24.		

Ásatás, vétel, ajándék, csere:	Elhelyezés:	Fényképnegatív:
Dr. Bóna I. 1969.		

Tárgy leírása, fénykép, irodalom:

Kovácsolási technikával készitett, lemezből visszahajlitott vas pajzsfogó és a pajzshoz tartozó vasalás töredékei. /lo db/

Péterfi

Balogh Ádám Múzeum 15000 74.636 Szekszárdi Ny.

Figure 21
Iron objects from the necropolis of
***Kajdacs-Homokbánya* and *Tamási-Csikólegelő* (Hungary).**
Metallographic analysis of fragmens of iron handle of a shield 75.22.2.
Pictures 19-20

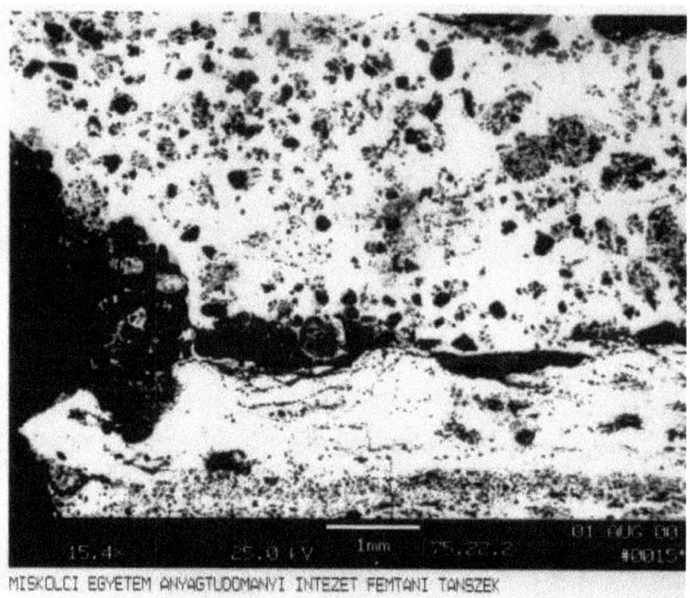

Picture 19

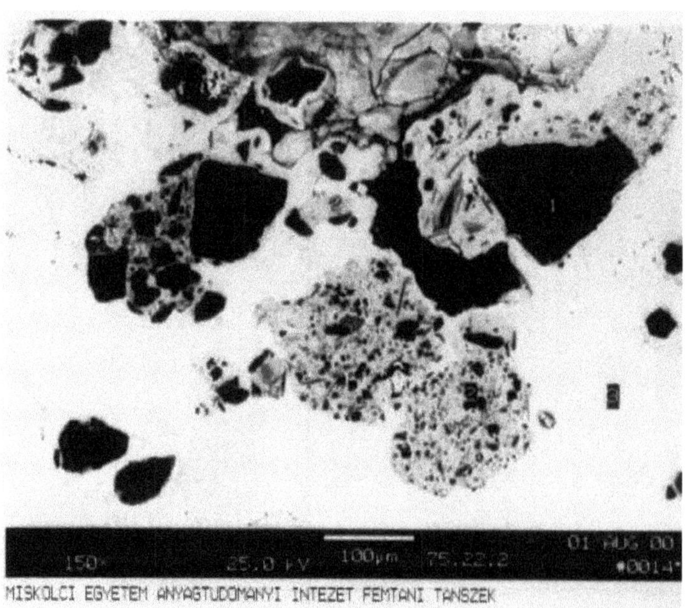

Picture 20

Figure 22
Iron objects from the necropolis of
***Kajdacs-Homokbánya* and *Tamási-Csikólegelő* (Hungary).**
Composition spectrum of spots 1, 2, 3, seen in Picture 20.
Fragmens of Iron Handle of a Shield 75.22.2

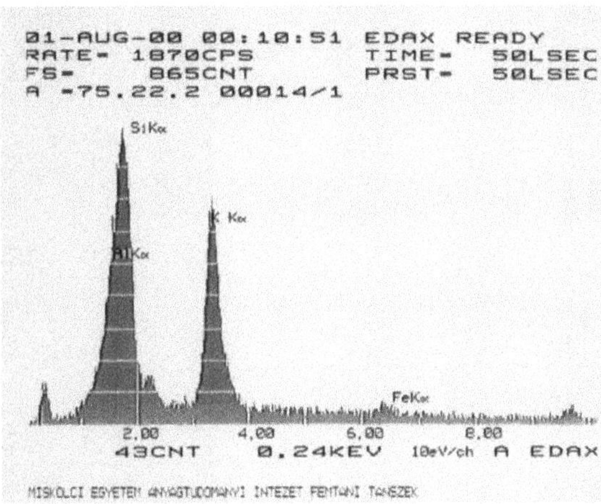

Spectrum 1

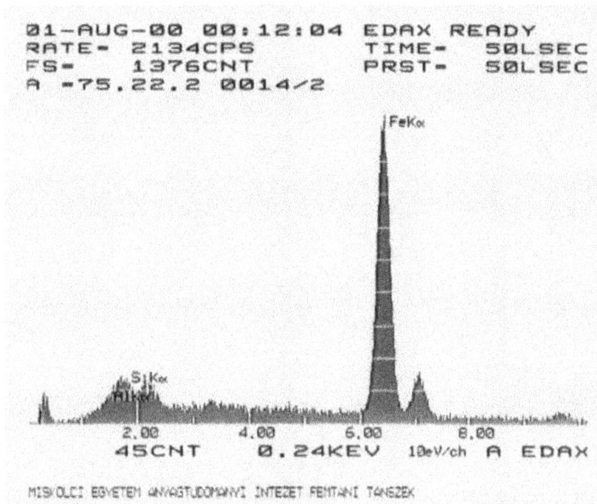

Spectrum 2

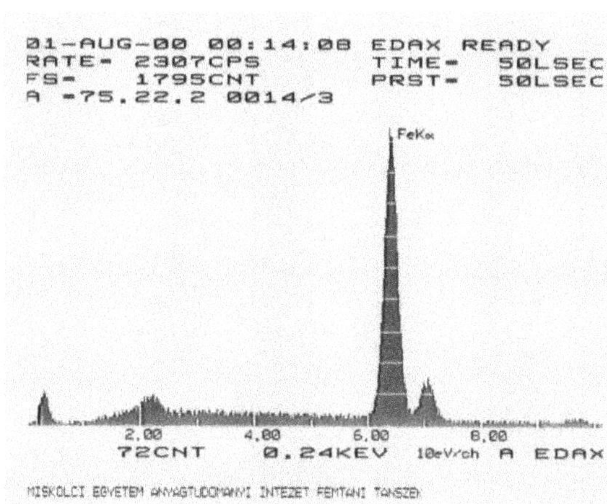

Spectrum 3

Figure 23
Agricultural tools from the site of *Belmonte* (Piedmont, Italy).
The spear like ploughshares

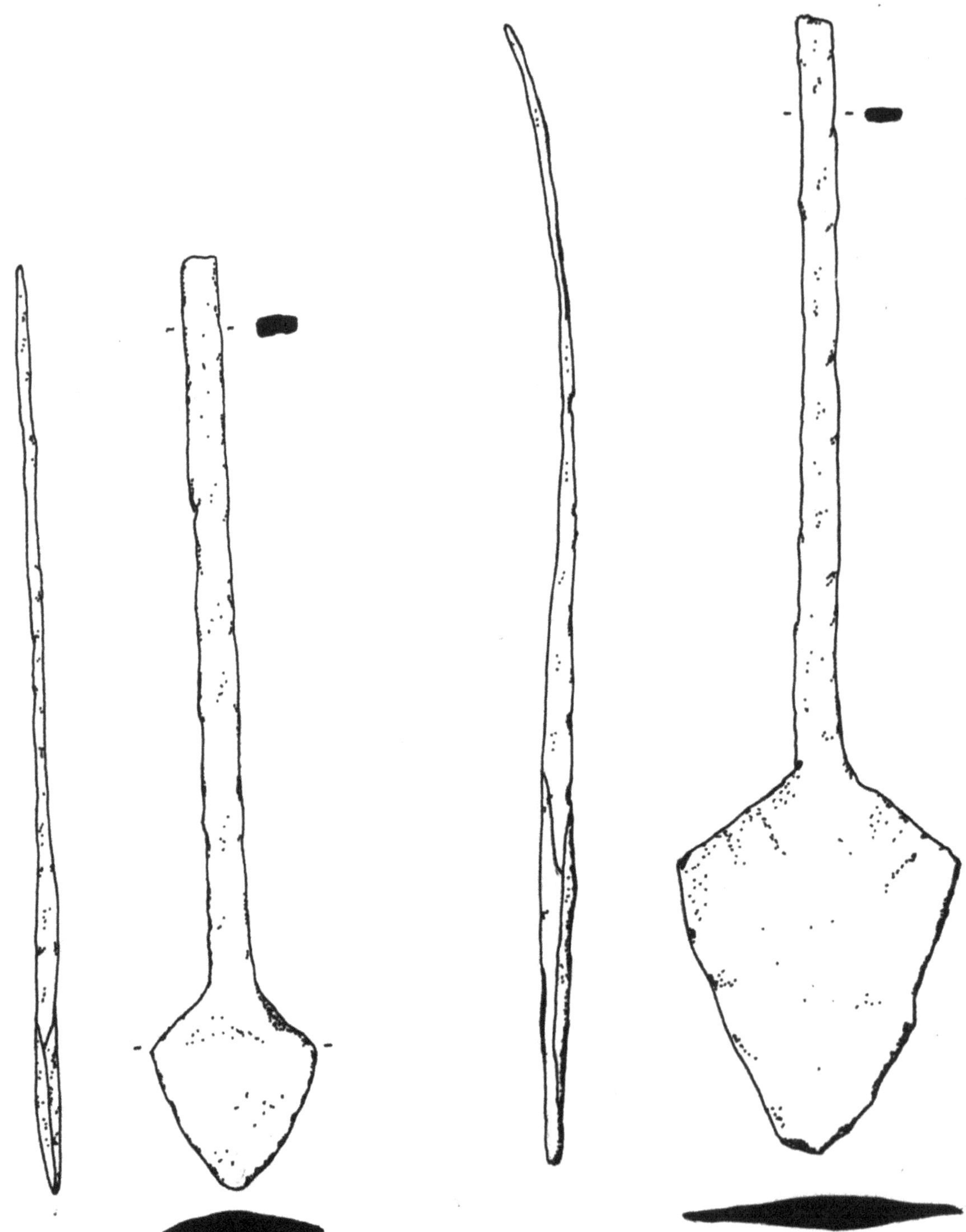

Figure 24
Agricultural tool from *Raetia.*
The spear like ploughshare

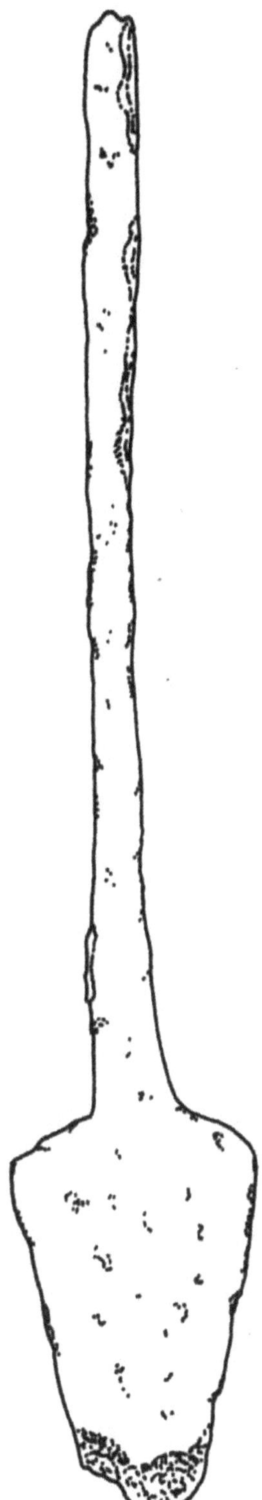

Figure 25
Distribution map of the finds of different ploughshares within Central Europe between Late Antiquity and Early Middle Ages.
From J. Henning 1986.

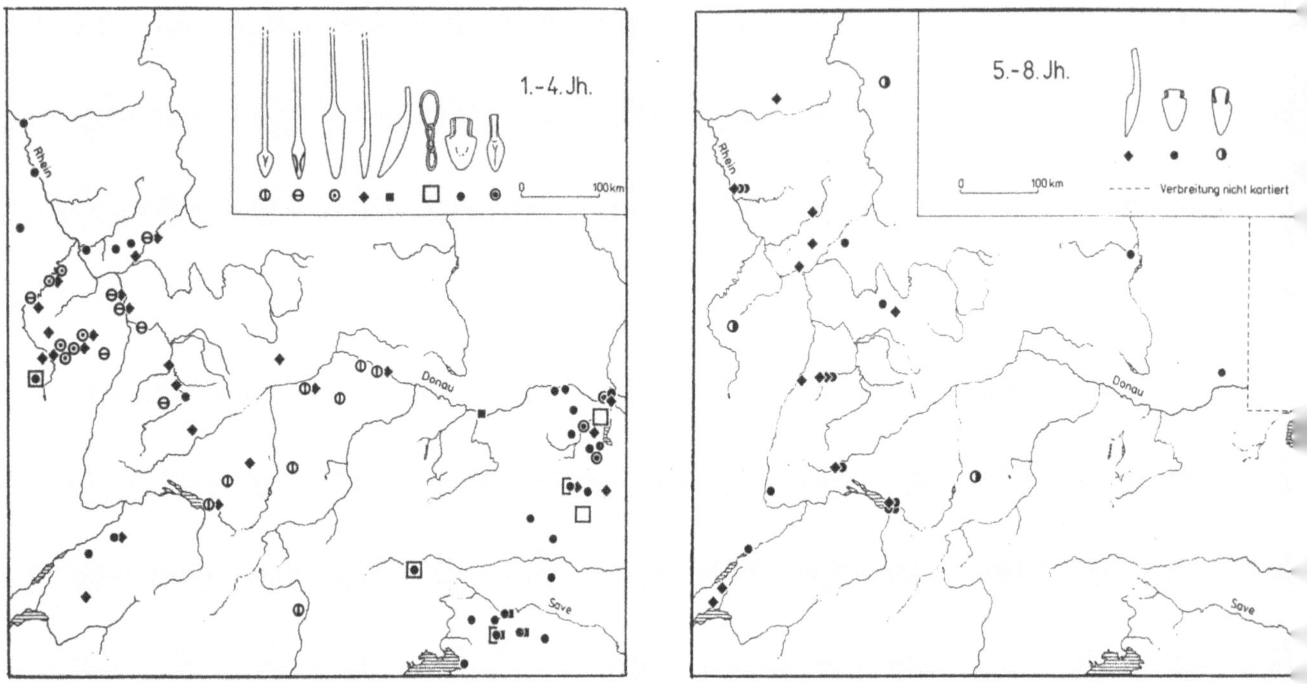

Figure 26
Distribution map of finds of different types of ploughshares between Late Antiquity and Early Middle Ages within south-eastern Europe. From J. Henning 1987

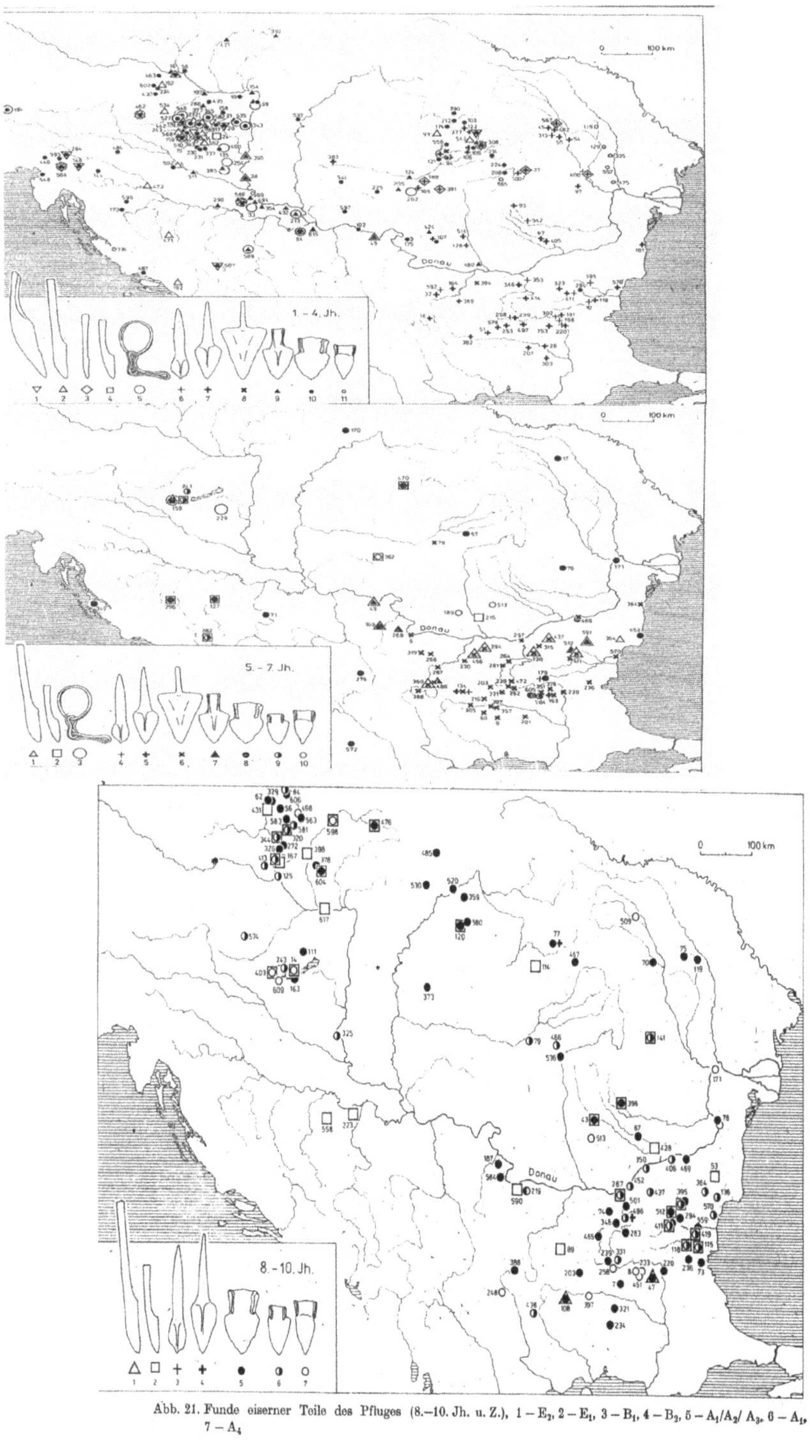

Abb. 21. Funde eiserner Teile des Pfluges (8.–10. Jh. u. Z.), 1 – E_2, 2 – E_1, 3 – B_1, 4 – B_2, 5 – $A_1/A_2/A_3$, 6 – A_1, 7 – A_4

Figure 27
Agricultural tools from the site of *Belmonte* (Piedmont, Italy).
The hoe and the spade

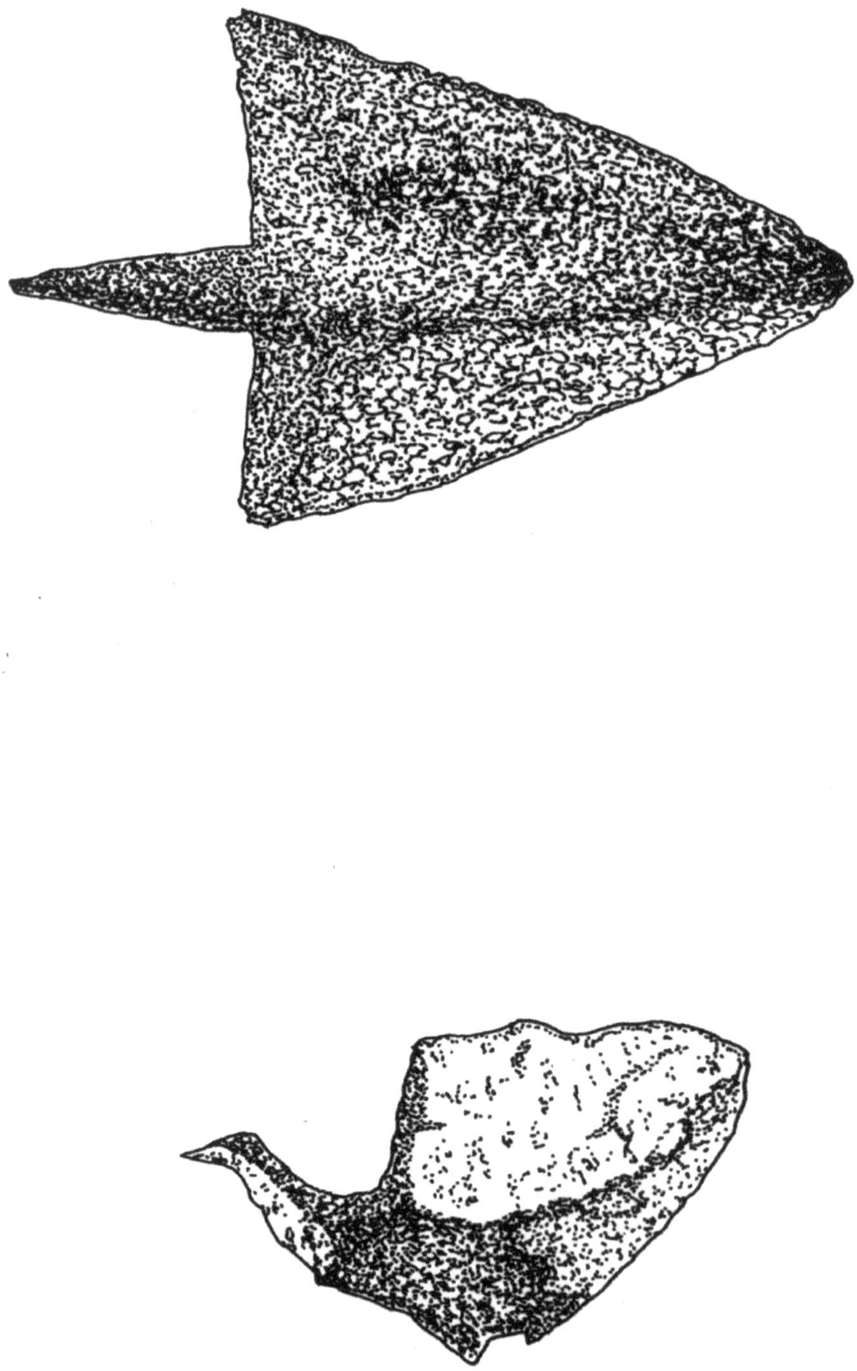

Figure 28
Agricultural tools. Picks from the site of *Bemonte* (Piedmont, Italy) and *Villa Clelia* (A) and from Late Roman *Pannonia* (B)

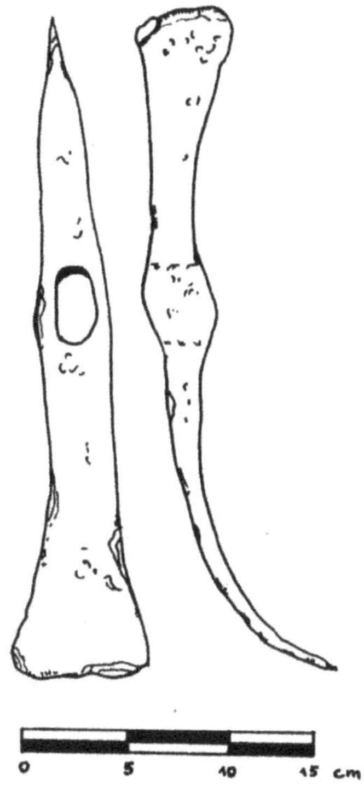

B

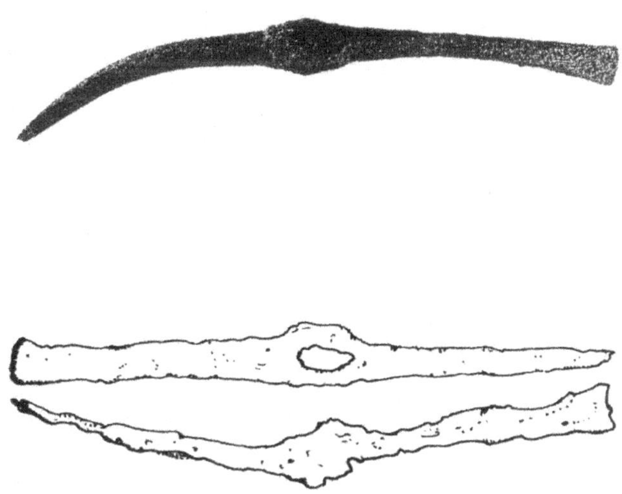

A

Figure 29
Development of agricultural tools between Late Antiquity and Early Middle Ages within south-eastern Europe.
From J. Henning 1987

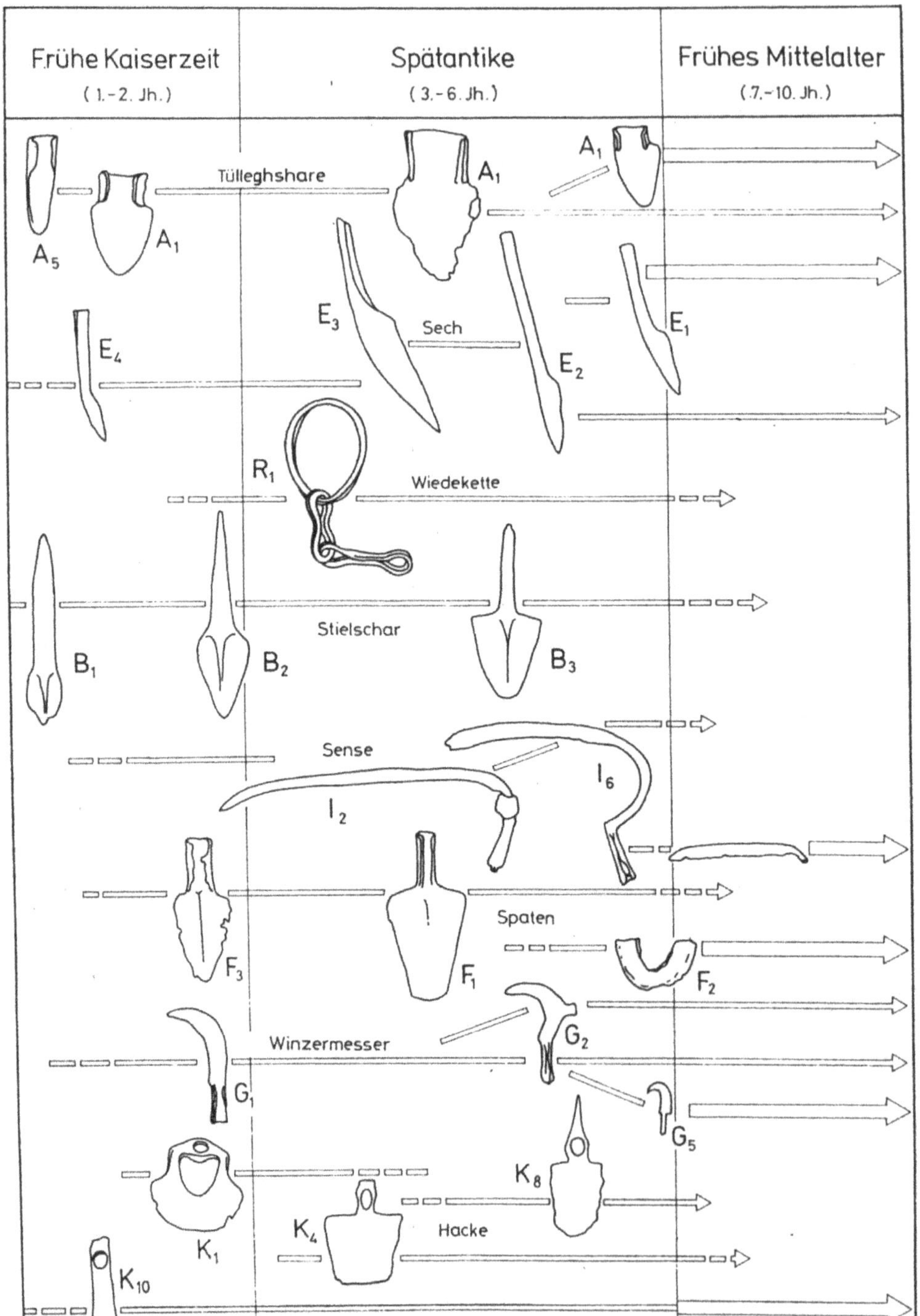

Figure 30
Development of real estates and villae betwen AD 1st and 4th within south-eastern Europe.
From J. Henning 1987

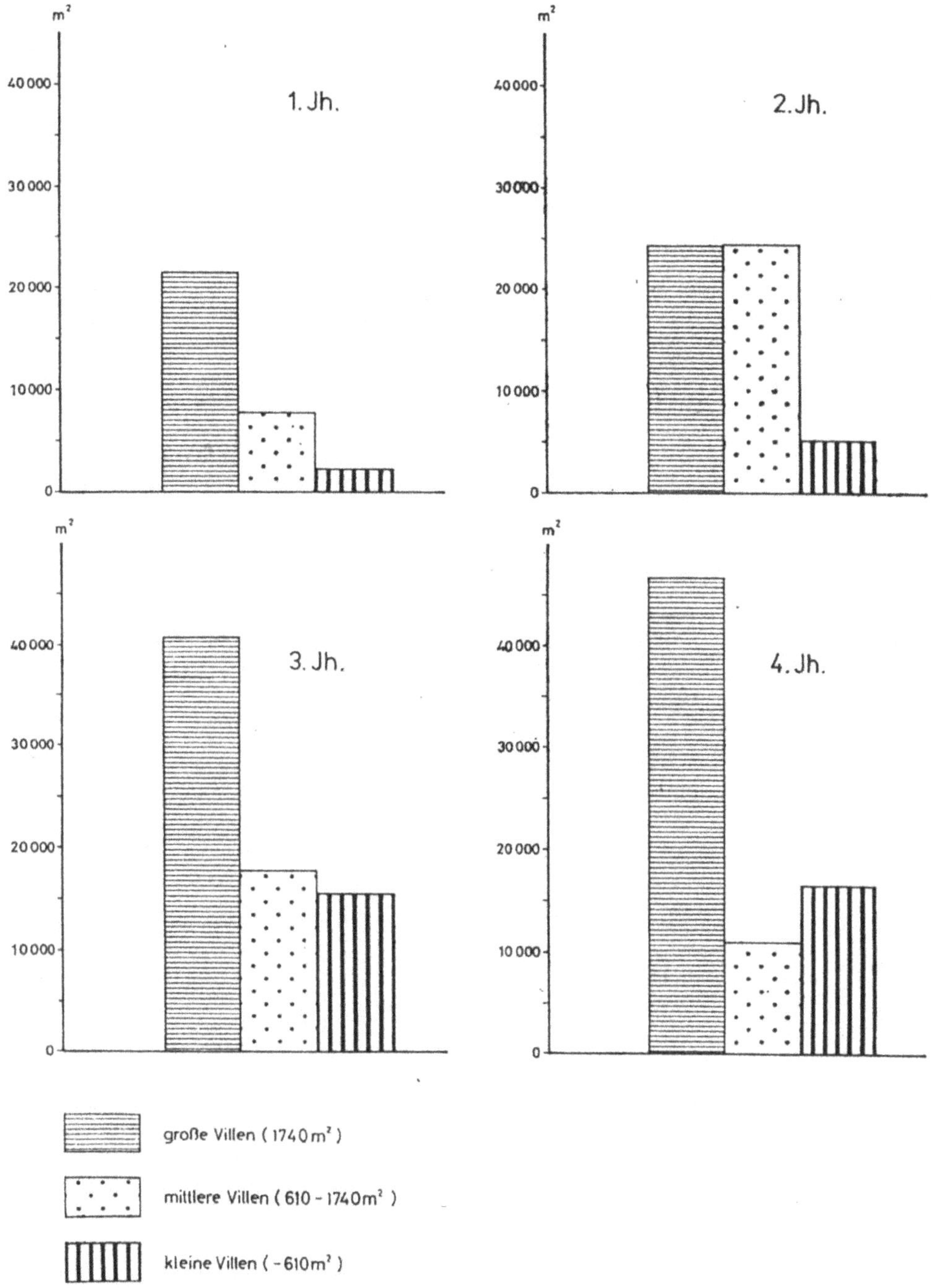

Figure 31
Patterns of the swords from
***Nocera Umbra* necropolis (Umbria, Italy).**
Courtesy of Dr. C. Rupp

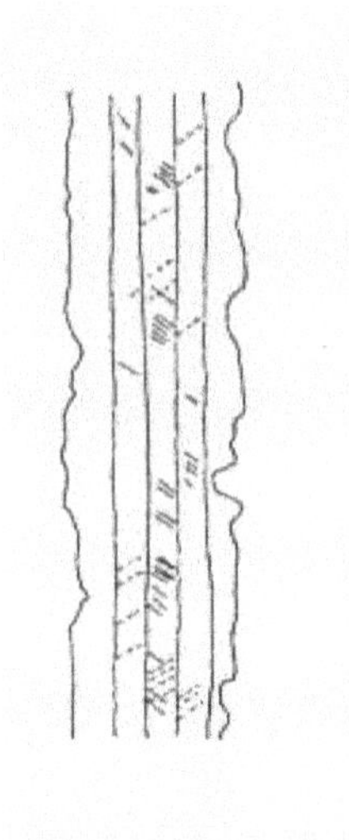

Group 1

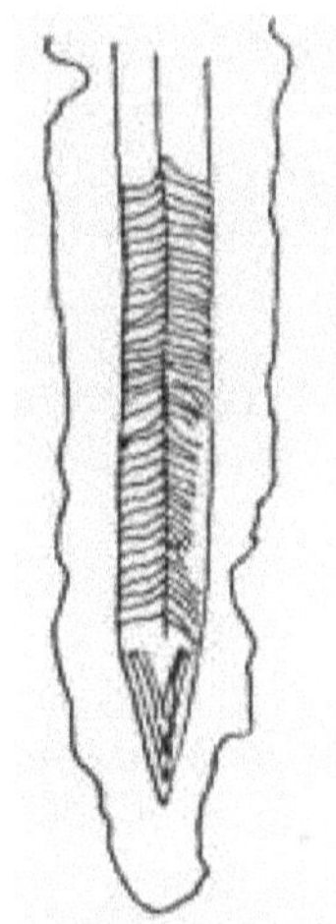

Group 2

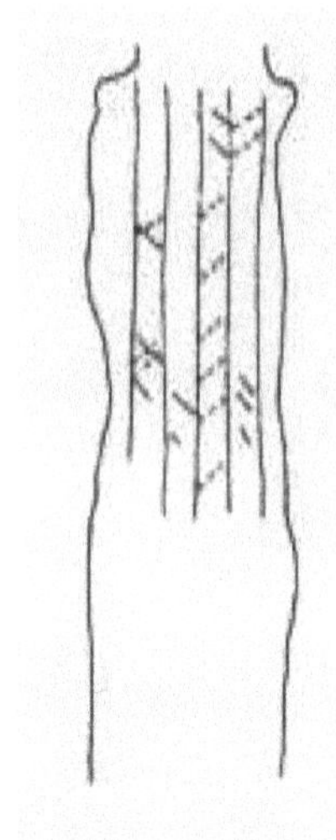

Group 3

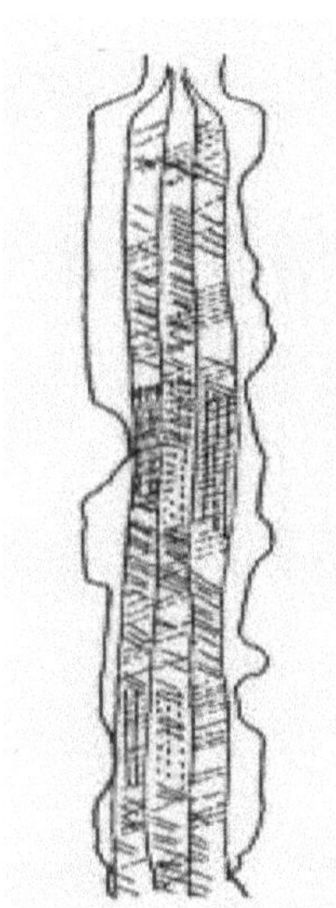

Group 4

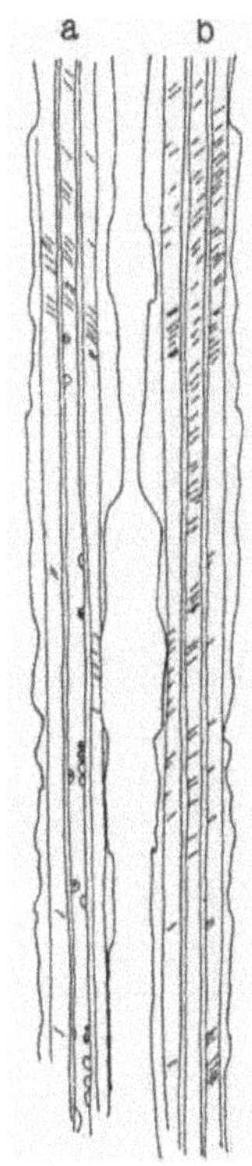

Group 5

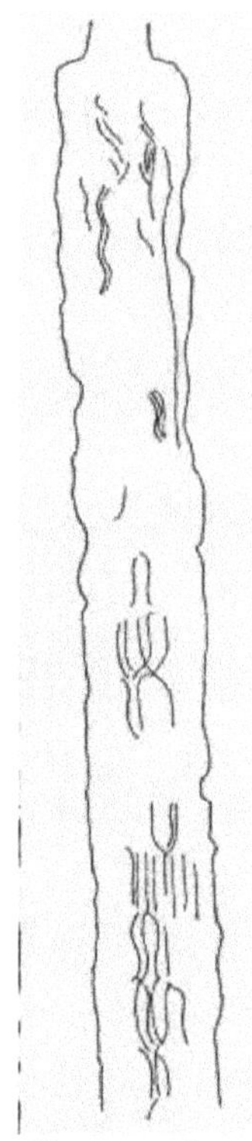

Group 6

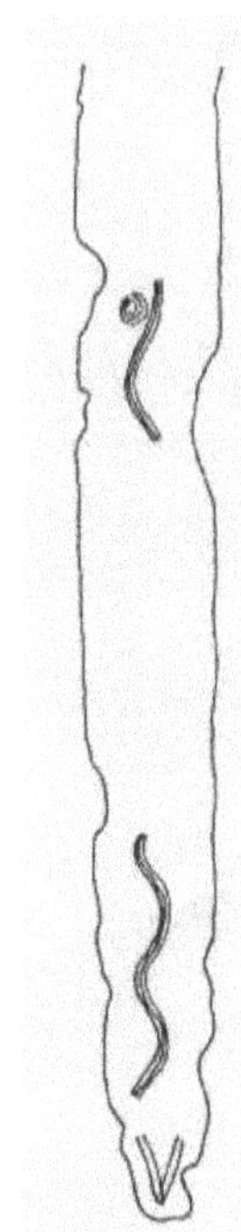

Group 7

Figure 32
"S" decorated swords.
Courtesy of Dr. C. Rupp

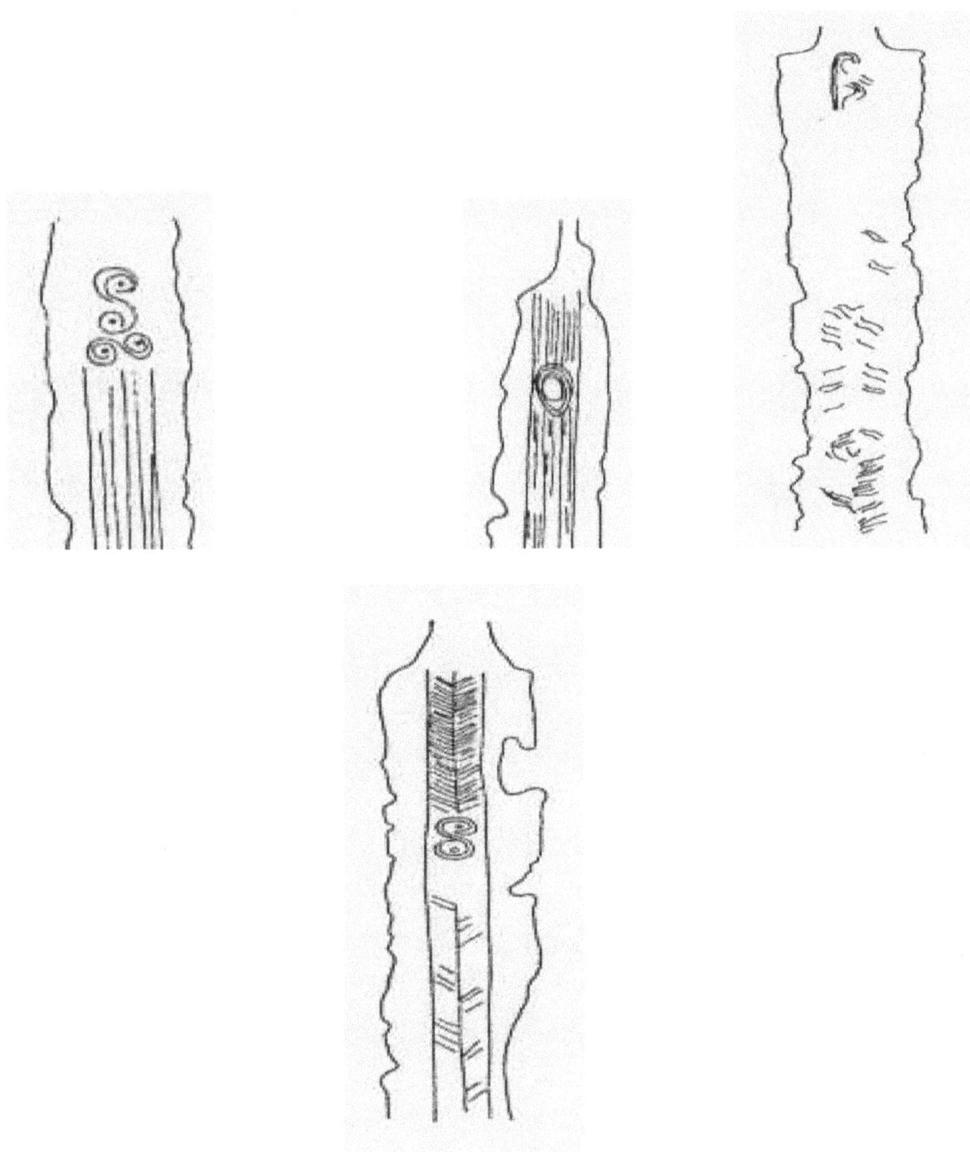

Figure 33
Main types of iron stirrups of early and middle Avar period.
From V. La Salvia and F. Zagari 2003

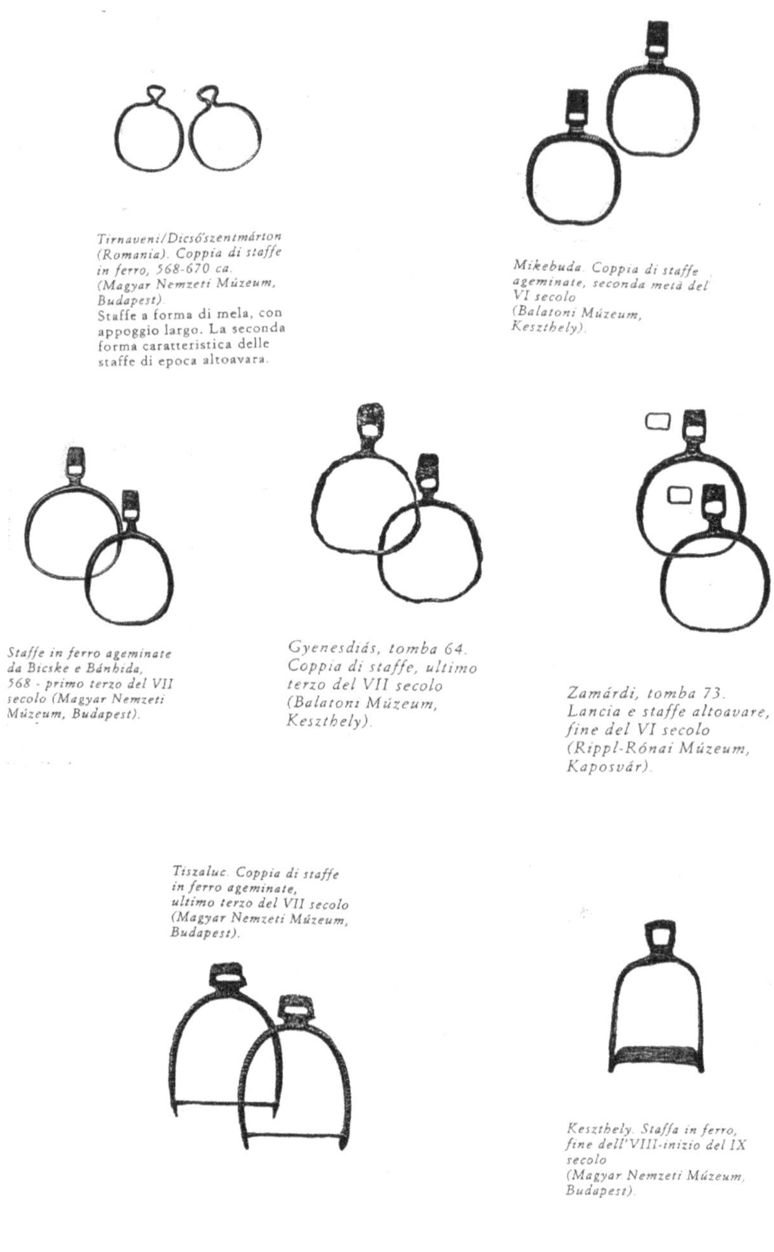

Figure 34
Belt fittings from grave 86/11, buckle from grave 86/9, stirrup from grave 86/2.
Necropolis of *Ischia di Castro* (Viterbo, Italy). From M. Incitti 1997

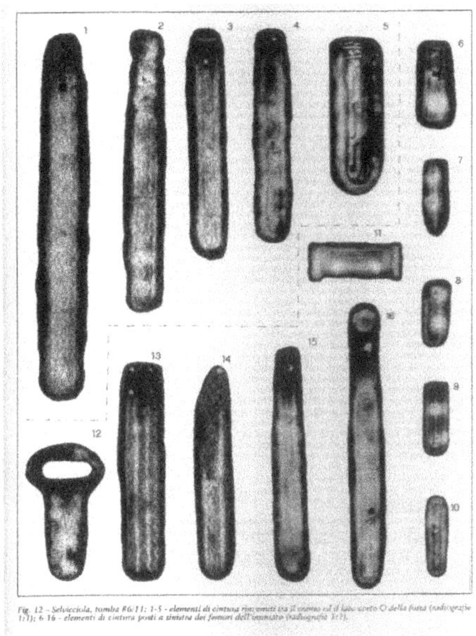

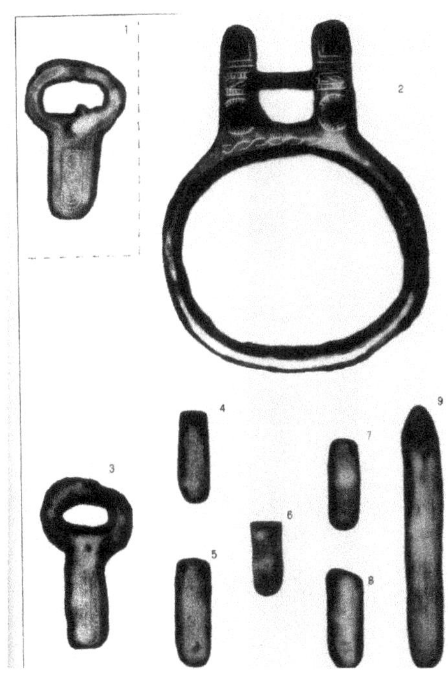

Figure 35
Stirrups like artefacts from A) *Caričin Grad* (Serbia) and B) from grave 1 of the early Avar necropolis of *Szegvár-Oromdülő* (Hungary)

A)

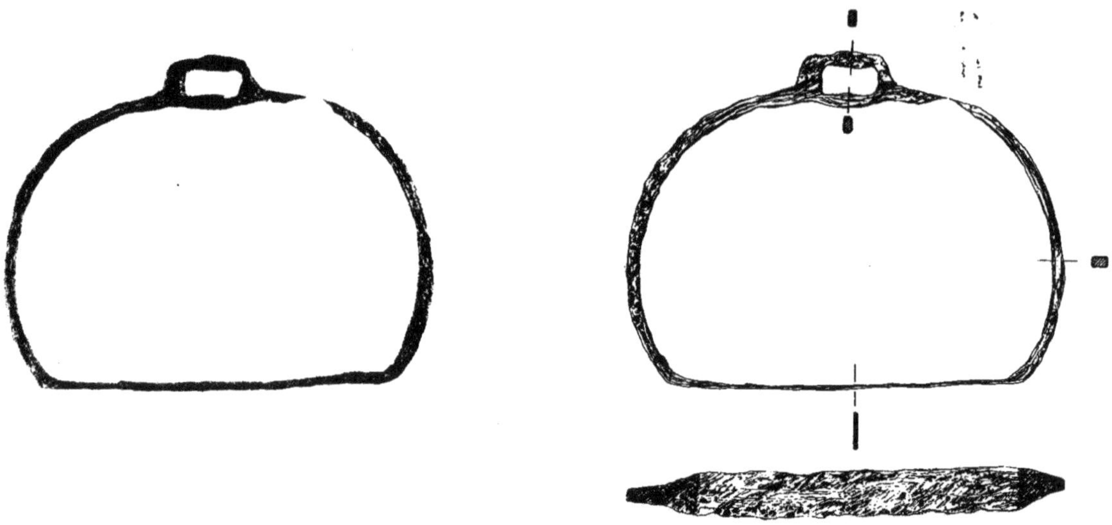

B)

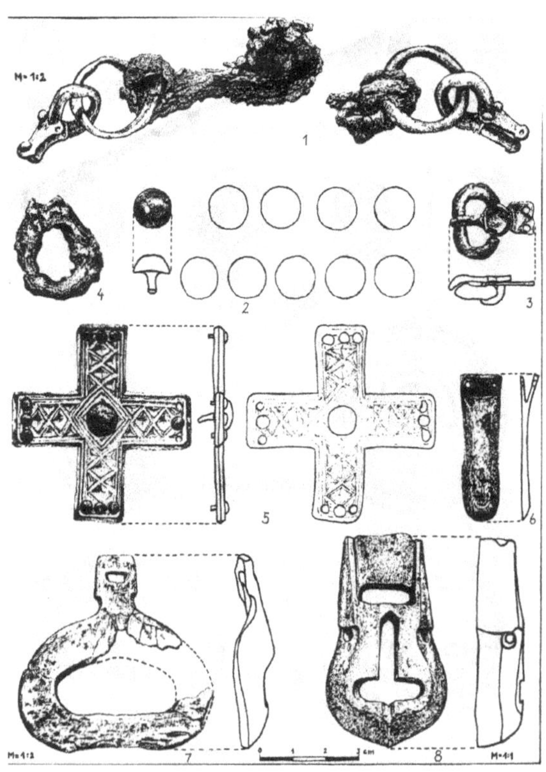

Figure 36
Early Avar iron stirrups and spear heads.
From I. Kovrig 1955

Figure 37
Working tools from the settlement of *Peveragno* (Piedmont, Italy) dated between the end of the AD 4th and the 6th century.
From E. Micheletto and L. Pejrani Baricco 1997

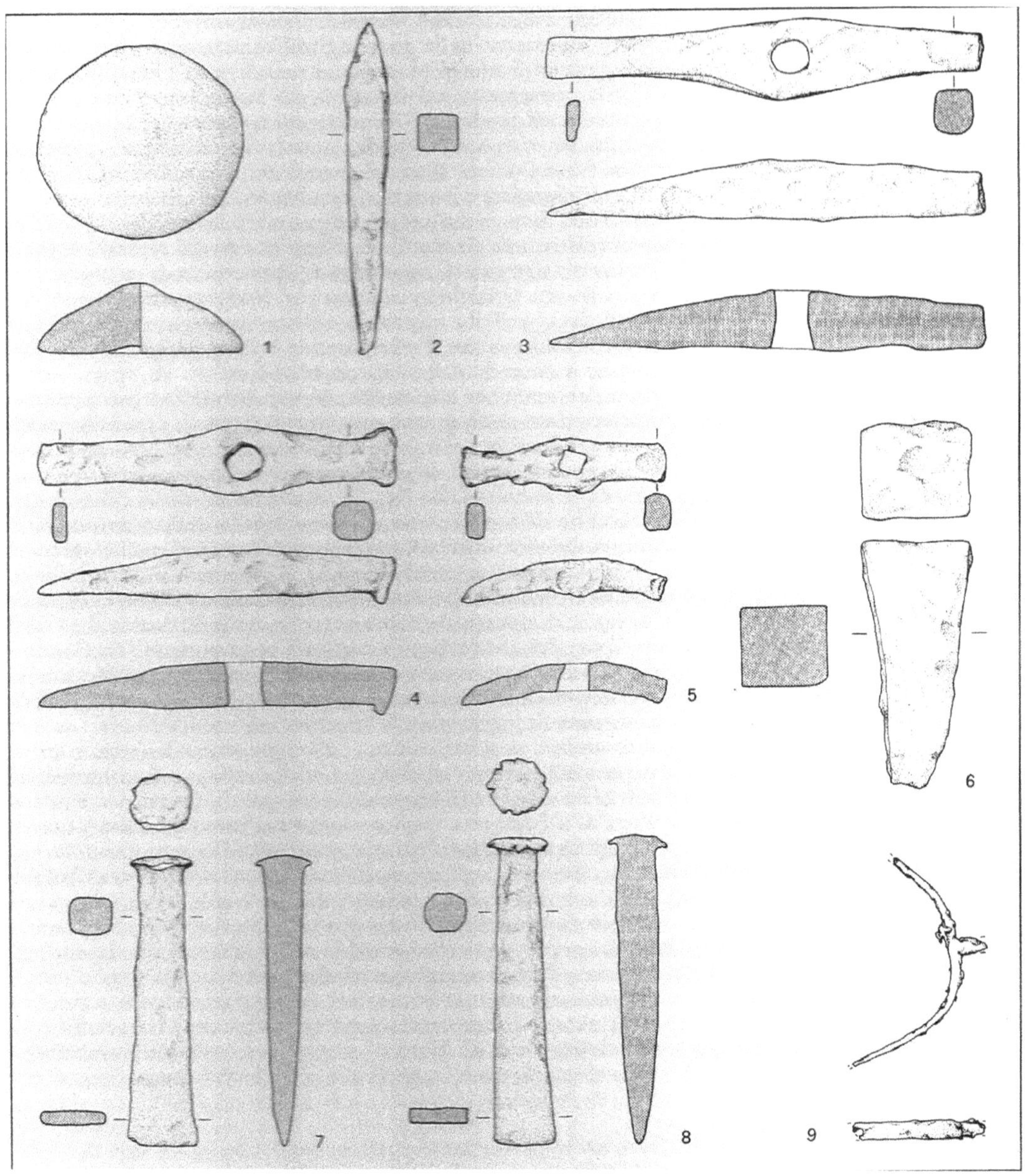

Figure 38
Stratigraphic features of the bowl furnace of *Misobolo* (Piedmont, Italy)
From M. Cima 1986

Fig. 6 – Profilo comprendente la struttura del basso fuoco a fine scavo.
Fig. 7 – Vista dallo zenith dei resti del basso fuoco a fine scavo.

Figure 39
Renditions of the stratigraphic features of the bowl furnace of *Misobolo* (Piedmont, Italy). From M. Cima 1986

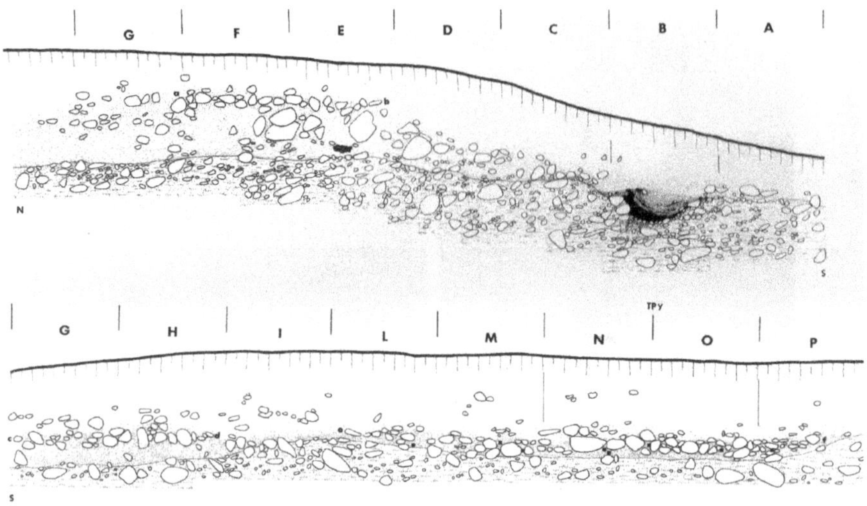

Fig. 4 – Rilievo stratigrafico della trincea del gasdotto nel tratto delle strutture. A. Lato E ove si collocano il basso fuoco (in corrispondenza del TPX) e la strada (a-b). B. Lato W ove è leggibile la continuazione della strada (c-d) e il fondo di abitazione (e-f) sul quale è stato realizzato il sondaggio (TPY).

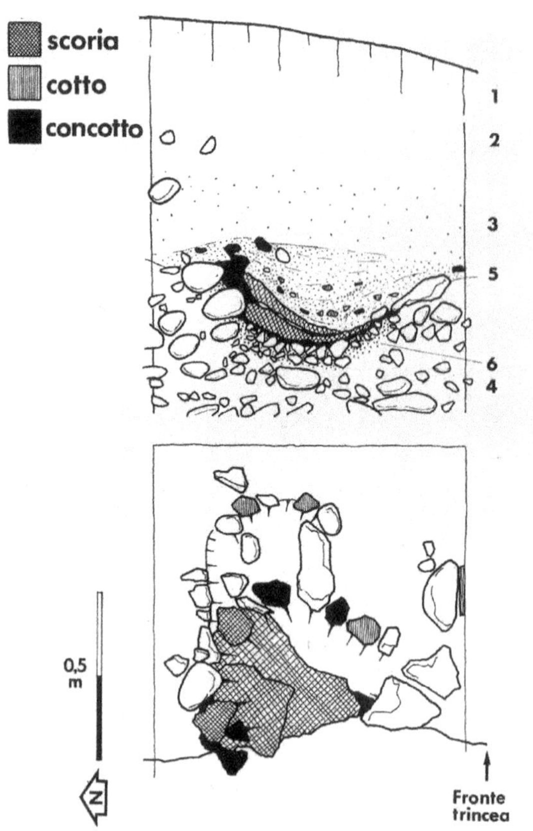

Fig. 5 – Sezione sui resti del basso fuoco e pianta di fine scavo (area del TPX).

Figure 40
Geographical location of the bowl furnace of *Misobolo* (Piedmont, Italy) together with indications of mines, archaeological evidence and towns/villages.
From M. Cima 1986

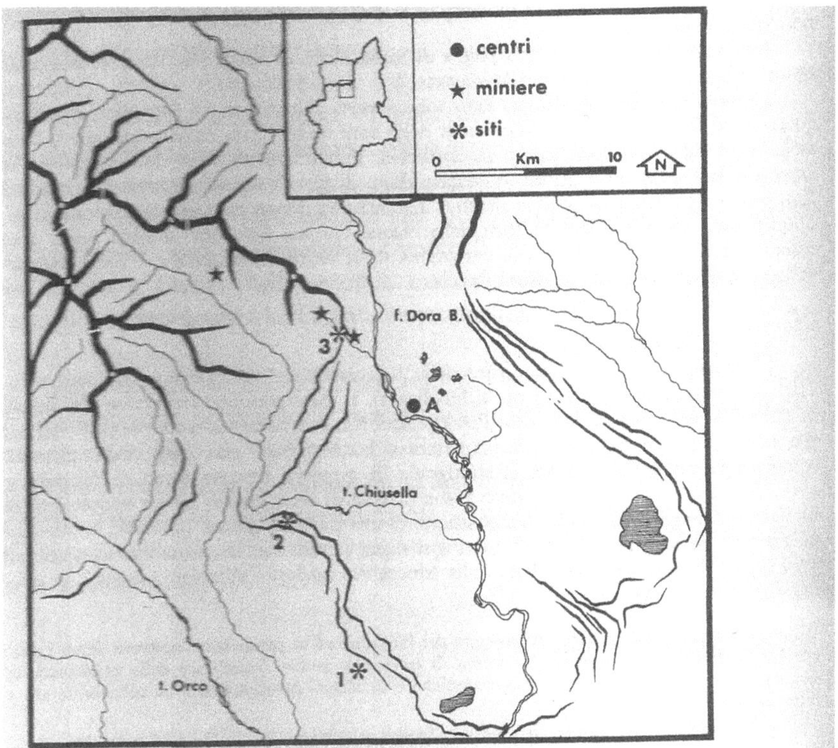

Piantina del Canavese orientale con la valle Chiusella e l'anfiteatro morenico d'Ivrea. A. Ivrea. 1. Misobolo, 2. Silva, 3. Assa.

Figure 41
Quantitative analysis of iron products of northern Italy from the period of Emperor Augustus until the AD 11th century. From C. Liborio, T. Marinig and A. Storti 1989.

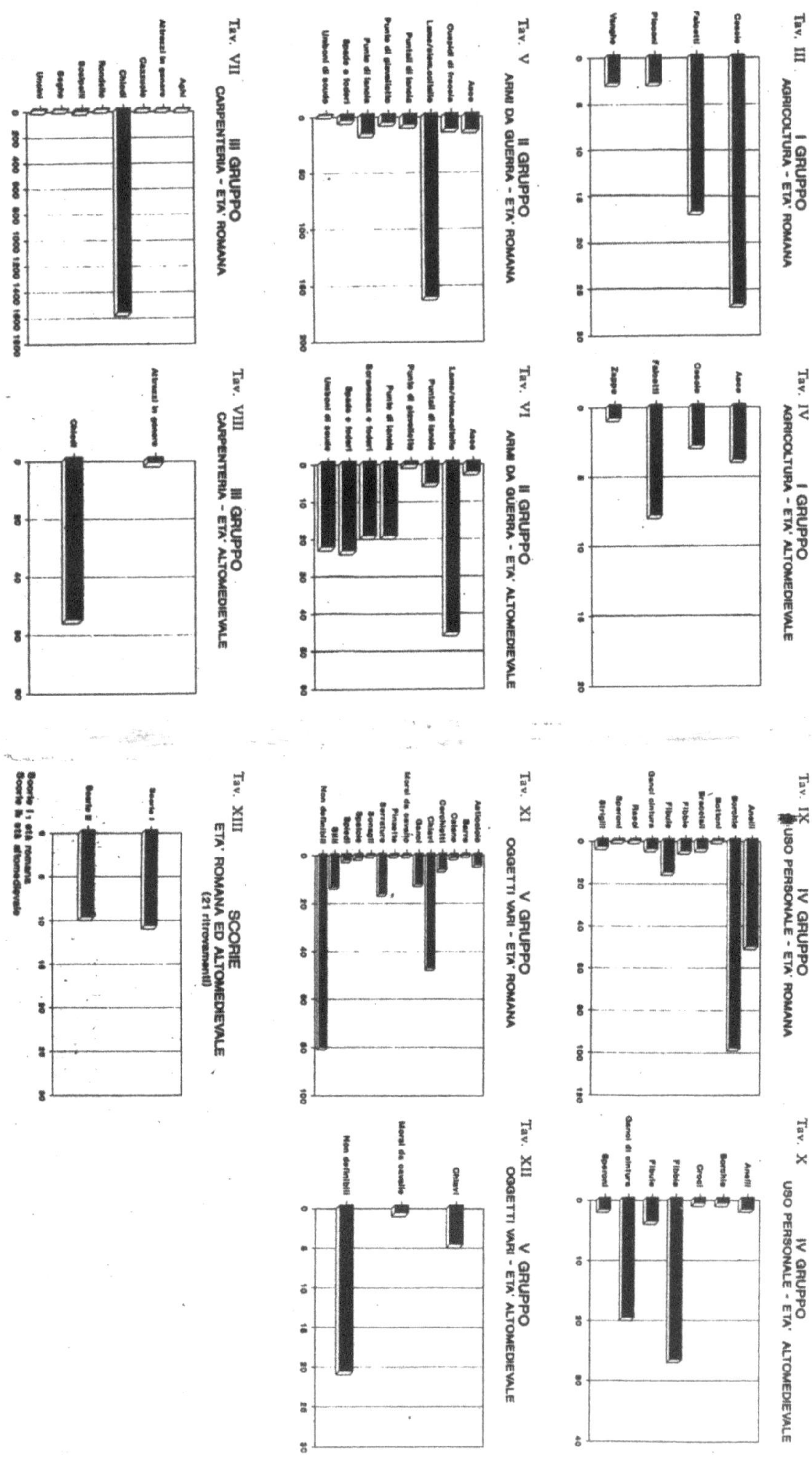

www.ingramcontent.com/pod-product-compliance
Lightning Source LLC
LaVergne TN
LVHW071216210926
848482LV00015B/829
* 9 7 8 1 4 0 7 3 0 1 5 9 4 *